Theory and Research in Behavioral Pediatrics

Volume 1

Edited by

Hiram E. Fitzgerald, Ph.D.

Professor of Psychology
Michigan State University
East Lansing, Michigan

Barry M. Lester, Ph.D.

Director of Developmental Research
Child Development Unit
Children's Hospital Medical Center
Assistant Professor of Pediatrics
Harvard Medical School
Boston, Massachusetts

and

Michael W. Yogman, M.D.

Associate Chief, Division of Child Development
Children's Hospital Medical Center
Assistant Professor of Pediatrics
Harvard Medical School
Boston, Massachusetts

PLENUM PRESS • NEW YORK AND LONDON

ISBN 0-306-40851-1

A Division of Plenum Publishing Corporation
233 Spring Street, New York, N.Y. 10013

Printed in the United States of America

Contributors

Heidelise Als, Ph.D. • Child Development Unit, Children's Hospital Medical Center, Boston, and Harvard Medical School, 333 Longwood Avenue, Boston, Massachusetts

Richard E. Behrman, M.D. • Department of Pediatrics, School of Medicine, Rainbow Babies and Childrens Hospital, Case Western Reserve University, Cleveland, Ohio

T. Berry Brazelton, M.D. • Child Development Unit, Children's Hospital Medical Center, Boston, and Harvard Medical School, 333 Longwood Avenue, Boston, Massachusetts

Katherine A. Hildebrandt, Ph.D. • Department of Psychology, State University of New York at Buffalo, 4230 Ridge Lea Road, Buffalo, New York

Barry M. Lester, Ph.D. • Child Development Unit, Children's Hospital Medical Center, Boston, and Harvard Medical School, 333 Longwood Avenue, Boston, Massachusetts

Edward Z. Tronick, Ph.D. • Department of Psychology, University of Massachusetts, Amherst, Massachusetts

Michael W. Yogman, M.D. • Department of Pediatrics, Harvard Medical School, and Child Development Unit, Children's Hospital Medical Center, Boston, 333 Longwood Avenue, Boston, Massachusetts

Philip Sanford Zeskind, Ph.D. • Department of Psychology, Virginia Polytechnic Institute and State University, Blacksburg, Virginia

Foreword

This volume initiates a series devoted to theory and research in behavioral pediatrics. Focusing attention on the limited scientific contributions to understanding the development of behavior will, we hope, stimulate further research in this vitally important area of human knowledge. The ability of an individual to achieve his or her full potential as an adult is to a significant degree determined by critical variables in this development within a given society. Study of this aspect of human biology, therefore, may have far-reaching effects on the evolving human species.

An awareness of the complexity of the behavioral patterns in infancy is of relatively recent origin and is an obvious essential starting point for this series of publications. The need to discriminate between objective observations and theoretical constructs and to design scientifically valid experiments is vital to progress in understanding early infant behavior. It is easy to misinterpret apparent responses to interventions, especially in infants at biological risk. Preterm infants and their caretakers are particularly challenging as subjects but very difficult to assess scientifically. The interactions between mother and newborn and father and newborn hold promise of substantial additional insights into the development of behavior. Thus, this volume provides interesting perspectives into the relationship of the evolving immature neurological system to complex behavior patterns in newborn infants which raise many new questions and exciting opportunities to extend our very limited knowledge about the newborn infant's psychosocial, emotional, and cognitive development.

RICHARD E. BEHRMAN, M.D.

Contents

Editors' Introduction

Publication of this biennial series on behavioral pediatrics reflects the historical evolution of the disciplines of pediatrics and child and developmental psychology. Around the turn of the century, when pediatrics first became an independent branch of medicine, the chief concern was the high infant mortality rate from infectious diseases such as smallpox. Research consisted primarily of the description and classification of childhood illnesses. By the early 1930s, when a specialty certifying board in pediatrics was established, infant mortality had decreased somewhat, and research began to focus on specific mechanisms of disease and therapy, particularly involving infant nutrition and fluid balance.

Around the same time, the Society for Research in Child Development was established, in part, to encourage interdisciplinary consideration of substantive and methodological issues in the field of child development. The spirit was reflected in the formation of a number of child development institutes in the 1920s and 1930s such as those at Iowa, Berkeley, Fels, Merrill–Palmer, Columbia, Minnesota, and Yale, and by the impact of people like Lawrence Frank, Arnold Gesell, Milton Senn, and Julius Richmond, to name just a few. Developmental psychology has become a specialty in psychology departments of American universities only within the last 20 years. It existed outside those departments precisely because, as G. Stanley Hall envisioned, the effort was toward a cross-disciplinary, practice-oriented enterprise. Interaction between the disciplines, however, was minimal. Perhaps these early attempts at cross-fertilization between child development and pediatrics were premature, coming at a time when the primitive level of development in both fields was not conducive to productive research in common areas of concern.

During World War II, pediatric concerns were refocused on the treatment of serious infections because the discovery of antibiotics offered dramatic uses. Together with immunizations and improved milk storage, these achievements permitted the control of most life-threatening infectious illness. More recently, technological advances have greatly improved the care of high-risk newborns and the diagnosis and treatment of immunological and genetic disorders. Because of the success of these contributions in treating serious disease and in decreasing infant mortality, pediatricians are now able to focus more attention on problems in child development and family adaptation.

The emergence of the field of behavioral pediatrics represents the increasing communication and collaboration between psychologists and pediatricians. Developmental psychology and pediatrics have evolved in parallel during the past century, and both are now firmly established and respected by their own and other disciplines. They tend to be eclectic and less committed to established authority and traditional methods and to share the common goal of fostering the optimal growth and development of the child and the family.

Interdisciplinary research and training in child development and pediatrics has increased dramatically during the past few years. Postresidency training programs in child development for pediatrics have been established, as have postdoctoral training programs for psychologists in clinical areas related to medicine. Much of this increased demand for communication between child development and pediatrics has come from the general acceptance of child development as the basic science of pediatrics, as stated by Julius Richmond (1967, 1975), "an idea whose time has arrived." Improved research and clinical training in child development and behavioral pediatrics have begun to be emphasized by the Task Force on Pediatric Education and by a special committee set up by the American Academy of Pediatrics. This new field of behavioral pediatrics faces the challenge of establishing itself as a science.

As with any new scientific field, behavioral pediatrics must offer a unique theoretical base relevant to the questions posed, define its assumptions, and create new research strategies. More specifically, it must support the study and management of complex psychosocial as well as biomedical problems. We hope that this series will further these goals.

The task of behavioral pediatrics as a newly emerging discipline is to develop a basic science that links biological science with behavioral science to provide a clinically relevant perspective on early human development. The purpose of this series is to provide a forum for the exposition of research findings in this new discipline. We predict that this approach to research through the binocular eyes of child development and pediatrics will generate multifaceted, dynamic theories and models of development and research methodologies that transcend and synthesize the strengths of the two disciplines, leading to a better understanding of development and the clinical care of children.

Perhaps because it is a new discipline, the term *behavioral pediatrics* has many uses, and it may be helpful to clarify how we plan to use the phrase. We are using the term *behavioral pediatrics* synonymously with the terms *developmental, psychosocial,* or *biosocial pediatrics.* For many purposes, behavioral pediatrics may be considered a subspecialty of the respective fields of developmental psychology and pediatrics. In developmental psychology, this would be akin to education, cognition, or physiology and related to but different from pediatric psychology, which emphasizes the clinical care of children in psychiatric settings. In pediatrics, this would parallel subspecialties such as cardiology or neurology, particularly with respect to research and training activities. Clinical skills in behavioral pediatrics, however, must be integrated into general pediatric training and practice, rather than viewed as another clinical subspecialty, since clinical problems in this area represent the core of general pediatrics.

Behavioral pediatrics is a research discipline in areas that intersect child development and pediatrics, with particular emphasis on the study of the integration of biological systems and behavioral systems. It includes basic and clinical research aimed at developing both interdisciplinary models that will lead to a better understanding of developmental processes themselves by studying the whole child. In seeking this goal, behavioral pediatrics begins with an attempt to understand normal development. This effort is designed to balance the traditional disease model of medicine by recognizing the strengths of normal development and integrating strengths and deficits into the whole-child concept. There is an emphasis on the study of intraindividual and interindividual differences and on their application to the individualized clinical care of the child and family. Studies of deviant populations, psychopathology, and at-risk populations broaden our understanding of normal developmental processes by forcing us to construct inductive systems of hypotheses that would not have been expressed had we not encountered these phenomena. At the same time, studying physical and behavioral deficits from the perspective of normal development should lead to clinical strategies for prevention and intervention that balance the deficits with recognition of the strengths of the child from a behavioral as well as a psychological perspective. The questioning of theoretical and methodological orthodoxy is enhanced as we are forced to deal with the whole child and the child as an individual.

In this series we will present substantive research findings that advance our understanding in areas relevant to child development and pediatrics. This series will emphasize the need for clinical practice in behavioral pediatrics to be based on sound theory supported by solid empirical research. With that goal in mind, empirical findings on a sound theoretical framework will be provided by contributors in chapters of sufficient length to enable a broad discussion of important issues and an opportunity to consider the clinical implications of this work. As such, the series should make useful contributions to developmental theory and research and to clinical practice concerning developmental issues. We expect this series to be useful to academic audiences in pediatrics, child psychiatry, psychology, nursing, anthropology, social work, and education. We hope through this series to facilitate communication and collaboration among disciplines and to share skills, insights, and perspectives toward the common goal of fostering the optimal growth, development, and welfare of all children.

The first volume of this biennial series focuses on the beginnings of human development during the infancy period. Each of the chapters begins by addressing an important theoretical question in early development, reports research findings, and attempts to relate the studies to clinical practice. Chapter 1 outlines a model for understanding plasticity in early development and relates this concept to the timing and quality of intervention by clinicians working with infants at biological risk. Chapter 2 focuses on the relationship between specific biological risk, prematurity, and newborn behavior. The authors postulate the existence of several behavior subsystems in newborn infants, provide a matrix for the careful description of these systems, and relate the behavior of these infants to their biological risk status. Appended is a manual for the assessment of preterm infants. Similarly, Chapter 3 describes crying, the most common behavior in young infants, and suggests its usefulness as both an index of and influence on the biological and social risk of the infant. All three of these chapters view the newborn and infant as embedded in a transactional relationship with caregivers right from birth.

The chapter on maternal perceptions of infants addresses one aspect of the caregiver's participation in this interactive process: How do infant physical characteristics influence the mother's attitudes toward her infant? The chapter on the father–infant relationship addresses the important issue of sex differences in caregiving. The author broadens our thinking about caregivers to include fathers and discusses both father–infant interactions and indirect influences of the father in his role as a member of the intrafamilial support system.

Chapters in future volumes in this series will address other topics in infancy and early childhood as well as the developmental issues of older school-age children and adolescents. We hope that over time this series will provide a useful forum for broad, general reviews of research areas. Typically, contributions to this series will be by invitation. However, readers and contributors who have ideas, suggestions, or manuscripts which they believe may be suitable for future volumes are encouraged to write the editors.

HIRAM E. FITZGERALD
BARRY M. LESTER
MICHAEL W. YOGMAN

REFERENCES

Richmond, J. B. Child development: A basic science for pediatrics. *Pediatrics,* 1967, *39*, 649.

Richmond, J. B. An idea whose time has arrived. *Pediatrics Clinics of North America,* 1975, *22*, 517–523.

Early Intervention

What Does It Mean?

T. BERRY BRAZELTON

1. INTRODUCTION

Coincident with our increasing capacity for conquering physical disease we have become sensitized to our capacity for (1) prevention of many physical and psychological disorders and (2) the mitigation of their effects on the quality of life for affected individuals. Our attitudes toward handicap and disease have been changing rapidly in the past two decades. *Prevention, intervention, quality of life* are becoming catchwords for the disciplines which are concerned with medical and psychological disorders. *Plasticity,* the capacity of a developing organism to find pathways around a deficit, has dawned upon many researchers and clinicians as if it were a new concept. Often this "new" awareness has come without a deep enough understanding of its mechanisms.

The remarkable ability of a child to recover from central nervous system (CNS) deficits is evidence of plasticity in medicine. Children with known insults and identified defects in CNS tissue seem to be able to compensate for these deficits over time (Neligan, Kolvin, Scott, & Garside, 1976; Sigman & Parmelee, 1979). With a defect such as blindness, for example, the child can learn to compensate with increased sensitivity of the other modalities—auditory, tactile, and vestibular. An instance of this is "radar vision" that was marked by heightened sensitivity to auditory, tactile, and vestibular cues in a 14-month-old blind baby, who vocalized as she walked and never ran into tables, rounded corners of doors by the differ-

---

T. Berry Brazelton, M.D. • Child Development Unit, Children's Hospital Medical Center, Boston, and Harvard Medical School, 333 Longwood Avenue, Boston, Massachusetts 02115. The author is supported by the Robert Wood Johnson Foundation, the Carnegie Corporation of New York, the National Institute of Mental Health, and the William T. Grant Foundation.

ences in reverberations as her vocalizations bounced off nearby objects (Als, Tronick, & Brazelton, 1980). Since CNS tissue is not thought to be regenerative, a capacity for recovery of function has always been poorly understood.

In this chapter, I would like to address the forces in development which can be harnessed to help the infant compensate for defects—either organic or behavioral—which might otherwise make him or her subject to failures in development. The timing of intervention is critical, not only in strengthening the infant's assets to overcome deficits, but also to reinforce a positive self-image from the first and to add motivation toward recovery or optimal function. Individual programs oriented to that particular baby are critical, and programs which reach out to grieving or anxious parents must take into account the reasons for the defenses around their anxiety. An understanding of the normal forces for development in the infant becomes critical, as does an understanding of the forces for attachment and for grieving in caregivers. All of these will be addressed in this chapter. The use of assessment becomes a way of identifying with the infant and with his parents as they work with him. The behavioral assessment of the neonate (NBAS) has taught us much about the uses of infant assessment. Using repeated assessments as recovery curves has taught us to understand the processes of recovery in early infancy from labor and delivery and gives us a way to assess and enhance the input from the environment, even in the face of CNS or autonomic nervous system (ANS) defects. In our face-to-face assessment of parent–infant interaction, we have come to realize the potency of violations of expectancy in distorting this interaction. By allowing parents to see this with us, they too can begin to understand and work for interactions appropriate to their child. In this way we have seen remarkable recovery in at-risk infants. Two cases will be cited which exemplify our use of assessments to follow the babies' development and to capture and encourage the parents' best efforts with them.

In small children, the efforts to overcome or compensate for handicapped pathways of function lead one to want to understand the process of learning to cope. An instance might be seen in a baby who learns to utilize compensatory musculature to make up for a paralyzed group of muscles. The need to learn to cope in the face of a deficit looks costly, but the developing child's determination to succeed in maturational steps seems to be equal to the task. At least two deficits seem to be involved—the localized neurological deficit and the reverberations of the interference in function around the deficit; a deficient function is likely to carry with it a spread of interference to other contingent systems. The autisms of a blind or deaf child reflect the cost to the organizational system of managing such a sensory deficit. When these deficits can be managed successfully, they reflect the relatively massive energy for coping which is necessary to compensate for such a real deficit (Fraiberg, 1977). Recent long-term follow-up studies indicate that at least two ingredients are critical: (1) timing of intervention and its quality, that is, it must be of a kind which fosters the child's sense of competence and fuels his own energies for learning to cope; and (2) the energy in the environment to back up the child's efforts at compensatory learning. The recent longitudinal study by Sigman and Parmelee (1979) is an example of this. Sigman and her associates have found that the best predictors for compensatory recovery of CNS deficits rest in (1) the energy of the infant to reach out for and interact with the environment (as reflected by visual behavior at one month) and (2) the richness of environmental input available at four months (Sigman & Parmelee, 1979). This points to a model which involves (1) maturation, (2) an inner source for fueling compensatory and rewarding developmental systems, and (3) a rewarding environment which reinforces both specific compensatory behaviors and a more general sense of achievement as each developmental step is reached and achieved. When these forces for development fail, they may add further to developmental failure in the infant. Unless we understand the processes which have contributed to the risk for failure, as well as the processes which will enhance plasticity or recovery from the deficit, we cannot begin to play an appropriate role in enhancing recovery in the infant. And unless we can understand the interactive processes between each individual parent and infant, we are not likely to enhance the environment's role in fueling that recovery. Oversimplified, nonindividuated programs of intervention may do more harm than good. Programs which are individuated have the possibility of reaching out to the target individual in a way that will allow him to feel special. Not only will the individuation be more likely to suit the baby's particular needs, but the very effort to understand them will create a kind of Hawthorne effect—a feeling of being special and important—to others, hence to oneself. Individuation of programs and an attempt to make the targeted individual feel that he is in control of his destiny are critical to optimal results.

"Stimulation" programs which are designed to correct identified deficits in development are widespread, and the lower socioeconomic status groups are targeted with middle-class therapists' goals. Since the risk for developmental failure seems often to be associated with lower socioeconomic status environments (Neligan *et al.*, 1976), this latter seems reasonable, but one wonders whether the goals of the therapist can match those of the recipient in such programs, and whether the risk of failure in basic communication may not enhance the sense of failure which already dogs the lower-class recipient. If the inability to "understand" it or to comply with the intervention is added to the sense of failure in those around him, the infant's own self-image is bound to be affected. Hence, the efforts of well-meaning professionals may compound the failure in caring parents who cannot reach the standards set for them.

2. WHY EARLY INTERVENTION?

As the potential for early intervention increases, it becomes more and more important that we be able to evaluate at-risk infants as early as possible with an eye to more sophisticated preventive and therapeutic approaches, before failure systems and the expectation to fail become established. Early intervention may prevent a compounding of problems which occur all too easily when the environment cannot adjust appropriately to the infant at risk. Premature and minimally brain-damaged infants seem less well able to compensate in disorganized, depriving environments than are well-equipped neonates, and their problems of organization in development are compounded early (Greenberg, 1971). Quiet, nondemanding infants do not elicit necessary mothering from already overstressed parents and are therefore susceptible because of their neonatal behavior for kwashiorkor and marasmus in poverty-ridden cultures such as those in Guatemala and Mexico (Cravioto, Delcardie, & Birch, 1966; Klein, Habicht, & Yarbrough, 1971). Hyperkinetic, hypersensitive neonates may press a mother and father into a kind of desperation which produces child-rearing responses from them that reinforce the problems of the child so that he grows up in an overreactive, hostile environment (Heider, 1966). Parents of children admitted to the wards of the Children's Hospital in Boston for such clinical syndromes as failure to thrive, child abuse, repeated accidents and ingestions, and infant autism are often successful parents for other children but not for the patient. By history, they associate their failure with this child to an inability to "understand" him from the neonatal period onward, and they claim a difference from the other children in his earliest reactions to them as parents. If we are to improve the outcome for such children, assessment of the risk in early infancy must mobilize preventive efforts and programs for intervention before the neonate's problems are compounded by an environment which cannot understand him without such help.

But we need more sophisticated methods for assessing neonates and for predicting their contribution to the likelihood of failure in the environment–infant interaction. The possibilities for synergism toward a failure in interaction between an infant who is not rewarding and an already stressed environment seem obvious. We also must be able to assess at-risk environments, for the impracticality of spreading resources too thin points to the necessity of selecting target populations for our efforts at early intervention. With better techniques for assessing strengths and weaknesses in infants and the environment to which they will be exposed, we might come to understand better the mechanisms for failures in development which result in some of the above syndromes. Even desperate socioeconomic conditions produce comparable stresses in many families whose children do not have to be salvaged from the clinical syndrome of child abuse, failure to thrive, and kwashiorkor. Minimally brain-damaged babies do make remarkable compensatory recoveries in a fostering environment. Understanding the infant and the problems he will present to his parents may enhance our value as supportive figures for them as they adjust to a difficult child.

In other words, there appear to be at least two sources of vulnerability which contribute to the risk of failure in developmental outcome: (1) in the baby's own organizational system and the capacity for growth—CNS and autonomic as well as physical—and (2) in the capacity of the environment (usually represented by the parents) to adjust to and nurture the at-risk infant in ways that are appropriate to individual needs. If the interaction between these two is positive, the opportunities are significantly increased for fueling feedback cycles necessary to the baby for developing energy for developmental progress.

2.1. Forces for Normal Development in the Infant

An understanding of the forces which work toward a child's development is critical to an understanding of his failure and toward any effort to prevent such failure. There are at least three forces that are constantly at work.

2.1.1. Maturation of the Central and Autonomic Nervous Systems

Maturation of the central and the autonomic nervous systems, which regulate the baby's capacity to control reactions to incoming stimuli, is one of these forces. If the baby is at the mercy of an overreaction of either a motor (Moro or startle) or autonomic reaction (as is seen in a overstressed pulmonary or cardiac system), he cannot learn to maintain attention or to react appropriately to a sensory stimulus or to other information necessary for development.

2.1.2. Realization of Competence in the Baby

A second force is that of competence within the child which is fed by a feedback system relying upon the completion of a task that he *himself* has done. White (1959) called this a "sense of competence," and one sees its power as a source of fuel when a toddler first learns to walk. His face glows, his body struts, his legs are driven to perform for long, exhausting periods, he chortles with delight in the achievement. The energy which has been mobilized to complete the task now fuels the realization of mastery and in turn is reinforced to press him on to the next step in development. In hospitalized or institutionalized children who do not have the opportunity for completing new developmental tasks, one sees the waning of this kind of inner excitement (Provence & Lipton, 1962).

2.1.3. Reinforcement from the Environment

The third force is the reinforcement from the environment which feeds the infant's affective and cognitive needs. The feedback cycles which are necessary for normal affective growth were pointed out by Spitz (1945), later by Harlow in monkeys (1959), and conceptualized as "attachment" by Bowlby (1969). That this affective base is critical to cognitive and motor development was well-documented in the institutionalized infants of Provence and Lipton (1962). We had not understood how critical was this environmental nurturance to all of the infant's development until the pathology was identified in environmentally deprived infants. The environmental forces can work powerfully to retard or to enhance the infant's progress. In our laboratory we are attempting to identify and conceptualize some of the ingredients of these interactive forces as they combine to fuel the child's recovery from severe deficits of prematurity, respiratory distress syndrome, and CNS dysfunction.

3. DEVELOPMENTAL MODEL

As we began in the early 1950s to attempt to document and understand neonatal behavior, very powerful mechanisms appeared to dominate the neonate's behavior (Brazelton, 1961a). In the tremendous physiological realignment that the changeover from intrauterine to extrauterine existence demands, it has always amazed us that there is any room for individualized responses, for alerting and stimulus-seeking, or for behavior which indicates a kind of processing of information in the neonate, and yet there is. Despite the fact that the newborn's major job is that of achieving homeostasis in the face of enormous onslaughts from the environment, we can see evidence of affective and cognitive responses in the period after delivery.

This very capacity to reach out for, to respond to, and to organize toward a response to social or environmental cues seems so powerful at birth that one can see that even as a newborn the infant is "programmed" to interact, as he wakes from sleep and is on his way to a disorganized crying state, to turn his head to one side to set off a tonic neck reflex, to adjust to this with a hand-to-mouth reflex and sucking on his fist. All of these can be called primitive reflex behaviors. But as soon as the newborn has completed this series, he sighs, looks around, and listens with real anticipation, as if to say, "This is what I'm really here for—to keep interfering motor activity under control so that I can look and listen and learn about my new world."

Our own model of infant behavior and of early infant learning goes like this: The infant is equipped with reflex behavioral responses which are established in rather primitive patterns at birth. He soon organizes them into more complex patterns of behavior which serve his goals for organization at a time when he is still prone to a costly disorganization of neuromotor and physiological systems and then for attention to and interaction with his world (Als, Lester, & Brazelton, 1979). Thus, he is set up to learn about himself, for as he achieves each of these goals, his feedback systems say to him "You've done it again! Now go on." In this way, each time he achieves a state of homeostatic control, he is fueled to go on to the next stage of disruption and reconstitution—a familiar model for energizing a developing system. We also believe that the infant's quest for social stimuli is in response to his need for fueling from the world outside. As he achieves a homeostatic state, and as he reaches out for a disruptive stimulus, the reward for each of these states of homeostasis and disruption is reinforced by social or external cues. Hence, he starts out with the behaviorally identifiable mechanisms of a bimodal fueling system (1) of attaining a state of homeostasis and a sense of achievement from within and (2) the energy or drive to reach out for and incorporate cues and reinforcing signals from the world around him, fueling him from without. He is set up with behavioral pathways for providing both of these for himself—for adaptation to his new world, even in the neonatal period. Since very little fueling from within or without may be necessary to set these patterns and press him onward, they are quickly organized and reproduced over and over until they are efficient and incorporated and can be utilized as the base for building later patterns. Greenacre's (1959) concept of early pathways for handling the stress and trauma of birth and delivery as precursors for stress patterns later on fits such a model. It is as if patterns or pathways which work were "greased up" for more efficient use later on. Our own concept is that other patterns are available too but that these are just readied by successful experience.

With this model of behavioral response systems which provide an increased availability to the outside world, one can then incorporate Sander's (1977) ideas of early entrainment of biobehavioral rhythms, Condon and Sander's (1974) propositions that the infant's movements match the rhythms of the adult's voice, Meltzoff and Moore's (1977) work on a kind of matching imitation of tongue protrusion in a three-week-old, and Bower's (1966) observations on early reach behavior to an attractive object in the first weeks of life. As each of these responsive behaviors to external stimuli fuels a feedback system within the baby toward a realization that he has "done it"—controlled himself in order to reach out for and respond appropriately to an external stimulus or toward a whole adult behavioral set—he becomes energized in such a powerful way that one can easily see the base for his entrainment. The matching of his responses to those in the external world must feel so rewarding that he quickly becomes available to whole sequential trains of behavioral displays in his environment and begins to entrain with them. He becomes energized to work toward inner controls and toward states of attention which maintain his availability to these external sequences. From the simpler form of attention to discrete stimuli, he is able to move toward prolonged periods of attention. In these periods, the prolonged attention is marked by sequential reactions to each stimulus. The sustained attention is modulated by brief but necessary periods of decreased attention. In this way, entrainment becomes a larger feedback system which adds a regulating and encompassing dimension to the two feedback systems of internalized control and externalized stimulus–response. Hence, entrainment becomes an envelope within which one can test and learn about both of his fueling systems. Thus, he can learn most about himself by making himself available to entertainment by the world around him. This explains

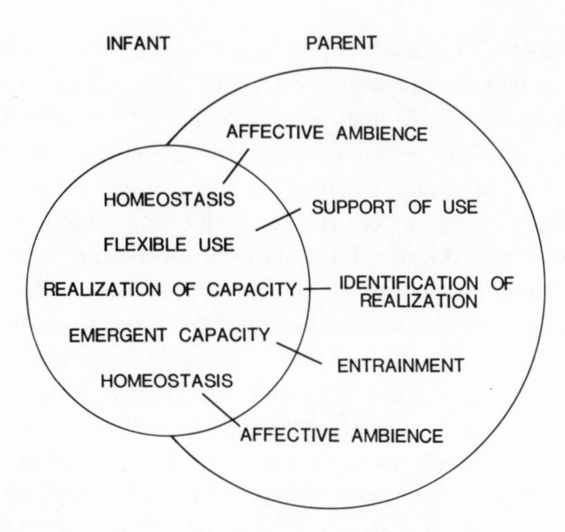

FIGURE 1. Process of interactive negotiation bringing about each stage of organization. From "Four Early Stages in the Development of Mother-Infant Interaction" by T. B. Brazelton and H. Als, *The Psychoanalytic Study of the Child,* 1979, *34*, 349–369. Copyright 1979 by Yale University Press. Reprinted by permission.

the observable drive on the part of the neonate to capture and interact with an adult interactant—and his need for social interaction. Figures 1 and 2 show a schematic presentation of this mutual fueling process (Brazelton & Als, 1979).

3.1. Forces for Normal Development in the Parent

The parents' own genetic potential is influenced by past experiences, and these form the most powerful base for their capacity to nurture a new baby. When this base is intact, as with healthy parents who have been nurtured themselves, one can expect them to mobilize resources to adapt to the individual baby. When, however, this base is stressed, for example, in parents who themselves are afflicted with physical or psychological deficits, their capacity to adapt to the needs of an individual baby may well be dominated by their own needs and their own past experience. In culturally deprived groups or in those whose energy (physical and psychological) is limited by the demands of poverty and its concomitants—disorganization, undernutrition, a sense of failure—it is no wonder that we find limitations of their ability to adapt to the individual child, especially if he is not rewarding or has special needs. It is not surprising, then, that one of the most important marker variables of this process is socioeconomic status in predicting the recovery of at-risk children—at risk for all conditions, physical and psychological (Drillien, 1964; Neligan *et al.,* 1976).

What is surprising is that there is energy in most parents to adapt to a new baby, as well as how readily available this energy can be. In order to understand the forces for adaptation to a new baby at delivery and in the immediate perinatal period, we studied a group of primiparous mothers and fathers in psychoanalytic interviews in the last months of pregnancy at the Putnam Children's Center in the 1950s. We found that the prenatal interviews with normal primiparas uncovered anxiety which seemed to be of almost pathological proportions. The unconscious material was so confused, so anxious, and so near the surface that before delivery one felt an ominous direction for making a prediction about the woman's

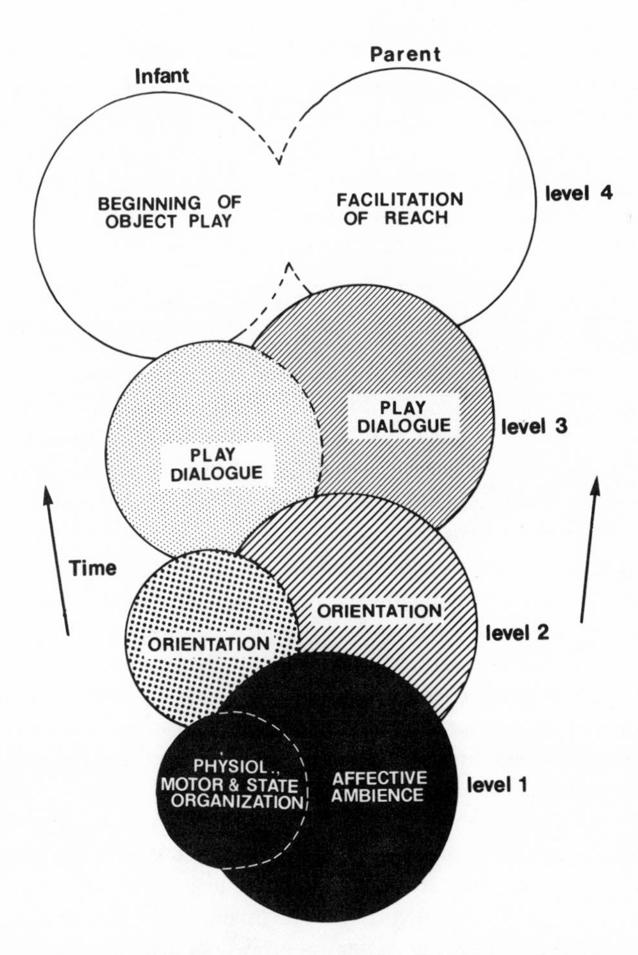

FIGURE 2. Stages of organization: Parent–infant interaction. From "Four Early Stages in the Development of Mother-Infant Interaction" by T. B. Brazelton and H. Als, *The Psychoanalytic Study of the Child,* 1979, *34,* 349–369. Copyright 1979 by Yale University Press. Reprinted by permission.

capacity to adjust to the role of mothering. Yet when we saw her in action in the postpartum month as a mother, this very anxiety and the distorted unconscious material seemed to become a force for reorganization, for readjustment to an important new role (Bibring, Dwyer, & Valenstein, 1961; Brazelton, 1963). I began to feel that much of the prenatal anxiety and distortion of fantasy was a healthy mechanism for bringing her out of the old homeostasis which she had achieved to a new level of adjustment. The "alarm reaction" we were tapping in on was serving as a kind of "shock" treatment for reorganization to her new role. I do certainly agree with Bowlby's concept of attachment and of the importance of providing opportunities for the mother to get to know her new infant early. I now see the shakeup in pregnancy as readying the circuits for new attachments; as preparation for the many choices which she must be ready to make in a very short, highly sensitive critical period (Klaus & Kennell, 1970); as a method of freeing her circuits for a kind of sensitivity to the infant and his individual requirements which might not have been easily or otherwise available from her earlier adjustment. Thus, this very emotional turmoil of pregnancy and of the neonatal period can be seen as a positive force for the mother's healthy adjustment and for the possibility of providing a more individualizing, flexible environment for the infant (Bibring *et al.,* 1961).

Prospective fathers must be going through a very similar kind of turmoil and readjustment (see Chapter 5). In an ideal situation we would be offering both parents much more support and fuel for their new roles than we do. So far, we in medicine have not done well in substituting for the extended family in this earliest period, but we surely are just on the brink of exercising our potential as supports for young parents. This energy is mobilized around the new infant. When the parents' needs are rewarded appropriately by responsiveness from the infant, the parents will develop in parallel with the infant. Because of the available energy, disrupted from old pathways in pregnancy—anxiety, if you will—the new parents are as ready to learn about themselves as is the neonate. Just as they learn about each new stage in their development, find the appropriate control system, and experience the excitement of the baby's responsiveness, the father and the mother are forced to learn about themselves. As each new stage in the infant's development presses them to adjust, they learn about the excitement *and the pain* of disruption and the gratification of homeostasis as they hit a plateau. In this way, we see mothers and fathers learning about themselves as developing people while they learn about their new baby. This is also the way in which the fueling for both nurturance and learning comes about at each new stage in the baby's development. Otherwise, nurturing a new infant would be too costly and too painful. In a reciprocal feedback system, the rewards are built in for the parents as well as the infant. The pain can be seen as a preparation for detachment later on when that becomes necessary to the baby's developing autonomy.

3.2. Forces for Failure in the Interaction

The pressures on parents, both internal and external, to succeed with their infant can work for failure. For when the feedback systems are not being completed in an expected way, we have been impressed with the power of violations of expectancy in the mother–infant interaction (Tronick, Als, Adamson, & Brazelton, 1978). The potential for withdrawal from each other and for ensuing failure in the interaction in caring people can best be understood by us in our "still-face condition," in which the mother violates the baby's expectation for interaction in a face-to-face play situation in our laboratory by remaining alert and unresponsive. She is instructed to follow a three-minute play period with a second three minutes in which she sits in front of the baby but remains unresponsive, staring at him with her face perfectly still. When this interactive system is violated by the parent's nonreciprocity, the infant will respond in an expectable manner indicating how powerfully he is affected by the violation of his expectation for playful, interactive responses in this situation.

Social interaction appears to be a rule-governed, goal-oriented system in which both partners actively share from the very beginning. The still face violates the rules of this system by simultaneously conveying contradictory information about one partner's goal or intent. The mother by her entrance and *en face* position is initiating and setting the stage for an interaction, but then her lack of response indicates a disengagement, withdrawal, or violation of the baby's expectancy.

An infant's recognition of the mother's violation of reciprocity in the still-face condition begins very early. Along with Stechler and Latz (1966) and Carpenter (1974), we have seen evidence of it as early as two to three weeks. The pattern described above is clearly established by four weeks and becomes increasingly complex. For instance, a three-month-old infant began reacting to the still face by showing the characteristic wary pattern of behavior. About a minute and a half into the interaction he looked at his mother and laughed briefly. After this brief tense laugh, he paused, looked at her soberly, and then laughed again, loud and long, throwing his head back as he did so. At this point, the mother became unable to maintain an unresponsive still face, broke into laughter, and proceeded to engage in normal interactional behavior. The intentions and emotions of the older infant are similar to those of a younger infant. The richness and skill in reestablishing a reciprocal interaction, however, are greater.

The strategies the infant employs to bring his mother out of her immobility demonstrate his growing confidence in his effectiveness as a social partner; the seriousness of the infant's reaction when the mother remains unresponsive despite his efforts demonstrates how critical reciprocity is to him. The final withdrawing of the young infant when he no longer seeks to pull his mother into the interaction reminds us of the withdrawn behavior and huddled postures of isolated monkeys (Harlow & Zimmerman, 1959) and of Bowlby's (1973) description of the withdrawn behavior of children separated from their caretakers.

The still-face mothers in our study remained unresponsive for only three minutes, yet their infants found even such a temporary violation greatly disturbing. Within 15 seconds the baby had recognized the violation and had begun to try to master it. His efforts included several programs aimed at eliciting her usual responses. This suggests that reciprocity and mutual achievement of the goals of social interaction form a necessary basis for the growth of affective well-being in early infancy.

The ability of the baby to precipitate and encourage the mother's attachment and caretaking behavior must be taken into account from the newborn period (Brazelton, 1961b). With an unresponsive neonate, the feedback mechanisms necessary to fuel mothering behavior are severely impaired. In a series of medicated newborns of normal mother–infant pairs, the effect on neonatal sucking coupled with the physiological effect of the medication on the mother's milk production delayed recovery of weight gain by 36–48 hours in a normal group (Brazelton, 1961a). Since this is an observation at a rather gross level, interferences in the interaction dyssynchrony and more subtle "lack of fit" in the earliest mother–infant attachment interaction should be carefully sought and observed over time.

The vulnerability of the parent to even mildly distorted cues from the infant can best be understood in the light of a "grief reaction" in the parent (adapted from Lindemann, 1944). Because of the heightened expectations at birth, the opportunity for grieving is enhanced by any minor violation of this expectancy. The very energy which has been mobilized to relate to a baby can turn inward into grieving in such cases.

The forces for grieving, as described by Lindeman in his studies of adults who experienced an unexpected loss of a beloved one at the time of the Cocoanut Grove fire, were those of overwhelming despair, self-incrimination, and guilt. Expected mechanisms were those of feeling guilty at not having "cared enough." The self-incrimination and depression were so intolerable that defenses were set up to preserve the adult survivor's emotional integrity. Expected defenses were denial, projection of guilt onto others, and detachment from the loved ones. The feelings and the defenses against them can be seen in all parents of a damaged or sick infant. *Denial* is used to handle the violations of expectation in behavior or in responsiveness. The parent denies that these matter, thereby covering up for herself and for others how deeply affected she really is by the inadequate responses she is receiving. *Projection* is also a common mechanism. The parent projects onto those around her the inadequacy she feels about having produced such a baby and about her inability to care for the baby properly. By projecting these feelings to others, she can tolerate them better in herself. But this very projection makes her less available for interaction with a helping person. *Detachment* is understandable in that the caring mother, who feels she has already damaged the infant, feels also that if she were out of the way or detached the infant might be less at risk. This serves to make her less available to the infant.

Since all of these forces of grief are expectable and are normal, those who want to intervene or help with a pair at risk for interactional failure must be aware of the power of these defenses as they operate to protect the emotional integrity of a caring parent who meets with a violation of expectation in the responses and the development of her infant. One must expect and work with these defenses if one is to capture the nurturant forces in the parent necessary to enhance the infant's recovery.

If the parents' resources are turned inward in the grieving process, they become encapsulated and the defenses strengthened to turn the available energy away from the baby. In the period of acute grieving, this self-protective mechanism may be necessary. At the point where the reorganization of the parents' ego has been accomplished, this energy can be made available to the baby and toward his recovery (Greenberg, 1979). It has constantly surprised us that, even in the face of a devastating diagnosis of retardation in the baby, a parent can have the energy available to search for and work with the baby's more hopeful, positive behavior. With this as a base, parents of babies with Down's syndrome work to achieve an interactive alert state in the baby, and, within this state, they can teach the afflicted infant to achieve remarkable developmental steps. Unless they can be captured to turn their grieving around, to turn its energy outward in the service of the child's best recovery, the chance is good that the parent can remain permanently withdrawn and unavailable to an at-risk child. The job of intervention is to *accept* the negative forces of grieving, but to work to free positive forces for interaction with the child as well. This work can best be done early and by utilizing the best behavior in the child as a demonstration to capture hope and reciprocity in the parents.

The danger of fixation in these defenses and of parents' developing insensitive, inappropriate patterns in dealing with the infant at risk makes it critical that we be ready to intervene early in their development together. Time is of the essence. The emotional availability and flexibility of parents of normal infants can be seen as a sign for capturing this availability to the less-than-expectably normal infant as soon as possible. Before fixation can occur, we must be ready to capture and reinforce the positives in the infant's behavior for the parents, and we must be ready to set up a working relationship to help them with their predictable grief to work around the violations in expectation. This energy becomes a force for the baby's recovery—not only as it fuels the parent, but as it serves as an external source of energy for the infant.

Anyone interested in intervention *must* understand in parents the process and defenses of grief as well as those of attachment. He must also understand the processes of behavioral reactions, their violation, and the potential for recovery in the baby. This is a large order.

4. IMPORTANCE OF ASSESSMENT OF THE INFANT

An ideal assessment for a clinician would have the elements of the past, present, and future which could become the base for an understanding of the infant and a window into understanding what he will do with his environment. An assessment of an infant must be seen as an opportunity for interaction with him, and through the infant's use of the examiner and his materials one might come to the best understanding of how this infant functions. By entering into a relationship and then reacting with him, the examiner is then provided with an opportunity to understand how his parents must react to him.

In other words, any scoring of an assessment of an infant should include the subjective and clinical insights of the examiner if it is to become a base for enhancing the infant's development. For it will be through these reactions that the examiner can enter into a working relationship with both the infant and his parents to further his development. We have learned the power of just such an assessment in our work with the Brazelton Neonatal Behavioral Assessment Scale (Brazelton, 1973). As we work to reach, to understand, and to envelop the newborn in an adult's containing and facilitating interaction, to obtain his "best performance" we have realized the power of such an envelope in providing him with an opportunity to show us his processes—of organization, of mastering his immature physiology and reflex nervous system, of achieving an optimal state of attention for interaction with his environment. As he shows us these processes, we can understand them by our identification with him. Through such an understanding we hope to predict fairly accurately what he will do to those caring adults around him who are also interested in helping him achieve his own best performance. If we are ready to do so, we can share these observations and these insights with his adults to join with them in setting goals for his best developmental outcome. An assessment of an infant is a multidimensional opportunity—for diagnosis, for prediction, and for entering the parent–infant interaction.

We have been struck with the power of modeling our own behavior to energize and teach parents. Because they care so much and the energy of caring makes them available to the baby, they can watch us produce his best performance and can model on our efforts to work to produce this performance themselves. As they produce his performance, the internal feedback systems in the baby couple with and are reinforced by delighted responses in the parents. These two systems (internal and external) signal and reinforce the achievement in the child and he is energized to reproduce this achievement again and again. The experience of completing such a circular process might act as an organizer for the CNS, and real recovery from deficits becomes more likely. That positive experience is a source of energy for the baby's development becomes most apparent in impaired infants.

The first evaluation of the neonate can provide an irreplaceable set of observations which point to his intrauterine experience and its effect on his fetal development. With the indications from animal research of the effects of intrauterine deprivation on the DNA content of the fetal brain and other vital organs, clinical assessment of depletion or underdevelopment in the neonate becomes of vital importance. The plasticity of the infant and his ability to recover from stress without external evidence of it make it more difficult to detect the signs in a viable neonate; there are not many ways to detect past injuries from which he has recovered. Neurological examination of reflex behavior may not detect minor damage to which he has made a functional adaptation in the uterus (Parmelee & Michaelis, 1971). We must look, therefore, for measures of intrauterine development which are sensitive to critical influences on his developing organs.

Signs of prematurity out of proportion to the infant's gestational age may point to a chronically stressed circulation in the fetus and an underdeveloped organism as a result.

An early examination of the neonate should include a detailed assessment of any signs of dysmaturity such as dried or peeling skin (Warkany, Monroe, & Sutherland, 1961). A small-for-date baby is more at risk than a well-nourished one. Evidences of extracellular depletion and speed of recovery may reflect the duration and depth of his deprivation *in utero*. Short-term signs such as skin texture and subcutaneous fat depletion may not be as significant as are long-term effects such as decreased linear bone growth, decreased head circumference, or minor congenital developmental defects; these may point to chronic intrauterine depletion which affected the fetus at critical periods of development.

An estimate of his stage of maturity using such physical signs as size of breast nodule, palm and sole creases, scalp and body hair distribution, earlobe, testes and scrotum (Lubchenco, Hansman, Dressler, & Boyd, 1963), and behavioral signs which measure reflex behavior (Dubowitz & Dubowitz, 1970) can be measured against a careful historical estimate of his gestational age from his mother's last menstrual period. When these physical signs do not coincide, the significance of relative immaturity becomes of predictive importance, suggesting long-term intrauterine deprivation which has affected cellular development in the fetus. A behavioral assessment of the maturation of reflex behavior and of external characteristics of the newborn, as outlined by Dubowitz and Dubowitz, should be coupled with an evaluation of the relation of gestational age to height, weight, and head circumference on normative intrauterine growth charts developed by Lubchenco *et al.* (1963). One can use the ponderal index as measured by Miller and Hassanein (1971), a relationship of weight to height at birth. These relationships are highly correlated with intrauterine growth and well-being, and we have found that they are highly correlated with neonatal behavior in babies who are small for their gestational age (Als, Tronick, Adamson, & Brazelton, 1976; Lester, 1979). The unexpected and hence disturbing violations in behavior in such neonates can easily shape the parent–infant interaction toward a failure system.

We have been working toward a more system-oriented method for presenting and scoring the Brazelton Neonatal Behavioral Assessment Scale as it is applied to at-risk infants (see Chapter 2). With this new method we are coming closer to a cost-effective look at organization in the infant, to see how the different developmental lines interact to create a picture of organization in the fragile premature or sick infant. With such a picture, the examiner can then identify the organizational processes which are already relatively stable and those which are easily overloaded and can more reliably identify the work for parents who must assist such babies toward optimal organization and function.

4.1. Neonatal Assessment

The behavioral assessment of the normal newborn infant presents a dual opportunity to the examiner. He can use the process of bringing the infant through his several states of consciousness to best performance as a measure of the infant's own capacity for social interaction. Through working to help the infant organize and produce his responses he can identify with parental responses. He can then use this experience to demonstrate to the parents what will be necessary for them to reach their baby. By the same token, while working to achieve the baby's best performance, the examiner must identify with the infant and will thus understand thoroughly the process of organization and social interaction in that infant. Thus, an examination fits the examiner to understand and analyze the organizational capacities of the baby in a powerful way and to be ready to understand the demands and limitations on the parents. He can thus help them to learn to interact with their baby for his best development. Since the examination not only identifies motor and CNS deficits in the newborn (Tronick & Brazelton, 1975) but over the 20 minutes produces responses to positive experiences as well, the opportunity for seeing the baby in nurturing situations and organizing around positive experiences gives the examiner an opportunity to visualize the infant as a whole person. In such demonstrations grieving parents are invariably struck with and relieved by the positive behavior of their high-risk or damaged infants. Since the nature of their grieving would otherwise fix their attention on the infants' deficits, such parents grasp hungrily at this evidence of responsiveness and are usually able to balance their perceptions of the children as damaged with the positive social responses which they have witnessed. In this way, the witnessed examination gives the grieving parents an opportunity to identify with the assets as well as the deficits of such a baby. We have seen parents immediately turn their grief reactions around and begin to work toward the baby's recovery. Thus, a shared demonstration of the baby's behavior and a description of the processes underlying it can become a powerful intervention in the neonatal period (Als, Tronick, & Brazelton, 1979).

Since the examination is a dynamic one and is clearly based on the baby's potential for organization with a nurturing interaction, the skills and sensitivity of the examiner are critical for bringing the baby to an optimal performance. The amount of effort required from the adult to produce organization and reactivity may be an important gauge of the infant's developmental status. Of course, examiners must be trained to a reliable level of awareness of the neonate's potential for organization and performance, and the examiners' performance and scoring must be uniform or the data collected will not be of value. The insight into the neonate's capacity for interaction with his environment will suffer as well.

We have been aware of the fact that one examination is not as fruitful as a series would be. Even in carefully controlled conditions there are far too many impinging variables to expect high test–retest reliability in the same neonate and with the same examiner. For example, a circumcision may affect a baby's performance for as much as 12 hours (Emde, Swedberg, & Suzuki, 1975). A blood test such as a PKU or bilirubin may well disrupt him for four to six hours. In addition, the neonatal period is expected to be a time of relative depletion and recovery from the effects of labor and delivery and exposure to his new environment. Hence his performance will be affected by these powerful experiences, and he will be behaving in expectable, relatively predictable ways which depend on these experiences. Each set of behaviors will recover differently over the first few days after delivery. The more basic responses, autonomic or reflex-motor, will be less affected by external factors and will be changing in a line affected primarily by maturation and time. State behavior and interactive behavior (orienting to face, voice, consolability, etc.) will be influenced by experience with the environment and will recover differently. In a study of the cumulative test–retest correlations over the first 10 days of life we found highest correlations for those behaviors that are expected to be most stable such as habituation and autonomous processes and lowest correlations for behaviors that represent newly emerging and rapidly developing processes such as state and interactive behaviors (Lester, 1980). Hence, day-to-day test–retest reliability will be poorer in these behaviors in an *expectable direction*. The recognition of the importance of the effects of the caregiving environment is reinforced by the accumulating evidence from longitudinal studies which show little relationship between early infant behavior and later developmental outcome (Sameroff & Chandler, 1975). We are treating these discontinuities as expectable results of the infant–caregiver system rather than as evidence of the infant as an isolated entity. The view of the newborn as a biobehavioral entity in transaction with his environment accounts for the modifications of his responsiveness over time and gives us a window into how he is utilizing his environment (Lester, 1979). As I have written:

> When the scale was conceptualized it was not intended to provide a 'one shot' assessment of the neonate. Rather it was designed to reflect the neonate's capacity to recover from the stress of events such as labor and delivery and the influences of the new environment. Since I see the neonate as an organism who is recovering from a stressful series of events, I think that the best test of his or her capacity to achieve physiological homeostasis in order to attend to social stimuli. Hence, a 'recovery curve'

> as demonstrated in improving behavioral responses becomes a test of physiological stress as well as of the infant's capacity to organize in spite of it to interact with his or her environment. I still believe that repeated examinations are necessary to evaluate the neonate's organizational adjustments and coping strategies in response to the demands of the new environment. Thus, I expect the maximum predictive validity of the scale in the pattern of recovery reflected in repeated assessments over the first few weeks of life rather than in any one assessment. By plotting each neonate's performance over time one can calculate individual 'recovery curves' which may in turn be used to predict the coping capacities of the neonate.
>
> This recovery curve approach implies that it would be foolish to expect more than moderate test–retest stability of the 27 scale items. Although some consistency in performance can be expected from day to day, it may be more appropriate to think of fluctuations in performance as indicative of how the neonate organizes around changing environmental demands. Moreover, those demands are likely to vary from infant to infant. There are probably individualized styles of recovery as well as coping with equal demands. Therefore, a recovery curve may be more meaningful than stability in performance. Particularly the more dynamic or interactive dimensions of the scale are likely to reflect change.
>
> The neurological reflex items would be least likely to be influenced by stress of recovery or by extrauterine events. Hence, they would be expected to be the most stable. By the same token, state items as well as items influenced by environmental stimuli might be expected to be the least stable in day-to-day test–retest scoring.
>
> In time, after data on many, many babies have been evaluated, perhaps all of us who are using the scale can set up a priori curves of recovery on all items against which individual babies can be measured. In this way, by predicting several expectable curves of recovery, we can see on which items the individual deviates and, thereby, understand the ingredients of his or her unique recovery curve. (Brazelton, 1978, p. 8)

Hence, in analyzing our data we are constructing recovery curves for an analysis of the data from repeated assessments to begin to understand the continuities *and the discontinuities* in the infant's behavioral responses to his new environment. The parents' responses to change over time can be seen as potentially bimodal—taking responsivity and pleasure in those which are improving and feeling less involved and responsible for those which are more static. We feel that by identifying these recovery curves over repeated assessments we can identify which behaviors will predictably reflect parental input and involvement and which will offer maximum predictive validity of the neonate's organization. We are convinced that a profile of recovery curves will identify patterns of individual differences that will discriminate among normal infants and will help identify the infant at risk (Lester, 1979).

The way the parents and the baby handle such discontinuity as their recovery from the stresses of labor and delivery may be one of the best predictors of their future. Bell (1971) and Kagan, Kearsley, and Zelazo (1978) point out the dearth of infant behavior which predicts the infant's future behavior. At a time when we all feel that there is continuity and even predictability from infancy, we are searching for better ways of assessing infants and their environments in order to predict their future competence. Perhaps we have been too simplistic in our search for continuities of behaviors from linear predictions. Emde, Gaensbauer, and Harmon (1976) point out that discontinuities and transformations of behaviors in development are much too powerful to make a linear prediction from behavior to behavior likely. I would go further and say that the way an infant and his parents handle discontinuities in behavioral responsiveness might lead us to better predictions than would the way they handle the continuities in behaviors. Using the discontinuities of neonatal performance to look at the way the neonate handles his recovery from labor and delivery, and his adjustment to his overwhelming new environment may certainly give us the best prediction about his capacities. The assessment of the parents' capacities to adjust to the changes in the neonate becomes a powerful prediction of how they will react to this particular infant. We can learn as much from the discontinuities as we can from more stable behavior.

Since we have had an opportunity to understand recovery mechanisms in the neonatal period through the organizational and interactive capabilities of the neonate uncovered by administering the Brazelton Neonatal Behavioral Assessment Scale, we have been struck with the power of sharing these processes with parents. Sharing the newborn's interactive and state organizational behavior with the parents of normal infants makes a significant difference in their future attachment—both to the infant (Eyler, 1979) and to the caregiver (Als & Brazelton, 1981). With the parents of high-risk infants this kind of sharing may be even more significant.

Our experience in demonstrating the behavior of the high-risk premature as we perform the behavioral assessment in the presence of the parents shows that they watch carefully as we work with the infant. The efforts to contain him, to adapt stimuli to him, to elicit responses which do not exhaust him are not overlooked. The parents observe in complete silence, ask us questions afterward, and often even make their own correct observations, such as, "I never knew he could see and would follow your face, but he can," "He gets exhausted if you do too much," "I see that you need to contain him so he won't get too excited," "You make your voice soft and insistent to get him to turn to you." All of these observations have come from uninstructed parents who then reproduce our techniques with their fragile baby to elicit his attention and to teach him how to maintain his own inner controls. Over time, they not only identify with us as we work with the baby to produce his best performance, but they begin to tell us how the baby functions best. In other words, in the face of an expectable reaction to his fragility and even in the case of identified CNS damage, parents have been able to lock onto the two aspects of these high-risk babies' behavior with which they can work: (1) the need for physiological control in order to maintain homeostasis while producing behavioral responses and (2) the attentional responses, which are often difficult to elicit. The fragile infant's long latency to response, his alternating high or low threshold for receiving stimuli, and then his tendency to overshoot with an unexpectedly total response or to overload from the cost of such a response—all of these make him difficult to work with. When he responds, he does so with responses which often violate a parent's expectancy. In other words, his behavioral responses are so extranormal that they set up the ingredients for a failure in their interaction—both because of the cost of responses to him and because of the grief reaction in the parent which is likely to be engendered by his unexpectedly deviant responses. We feel that if *we* can understand the mechanisms behind the deviant response, the parent can begin to understand them by observing us. And we have come to realize that the most powerful therapy for grieving parents is to set them to work in a appropriately sensitive way with such an infant. As they see him learn about his own inner organization, learn to control his overreactions; as they see him coming to accept and respond to their social stimuli, geared especially to him, they can begin to find techniques for stimulation and for helping him toward recovery that we have not thought of ourselves. We feel that our therapy is that of allowing the parents to see and understand the baby's observable and positive behavior, thereby accepting his negative or deviant responses. By modelling on our techniques for eliciting his best behavior, they can begin to understand him and to see his progress. By harnessing the overloading stimulation and by helping him learn containment for himself and achieve homeostasis, they free him to begin to interact with his environment. The energy for plasticity for recovery comes from two sources: (1) from within, as he learns to achieve control and maintain an alert state; then he can achieve the inner feedback cycle of completing an attentional or motor task which gives him a "sense of competence"; and (2) from without, as his parents get to know him, to understand his need for containment and homeostasis, as well as his need for social and motor stimuli which are appropriately and individually geared to his capacity to utilize them. Of course, this kind of fueling is most appropriate to early infancy and to the early parent–infant interaction. Whether it serves a comparable purpose in older children remains to be tested.

5. FACE-TO-FACE SESSIONS

As the infant's organization becomes differentiated and begins to be regulated, usually in the course of the first month, the next emerging capacity is that of increasing differentiation of his alert state. The infant's social capacities begin to unfold. His ability to communicate becomes increasingly sophisticated. The repertoire of facial expressions and their use, the range and use of vocalizations, cries, gestures, and postures in interaction with a social partner begin to expand. On the basis of well-modulated state organization he can negotiate social interaction with his caregiver (Als, Tronick, & Brazelton, 1979).

We have designed a face-to-face interaction paradigm in the laboratory to serve as an opportunity to maximize the display of social interaction skills between adult and infant (Brazelton, Tronick, Adamson, Als, & Wise, 1975). The infant is seated semi-upright in an infant seat. The mother is instructed to play with the infant without picking him up. The face-to-face situation in the laboratory is a stressful situation in that it provides no functional goals other than a situation for playful interaction. In that way is presses both partners to use skills and resources to engage one another within a mutually satisfying cyclical interchange, and most of all it demands all of the newly emergent capacities of the infant in the first new months of life.

In this situation the infant must (1) control his posture, sitting in an upright position in an infant seat, (2) maintain an alert state, (3) control his motor arousal, (4) control the autonomic demands this engenders, (5) process complex social information, and (6) simultaneously enact his own goals for the interaction. This situation, then, taps the important organization systems available to the very young infant (Brazelton, Tronick, Adamson, Als, & Wise, 1975). The mother, in turn, must be aware of the infant's needs for recovery of a homeostatic balance in such an exciting situation, and her sensitivity to him becomes measurable as we quantify her behavior.

We have found that a three-minute face-to-face session with healthy infants up to five months of age produces a typical sequence of engagement, acceleration of attention by each partner, a period of reciprocal play and cyclical attention, followed by deceleration of attention (Tronick, Als, & Brazelton, 1977). This then has become the target sequence for our analysis of reciprocity. Mothers who can allow for this sequence in their infants, without overloading them, are already aware of and in tune with their infants' stage of development.

This technique has been a powerful window for us into the stage of development of their interactive capacities within this system. We have watched them grow together and have identified four stages of development in the mother–infant interaction over the first four months (see Figure 2) (Als, 1979; Brazelton & Als, 1979). The differences in the amount of maternal input which is necessary to organize and enhance reciprocal interaction in a high-risk infant are easily identifiable in this system.

The most valuable use of the paradigm has been sharing it with parents. As they watch the videotaped interactional sequence of them with their babies, they have an opportunity to observe more objectively their own reactions and skills and to chart the development of these in their infants. They share questions and observations with us as they observe this recorded session. We have found that they use the observation as a powerful opportunity for self-evaluation as well as for noting progress in the infant. In this observational paradigm, they have brought their concerns about their babies' developmental progress and their child-rearing questions. In the case of delayed infants, mothers have identified the positive behaviors and have appeared to cling to them as they work to organize the baby within this social paradigm.

We are convinced that the timing of our observations to meet the critical periods of early adjustment to the baby have made us more useful to the mother in that adjustment. Our capacity to produce the baby's best behavioral responses bonds us to the mother, as she also learns from us how to produce these responses. Within the context of such a relationship, then, she can begin to model her behavior on ours. Her grief over her baby's deficits begins to turn into more productive channels, for instance, adopting our model for organizing him and searching for appropriate ways of interpreting his behavior in order to interact with him. The parents who are able to model on the information gained in our assessments were examples of the powerful energies present in caring parents for shaping their at-risk babies toward organization and eventual recovery of function.

Behavioral assessments were utilized according to the schedule in Table I for both 21 preterm and 20 term infants. The term infants were originally seen as a comparison group for the preterm infants and were followed with all of the same techniques. The mothers, fathers, and infants were involved in the study with the same techniques as the preterm infants. They were enlisted in the perinatal period. After Day 3, the mothers (and, less often, fathers) observed each of the assessments. Although we had originally conceived of this as a study of the developmental processes of these babies, after our experience with the modeling effect

TABLE I. Overview of Assessments by Time Period

Type of assessment	Birth to 1 month postterm	3–5 months postterm	9 & 18 months postterm
Behavioral organization	Neonatal assessment	Face to face interaction	BASIC Kangaroo Box Paradigm
Medical factor	Dubowitz Ponderal Index Obstetrical Complications Scale Perinatal Complications Scale Neurological	Perinatal Complications Scale Pediatric	Perinatal Complications Scale Pediatric
Sociocultural	Parent interview Caldwell Life Events Socioeconomic status	Parent interview Life Events	Parent interview Caldwell Life Events
Psychometric		Bayley	Bayley Uzgiris–Hunt

on parents observing our assessments, we began to realize that to some extent this was indeed an intervention study for all babies in our sample.

Two sample cases will be described to exemplify the ways in which each baby was assessed and to attempt to underline the opportunities for parental interpretation and modeling for our understanding of their babies and how each assessment functioned as an opportunity for communication with the parents and as an intervention.

5.1. Case 1

Alice B. was a full-term, 40-week infant whose birth weight was 3.2 kg, Apgars 9 at 1 minute, 9 at 5 minutes. Her obstetrical, perinatal, and postnatal complication scores were optimal, and there were no events which complicated her hospital course. The mother was an intelligent young woman in her late twenties, interested in the details of our study. She told our researchers (Heidelise Als and Emily Burrows) that she felt very special at being recruited. She was a doctoral student in special education and had taught blind children before her pregnancy, and she had taken courses in prenatal influences on development in the first trimester to avoid damage to Alice. She admitted that the thought of fetal damage had been frightening and that she was reassured to see how well her baby performed on the neonatal assessment. She watched her baby being examined with avidity. When Alice got upset during the exam, the mother took her quickly to put her to breast. After the examination Dr. Als discussed the only unusual aspects, her sensitivity to stimuli with resulting irritability and her variable, expressive facial responses. Her mother responded to her crying by cuddling and breastfeeding the baby, looking down at her to say, "You're not so dumb after all." She thanked the examiners for "letting her see 'her' baby."

At ten days, the examination was done by Dr. Als and Ms. Burrows at home. The parents lived in a quiet, upper-middle-class home rich in stimuli for the baby. The father worked at home and participated in the baby's care and was a quiet but watchful onlooker as we played with his baby. By this time, the infant's irritability had decreased, and she scored very high on orientation items. She was a very appealing baby. Mrs. B. told us she carried Alice in a canvas carrier as she did her housework. Some of her comments seemed to be couched within the question to us of "whether she was doing right."

At one month, Alice was examined while visiting her grandparents. Her mother appeared to be entirely adequate in handling Alice and had few questions for us. The parents were already trying to find a way to balance their caregiving tasks between them, but because of breastfeeding, the mother dominated the caregiving scene. The mother's play was noisy and active; her father was quieter and more observant. Alice was optimal in her performance in all respects on the neonatal assessment, and her mother nodded approvingly as she performed. In addition, she elicited the baby's responses to her voice and face and was already convinced that Alice "knew her." On each visit, our team asked whether there were any problems. None were suggested by mother or grandmother.

In the face-to-face situation at three and five months, Alice and her mother seemed at first to be harmonious as before. However, on analysis of the tapes, we began to realize that the mother was actively dominating the interaction. There were surprisingly few episodes in the three-minute interaction in which Alice was given the lead in their dialogue. Her mother continued to be overwhelmingly active in their interaction at the five-month session. Although Alice handled her mother's active, repeatedly changing bids to play in an amazingly competent way (scoring at optimal in all parameters), we sensed that she was having to manage her reactions to being overwhelmed when she was in the still-face condition. At this point, Alice became significantly more active herself, eliciting her unusually quiet mother repeatedly to join her in play. Not at all daunted by three minutes of an unresponsive mother she continued to elicit her with chortling playful faces and vocalizations. We have seen a side of Alice's ability to interact which tended to be somewhat submerged by her active mother. Although she was quietly responsive under the usual circumstances, she demonstrated her own vigor and resourcefulness in coping when she was given the necessary space of the inactive mother. Since our laboratory routine included showing the film to the mother, we set the stage for the next step.

Our research team felt that Mrs. B. was still actively dominating Alice's responses at nine months, more than other parents whom we were seeing. But Alice was delightful, responsive, and in no way subdued. The parents reported that she would not eat "for them" any longer and wondered whether they must now let Alice feed herself. The mother told us that she had "learned from watching our videotapes how to interact with Alice." In fact, she had returned home to teach her husband what she had learned and they both felt very grateful.

At their return at 18 months, Mrs. B. reported that she had heard from friends who attended a lecture that "Brazelton's group felt that mothers could be too active." She has tried to calm down in her play with Alice. She had also turned more of the care of her over to her quietly gentle father and a babysitter. She had gone back to school and had weaned Alice to a bottle at around seven months.

At 18 months, Alice was already a gentle but strong-minded little girl. She seemed more dependent on her mother. Her mother seemed gentle, softer, less driving. Alice's Bayley scores were excellent in the mental performance scale and average in motor performance. She was rather quiet motorically by now but was very competent when she was presented with test objects on the examination. Her mother seemed somewhat strained, eager for her to perform optimally, and kept questioning the examiner about whether Alice was doing well. She told us that she was already teaching her to communicate in sign language since she could not speak well yet. She asked repeatedly whether she might be expecting too much of

Alice and felt reassured when we suggested that Alice was very competent and that she herself had done an excellent job.

At the close of the 18-month session, we asked Mrs. B. for her evaluation of the study. She was quite honest in her responses, as have been most of our subjects. She told us that she had learned a great deal about herself as she watched the taped face-to-face sessions. She assured us that she had always been happy with Alice but valued specific reassurances. She gave our study the credit for having alerted her at times to being too pushy "without criticizing or even pointing it out to her." In fact, she felt nothing but support from the study and felt it had helped her to be the best mother to Alice that she could be. Her praise of the study as "supportive" was repeated several times. Indeed, the fact that we had not given her advice along the way was interpreted as support. Since she and Alice were both such competent people, our "intervention" could really only be interpreted as supportive and as an opportunity for the mother to see herself as pressuring Alice in a more realistic way than she might have done otherwise. We had seen them as an optimal pair all along, and our belief in their competence certainly was transmitted to them over the course of the study.

5.2. Case 2

Clarissa D. was the 27-week baby of a 31-year-old mother who had had several miscarriages. She had had a sutured cervix in order to carry this baby. When she went into labor at 27 weeks, the sutures were removed and the 950 gm baby was delivered. She survived despite many complications. She was small for gestational age, was in severe fetal distress, Apgars of 5 at 1 minute, 7 at 5 minutes, requiring bagging and masking with oxygen for resuscitation. Severe respiratory distress ensued and for seven weeks, Clarissa was intubated and in oxygen, being given constant artificial respiratory support and numerous manipulations including aminophylline treatment for apnea and bradycardia. Jaundice ensued and on the second day an exchange transfusion was necessary. She received phototherapy for five more days. She had a pulmonary ductus arteriosus (PDA) which was ligated surgically and a sepsis and pneumonia (for which she received antibiotics for 14 days). Despite all of these complications, Clarissa survived and was placed in room air at the age of 35 weeks. At this time, we were able to start following her with our assessment techniques. Her weight was 1.56 kg, length 44 cm, head circumference 31 cm. She still had bronchopulmonary dysplasia and evidence of a CNS hemorrhage which had left her with suspected mild neurological damage.

Her mother had visited her 51 times, her father 49, in that period. When Dr. Als asked them to join the study, Mrs. D. was very happy because she hoped we would find that, as she firmly believed, Clarissa "could do things for which no one seemed to credit her." She quickly added that she had seen a normal full-term baby and realized how different one like that was from Clarissa. She knew it would take a long time, but the fact that the infant had survived and was already much better made her feel that she would get better over time. Mrs. D. was back at work as an editor but planned to take time off when Clarissa came home.

Like the other participants in our study, Clarissa's parents formed a very intense relationship with our research team. One research assistant, Joan Gilman, became a close friend and confidante and her relationship with the parents approached that of a social worker. The confidences which were exchanged and the eagerness with which the parents relied on her and the other members of our team bespoke their intense need for professional backup.

This was true of all the parents in our study. All of them developed close relationships with our workers and looked to them for support and advice at each visit. As a result, we have lost only one participant (out of 43) to follow-up. None of our team members was trained for this kind of dependent relationship, and we did not originally expect to intervene in such a therapeutic fashion. But their availability and supportive approach undoubtedly served a purpose.

Dr. Als and Dr. Lester administered the Brazelton Neonatal Behavioral Assessment every two weeks from 35 to 44 weeks of gestational age. Both mother and father took off from work in order to be present on each occasion. The nurses had taught them how to administer necessary oxygen when Clarissa became too stressed, and they appeared to know when and how to handle her. But when she became stressed and agitated with our stimuli as we tested her, both parents would look discouraged and depressed. On one examination at 36 weeks, Clarissa became aroused, and as her color worsened the exam had to be discontinued. However, the baby brought her right hand to her mouth and calmed herself a bit as Dr. Als held onto her legs. Mrs. D. noticed Clarissa's efforts to calm herself and virtually beamed with delight. As Clarissa briefly watched the examiner's face to follow it once, her mother could hardly contain herself. Despite the extreme fragility of this baby, despite her marginal respiratory status, which caused her to become cyanotic and hyperpneic whenever she was stimulated, despite poor motor tone of her arms and increased tone in her legs (raising the probability of a CNS deficit), despite inadequate state control, which caused her to shoot from a quiet state to an agitated crying state in a very short time, Clarissa's attempts to control herself and her brief transient response to auditory and visual stimuli were beginning and were noted by the examiners. Her parents seemed to be aware of all of these deficits but were much more involved with her attempts to control herself and concerned about the limits of her responses—the narrow threshold for her ability to respond and above which she would be overwhelmed.

At 40 weeks, the first examination after her discharge home, both parents watched the examination and brought many questions and observations to the examiners about how to distinguish hunger from agitation, when to carry her around, when and how to try to help her to control herself, and when to leave her alone. She was quite difficult to reach at this time, consoling *herself* more easily than Dr. Als could. The parents confimed this, calling her a demanding baby whom they just had to "wait to reach." On this visit Mrs. D. told the research team about their trouble with infertility and their determination to have a baby. Clarissa was a "special" child as a result. They were determined to see that Clarissa came along as well as possible now, "no matter what." The mother maintained her breast milk for 12 weeks and was trying hard to feed Clarissa despite all of these odds. The baby's neurological score was 5 out of 14 on this visit, with inadequate reflex behavior throughout, mixed hypotonia and hypertonia, at times gross abnormalities of eye movements, and less than adequate state behavior. Her cry was very high and strained, with a piercing quality. On a recovery curve, she showed little improvement over the five weeks of exams, and we felt she was still potentially worrisome from a neurological standpoint. Her parents were bravely cheerful, and found it difficult to admit that they had real worries about Clarissa's outcome. They reported, however, that the neurologist was worried about Clarissa and possible "brain damage." At each visit they reiterated this kind of statement. They watched the difficulties of the examiners in trying to help her organize herself, for simply being looked at aroused the baby to agitated crying. The efforts of the research team gradually met with a little more success, and by 44 weeks she was less fragile, less easily overloaded, and a bit more reachable for auditory and visual responses—as long as she was carefully swaddled and the stimuli were offered very quietly and slowly.

The parents talked uncomplainingly about the fact that they could never take her out, had not left her for five weeks, watched her almost day and night, and met her frequent, unconsolable crying periods with swaddling, frequent feedings, and all the techniques they had seen us apply. They used our visits as opportunities for support and reorganization for themselves as well as for opportunities to catch up with Clarissa's progress. The mother decided not to go back to work at all, and the father came home each night to relieve her. They supported each other through it all, each praising the other to us, but they openly wished for an extended family which was nearby and could support them.

At the five-month visits, Clarissa's postural performance was worrisome enough that the research team decided to refer her to a nearby United Cerebral Palsy intervention program for evaluation and treatment. With the efforts of a psychiatrically trained social worker who was a member of our team, the parents voiced many questions regarding our referral and indicated their anger and frustration over the lack of definitive answers to their immediate questions about the child's long-term outcome. Nevertheless, they followed through on our recommendation. By nine months, Clarissa's ability to maintain an active playing period was increasing, and her responses to those around her were better; yet she had developed a severe esotropia (crossed eyes) and bilateral vision was questionable. She no longer cried as intensely as at five months, but she was still difficult to regulate in her play and in sleep-wake transitions and was awake and needed to be fed at least every three hours day and night. Her parents looked exhausted, and our team quietly but firmly insisted that they should continue to use the help of an experienced agency. Physiotherapy and a massive program for Clarissa's recovery had been instituted at eight months. Weekly visits with both parents present resulted in slow but sure improvement in Clarissa's motor and neurological status. For her visual problems, glasses had been prescribed and later on an operation for surgical correction of her internal strabismus was performed successfully.

At the pediatric examination at nine months, Dr. Yogman spent a large part of the time reassuring the parents that they were on the right track in pursuing Clarissa's intervention program. The parents insisted that Clarissa did not need "intervention" and that they preferred our kind of "observation" which they felt was "a lot more helpful." Mrs. D. by now was taking a psychology course in child development to understand better how best to help Clarissa, and she was convinced that she learned much from our techniques to help Clarissa to organize herself. She began to derive more confidence from her observations of the therapists and from the play sessions they attended. We praised her resourcefulness and pressed her to continue.

After a long discussion of Clarissa's feeding and sleeping problems with Dr. Yogman, the parents asked openly about her prematurity and the questions they had stored up about brain damage. "Would she completely recover?" was foremost on their minds, but they had not dared to admit it to each other, they said, until we "forced them to." It seemed a relief to be able to talk about it. They quickly added that they were enjoying Clarissa and were aware of all the emerging skills that she showed. Indeed, on her Bayley assessments at nine months she was performing at her age (mental 111, motor 100), although the administration of the exam demanded a low-keyed, patient examiner who was willing to spend twice as long as usual with her. On our two other assessments, Clarissa performed competently despite relative visual and motor handicaps (her mental and motor Bayley exam scores at 3 months were 83 and 83 respectively, whereas those at 5 months were 89 and 106). I quote from Dr. Als's letter to the United Cerebral Palsy Foundation of May 1, 1979:

> Clarissa's father wanted to be involved in the "kangaroo box" episode with her. [The child] showed good gross motor ability, crawling to the box, pulling herself up on it, and also letting herself down again. She understood the demands of the task quickly and attempted to hold the porthole of the transparent box open in order to reach in to retrieve the kangaroo. She refused her father's help despite her own frustration at her inability to get the kangaroo through the porthole. She recovered her composure repeatedly and stayed well focused on the task.
>
> When her father was instructed not to interact with her around the kangaroo box (our experimental condition), Clarissa showed even greater determination to obtain the object; she looked to her father for help occasionally, then got fussy but recovered and tried again. As her efforts were to no avail, she took a crawling tour

through the room as if to give herself time out, showing active self-regulation in the face of frustration. She eventually returned to the box to renew her efforts.

The Bayley Scales and Uzgiris–Hunt Scales confirmed our impression of the baby's competence. Throughout the testing, Clarissa showed continued interest in each new toy presented. She approached each toy with a serious intent expression on her face and examined each by fine finger exploration and by delicate exploration with tongue and lips. She seemed determined to find out about the details of each object. On the rubber whistle doll she discovered that the little air hole at the top would blow a stream of air into her face; she repeated this event several times and took pleasure in it. Presumably because of her poor vision, she held each object close to her right eye, then explored it with her mouth and often went back and forth between these two modes of exploration.

Clarissa's affective facial displays were somewhat narrow. She was predominantly sober and serious, although she did smile repeatedly, e.g., to the bell and the mirror. Her excitement over success with certain toys was expressed motorically with hitting, bouncing, shaking, and waving her arms. Her excitement was also shown in her persistent, determined involvement and in her readiness for new toys.

On the Bayley Scales of Infant Development she achieved an overall Mental Development Index of 111, and a Psychomotor Developmental Index of 100, when her age is corrected for her prematurity. This reflects quite well her increasing organizational expansion and the use she is able to make of her very supportive and facilitative home environment.

Upon the return visit at 18 months, we were amazed at the continued recovery of competence in this baby. She was still a potentially disorganized child, but she appeared to know her own capabilities and could defend herself from "falling apart." She now slept through the night, fed herself, and was talking in phrases. She played by herself creatively. Her mother talked at some length about the major problem she had had of letting Clarissa learn to play alone and to ignore her constant demands. At night she had first let Clarissa cry for a while and then, to her surprise, Clarissa had begun to sleep through the night. In the daytime, she had found that Clarissa could be independent and resourceful if her initial whines were ignored. This was hard for Mrs. D., but she had seen at the center that Clarissa could be more outgoing and independent than she had realized. Also, she had noticed here in the nine-month examination that Clarissa had clung to and teased her parents more than they had liked. Dr. Yogman had also helped her by his advice about night waking. (He had not specifically advised her but had allowed her to ventilate her concerns, and she had taken it as advice.) Through all these difficult months, the parents had turned to their religion for support, but they insisted that our program had given them "what they needed to see Clarissa through." They had continued with the United Cerebral Palsy program and found it supportive and very helpful. The child's vision had improved markedly with the aid of her glasses. She spoke now in three- and four-word phrases and her receptive language was entirely adequate. When she was examined physically, she had an extended temper tantrum which ended when her mother picked her up to comfort her. On the pediatric exam, tone and reflexes in her lower extremities were only slightly increased, as was her sensitivity to auditory stimuli. She walked with a slightly wide base. Her eyes still drifted inward at times despite the glasses, but her vision appeared to be entirely adequate.

On the Bayley exam she performed slightly above her age (mental 107, motor 113) with above-average scores on energy level and coordination of fine and gross motor skills, and no deviant behaviors.

Her parents described her as "fun, talking all the time, and rewarding." Indeed, she was delightful, determinedly stubborn, and charming in a social situation. When she began to fail at a task, she seemed to sense that it was expected of her, and she kept at it, repeating it over and over until she completed it. When she finally failed at a task, she quickly looked to her mother or father for support, as if failure could be very disappointing for her. We felt that her parents' determination to help this child recover was now reflected in her determination to succeed.

The parents were grateful for our study but felt that we had not "told them enough." Although they had no real concern about her recovery, they would have liked to have been kept better apprised of each step in that recovery and what to expect. After all, they said, they were both scientists and could have been told more. Despite our assurance that neither we nor the United Cerebral Palsy program could have disclosed any more than we did and that none of us could have predicted such recovery, they seemed to feel that they had been struggling in a rather lonely way. But after this statement they began to recount in detail their memories of each of our examinations and how much they'd learned from each one. "The hospital team and their exams, as well as those at United Cerebral Palsy, made parents feel defensive, whereas watching Clarissa perform on behavioral exams was rewarding. You could see and feel that she'd learned from one time to another." They wished our study could have gone on longer.

6. SUMMARY

The opportunity in early infancy for joining the developmental processes of the infant and the energy in the parent for bringing the infant to his best potential has been out of proportion to our expectations. By visualizing the infant's capacity to organize around the positive experiences of interaction with a nurturing adult in our neonatal assessments, we have better understood normal recovery processes from labor and delivery and plasticity for recovery from CNS insults in the neonate. As we observed and identified with them to produce best behavior in the infant, the parents were able to believe in our methods and to work toward their infants' optimal recovery. Hence, our assessments of premature and normal infants became a window for us and for parents into organization and ongoing development. In later infancy, the face-to-face procedure served a similar purpose.

In other words, within the observational assessment of a small baby is contained not only the organizational capacities of his present status, but also an opportunity to see what he will do to his parents and what they will have to do to organize him. By sharing this observation with them, we share these processes and give them an opportunity to identify his positive potential in addition to his deficits. As such, they can utilize the forces for nurturance in their relationship to produce his optimal recovery. Serial observations over time provide an opportunity to observe the baby's capacity to organize himself as he recovers from a known set of experiences surrounding labor, delivery, and his new environment. In addition, they provide insight into his capacity to utilize the nurturant stimuli from his environment. Further, these observations over the first year offer us a chance to share with the parents the baby's progress and emphasize their effect on it. By repeated observations shared with parents, we have seen their powerful effects as interventions. They have fueled the parent's perceptions of the infants as organizable and as potentially recoverable even after CNS insults or prematurity. We are impressed with the potential energy for recovery and plasticity after identified damage in both the infant and the parent. The timing in early infancy and the sharing quality of the intervention seem to be important aspects of this potential.

Acknowledgments

The work for this chapter was done in close collaboration with Heidelise Als, Barry Lester, and Michael Yogman. As a result, they are indeed valued associates and their intuitions and artistry have been a delight to work with. The invaluable assistance of Ann Berger, Emily Burrows, Joan Gilman, Carey Halsey, Sylvia Howe, Nancy Kozak, Cassie Landers, Anna Lombroso, Kate Neff, and Carol Sepkoski made this study possible in the first place. Their sensitivity to our study subjects, their ability to make the parents feel partners in the study, and their hard work in organizing the new instruments have given us the base for the assertions made in this chapter. Last of all, Karin Madrid has made all of this a joy, thanks to her constant help and industry.

7. REFERENCES

Als, H. Social interaction: Dynamic matrix for developing behavioral organization. In I. C. Uzgiris (Ed.), *Social interaction and communication in infancy. New Directions for child development* (Vol 4). San Francisco, Calif.: Josse Bass, 1979, pp. 21–41.

Als, H., & Brazelton, T. B. Assessment of behavioral organization in a preterm and a full term infant. *Journal of the American Academy of Child Psychiatry,* 1981, *20,* 239–263.

Als, H., Tronick, E., Adamson, L., & Brazelton, T. B. The behavior of the full term but underweight newborn infant. *Developmental Medicine and Child Neurology,* 1976, *18*(5), 590–602.

Als, H., Lester, B. M., & Brazelton, T. B. Dynamics of the behavioral organization of the premature infant. In T. M. Field, A. M. Sostek, S. Goldberg, & H. H. Shuman (Eds.), *Infants born at risk.* New York: Spectrum, 1979, pp. 173–192.

Als, H., Tronick, E., & Brazelton, T. B. Analysis of face-to-face interaction in infant–adult dyads. In S. J. Suomi & G. R. Stephenson (Eds.), *Social interactional analysis: Methodological issues.* Madison: University of Wisconsin Press, 1979, pp. 33–76.

Als, H., Tronick, E., & Brazelton, T. B. Affective reciprocity and the development of autonomy: The study of a blind infant. *Journal of the American Academy of Child Psychiatry,* 1980, *19*, 22–40.

Bell, R. Stimulus control of parent or caretaker behavior by offspring. *Developmental Psychology,* 1971, *1*(4), 63–72.

Bibring, G. L., Dwyer, T. F., & Valenstein, A. F. A study of the psychological processes in pregnancy and of the earliest mother–child relationship. *The Psychoanalytic Study of the Child,* 1961, *16*, 9–72.

Bower, T. G. R. Visual world of infants. *Scientific American,* 1966, *215*, 80–92.

Bowlby, J. *Attachment and loss, Vol I: Attachment.* New York: Basic Books, 1969.

Bowlby, J. *Attachment and loss, Vol II: Separation.* New York: Basic Books, 1973.

Brazelton, T. B. Psychophysiological reactions in the neonate. No. 1: The value of observation of the newborn. *Journal of Pediatrics,* 1961, *58*, 508. (a)

Brazelton, T. B. Psychophysiological reactions in the neonate. No. 2: Effect of maternal medication. *Journal of Pediatrics,* 1961, *58*, 513. (b)

Brazelton, T. B. The early mother–infant adjustment. *Pediatrics,* 1963, *32*, 931.

Brazelton, T. B. *Neonatal Behavioral Assessment Scale.* Spastics International Medical Publications, Monograph #50. London: William Heinemann; Philadelphia: J. B. Lippincott, 1973.

Brazelton, T. B. Introduction. In A. Sameroff (Ed.), Organization and stability of newborn behavior: A commentary on the Brazelton Neonatal Behavioral Assessment Scale. *Monographs of the Society for Research in Child Development,* 1978, *43*, 14–29.

Brazelton, T. B., & Als, H. Four early stages in the development of mother–infant interaction. Presented as the Helen Ross Lecture, Chicago Psychoanalytic Society, Chicago, April 1978.

Brazelton, T. B., & Als, H. Four early stages in the development of mother–infant interaction. *The Psychoanalytic Study of the Child,* 1979, *34*, 349–369.

Brazelton, T. B., Tronick, E., Adamson, L., Als, H., & Wise, S. Early mother–infant reciprocity. In M. A. Hofer (Ed.), *Parent–infant interaction.* London: Ciba, 1975, pp. 137–154.

Carpenter, G. C. Visual regard of moving and stationary faces in early infancy. *Merrill–Palmer Quarterly,* 1974, *20*, 181–195.

Condon, W. S., & Sander, L. W. Neonate movement is synchronized with adult speech. *Science,* 1974, *183*, 99–101.

Cravioto, J., Delcardie, E., & Birch, H. G. Nutrition, growth and neurointegrative development. *Pediatrics Supplement,* 1966, *38*, 319.

Drillien, C. M. *The growth and development of the prematurely born infant.* Baltimore: Williams & Wilkins, 1964.

Dubowitz, L. M., & Dubowitz, V. Clinical assessment of gestational age in the newborn infant. *Journal of Pediatrics,* 1970, *77*, 1.

Emde, R., Swedberg, J., & Suzuki, B. Human wakefulness and biological rhythms during the first postnatal hours. *Archives of General Psychiatry,* 1975, *35*, 780–783.

Emde, R., Gaensbauer, T. J., & Harmon, R. J. Emotional expression in infancy. *Psychological Issues X: Monograph 37,* 1976.

Eyler, F. *Demonstration of premature infants' capabilities to improve maternal attitude and facilitate mother–infant interaction.* Doctoral dissertation, University of Florida, 1979.

Fraiberg, S. *Insights from the blind.* New York: Basic Books, 1977.

Greenacre, P. On focal symbiosis. In L. Jessner & E. Pavenstedt (Eds.), *Dynamic psychopathology in childhood.* New York: Grune & Stratton, 1959, pp. 243–256.

Greenberg, D. *Parental reactions to an infant with a birth defect.* Doctoral dissertation, Smith College, School of Social Work, 1979.

Greenberg, N. H. A comparison of infant–mother interactional behavior in infants with atypical behavior and normal infants. In J. Hellmuth (Ed.), *Exceptional infant: Studies in abnormalities* (Vol. II). New York: Brunner/Mazel, 1971, p. 390.

Harlow, H. F. Love in infant monkeys. *Scientific American,* 1959, *200*, 68–74.

Harlow, H. F., & Zimmerman, R. R. Affectional responses in infant monkeys. *Science,* 1959, *130*, 421–432.

Heider, G. M. Vulnerability in infants and young children. *Genetic Psychology Monographs,* 1966, *73*, 1.

Kagan, J., Kearsley, R., & Zelazo, P. *Infancy: Its place in human development.* Cambridge, Mass.: Harvard University Press, 1978.

Klaus, M. H., & Kennell, J. H. Mothers separated from their newborn infants. *Pediatric Clinics of North America,* 1970, *17*, 1015.

Klein, R. E., Habicht, J. P., & Yarbrough, C. Effect of protein calorie malnutrition on mental development. *Incap Publication,* 1971, No. 1, p. 571.

Lester, B. M. *Assessment of the behavioral organization of the neonate: Research and clinical perspectives.* Paper presented as part of the Joint Colloquium Series between the School of Education and the Program of Developmental Psychology, University of Michigan, Ann Arbor, November 6, 1979.

Lester, B. M. Issues in neonatal behavioral assessment. In E. Sell (Ed.), *Follow-up of the high risk newborn.* Springfield, Ill.: Charles C Thomas, 1980, pp. 291–295.

Lindemann, E. Grief. *American Journal of Psychiatry,* 1944, *101*, 141.

Lubchenco, L. O., Hansman, C., Dressler, M., & Boyd, E. Intrauterine growth estimated from liveborn, birthweight data at 24 to 42 weeks of gestation. *Pediatrics,* 1963, *32*, 793.

Meltzoff, A. N., & Moore, M. K. Imitation of facial and manual gestures by human neonates. *Science,* 1977, *198*, 75–78.

Miller, H. C., & Hassanien, K. Diagnosis of impaired fetal growth in newborn infants. *Pediatrics,* 1971, *43*, 511–515.

Neligan, G. A., Kolvin, I., Scott, D., & Garside, R. F. *Born too soon or born too small.* Spastics International Medical Publications. London: William Heinemann, 1976.

Parmelee, A. J., & Michaelis, R. Neurological examination of the newborn. In J. Hellmuth (Ed.), *Exceptional infant: Studies in abnormalities* (Vol. II). New York: Brunner/Mazel, 1971.

Provence, S., & Lipton, R. C. *Infants in institutions.* New York: International Universities Press, 1962.

Sameroff, A. J., & Chandler, M. Reproductive risk and the continuum of caretaking casualty. In F. D. Horowitz, M. Hetherington, S. Scarr-Salapatek, & G. Siegel (Eds.), *Review of child development research* (Vol. 4). Chicago: University Press, 1975, pp. 187–244.

Sander, L. W. The regulation of exchange in the infant–caregiver system and some aspects of the context–content relationship. In M. Lewis & L. A. Rosenblum (Eds.), *Interaction, conversation, and the development of language.* New York: Wiley, 1977, pp. 137–156.

Sigman, M., & Parmelee, A. H. Longitudinal evaluation of the preterm infant. In T. M. Field, A. M. Sostek, S. Goldberg, & H. H. Shuman (Eds.), *Infants born at risk.* New York: Spectrum, 1979, pp. 193–219.

Spitz, R. Hospitalization: An inquiry into the genesis of psychiatric conditions in early childhood. *The Psychoanalytic Study of the Child,* 1945, *1*, 53–74.

Stechler, G., & Latz, M. A. Some observations on attention and arousal in the human neonate. *American Academy of Child Psychiatry,* 1966, *5*, 517–525.

Tronick, E., & Brazelton, T. B. Clinical uses of the Brazelton Neonatal Behavioral Assessment. In B. Z. Friedlander, G. M. Sterritt, & G. E. Kirk (Eds.), *Exceptional infant: Assessment and intervention* (Vol III). New York: Brunner/Mazel, 1975, p. 137.

Tronick, E., Als, H., & Brazelton, T. B. Mutuality in mother–infant interaction. *Journal of Communication,* 1977, *7,* 74–79.

Tronick, E., Als, H., Adamson, L., & Brazelton, T. B. The infant's response to entrapment between contradictory messages in face-to-face interaction. *Journal of American Academy of Child Psychiatry,* 1978, *17,* 1–13.

Warkany, J., Monroe, B., & Sutherland, B. Intrauterine growth retardation. *American Journal of Diseases of the Child,* 1961, *102,* 127.

White, R. W. Motivation reconsidered: The concept of competence. *Psychological Review,* 1959, *66,* 297–333.

2

Toward a Research Instrument for the Assessment of Preterm Infants' Behavior (APIB)

HEIDELISE ALS, BARRY M. LESTER,

EDWARD Z. TRONICK, AND T. BERRY BRAZELTON

1. INTRODUCTION

Recent advances in the care of premature infants have led to rapid increases in survival rates. This in turn has led to a growing demand for intervention at earlier and earlier stages aimed at preventing developmental dysfunction and supporting optimal development. The demand for support and intervention necessitates early assessment procedures on which to base appropriate intervention and with which to assess individual progress. This demand, as Thoman and Becker (1979) point out, reflects a profound change in the assumptions about early infancy, namely that assessment and support at this early stage of life can effect significant changes

Heidelise Als, Ph.D. • Child Development Unit, Children's Hospital Medical Center, Boston, and Harvard Medical School, 333 Longwood Avenue, Boston, Massachusetts 02115. **Barry M. Lester, Ph.D.** • Child Development Unit, Children's Hospital Medical Center, Boston, and Harvard Medical School, 333 Longwood Avenue, Boston, Massachusetts 02115. **Edward Z. Tronick, Ph.D.** • Department of Psychology, University of Massachusetts, Amherst, Massachusetts 01002. **T. Berry Brazelton, M.D.** • Child Development Unit, Children's Hospital Medical Center, Boston, and Harvard Medical School, 333 Longwood Avenue, Boston, Massachusetts 02115. This work was supported by grant #3122 from the Grant Foundation, New York, Grant #HD 10899 from NICHD, and Contract #278-78-0558 from NIMH. Parts of this work were carried out at the facilities of the Mental Retardation Research Center, Children's Hospital Medical Center, Boston, Massachusetts.

in the developmental course of the infant. This assumption is confronted at present with a remarkable lack of predictive success associated with many of the present assessment procedures (see McCall's review, 1976; Lewis's review, 1973). As one might expect, prediction at the extreme low end of the continuum is most reliable (Honzik, 1976); yet even with complex statistical procedures applied to comprehensive and cumulative batteries of assessments the diagnostic measures at best have been able to account for one tenth of the variance in the major outcome measures at two years, as, for instance, reported in the UCLA Infant Project (Sigman & Parmelee, 1979). Sigman and Parmelee conclude from their extensive study of preterm infants that the "nature of the outcome measurements should be broadened to include social, motivational, and personal qualities of the infant as well as more stable intellectual assessment" (p. 215), and they conclude that predictions which do not take into account the ongoing transactions between child and environment are bound to be weak, since early diagnosis is complicated by the responsiveness of the environment and the adaptability of the human infant.

It appears that Sroufe and his associates (1979; Arend, Gove, & Sroufe, 1979) recently have made headway concerning older infants in just that direction. They are stressing the fact that although qualitative advances in developmental level and dramatic changes in the behavioral repertoire occur during the early years, they can demonstrate good continuity in their studies of infants from 18 months to 4 and 5 years, when they use broadly based assessments of competence with respect to salient developmental issues. In order to develop assessments of competence, they have drawn on the theoretical constructs of *ego resiliency,* defined as the ability to respond flexibly, persistently, and resourcefully, especially in problem situations; *ego brittleness,* defined as inflexibility, an inability to respond to changing requirements of a situation and a tendency to become disorganized in the face of novelty or stress; and *ego control,* referring to "the disposition or threshold of an individual with regard to the expression or containment of impulses, feelings, and desires (Block & Block, 1977, p. 2). Thus, when broadly conceived constructs reflecting developmentally appropriate organizational issues are assessed, a picture of continuity in the quality of the infant's and young child's adaptation emerges. Arend *et al.* (1979) state:

> The importance of the results is enhanced because theoretical derivation of outcome measures permitted a priori statements concerning expected relationships. This precluded the possibility of capitalizing on chance relationships which exist in multivariate exploratory studies.

The recognition of the need to develop measures of such theoretical and, needless to say, complex constructs as competence is also reflected in the work of L. Yarrow and his group (Harmon, 1979; Jennings, Harmon, Morgan, Gaiter, & Yarrow, 1979; Morgan, Harmon, Gaiter, Jennings, Gist, & Yarrow, 1977). Our own work has focused on the development of measures of competence in the newborn and very young infant. On the basis of conceptualizations of the Brazelton Neonatal Behavioral Assessment Scale (1973), we have sought to address the issues of understanding and assessing a prematurely born newborn's competence and adaptations. We concur with the implications of Sroufe's and Yarrow's work that comprehensive constructs of competence may allow us to study developmental processes in the individual organism. What constitutes being a competent newborn—being in control of his own intraorganismic negotiations and environmental interactions? What are the behavioral analogues of *ego resilience* and *ego strength* in the newborn? How can we assess competence, and how can we foster it? And, if we are successful, can we begin to see continuity of development? Clinicians and parents have always known that it exists. Researchers have fallen short of being able to articulate and measure wherein the process of continuity lies.

We would like first to sketch some theoretical notions about development and certain ideas about assessment aimed at possibly bringing us closer to a clinical *and* quantifiable understanding of the individual. We would then like to describe a new research instrument based on the Brazelton Neonatal Behavioral Assessment Scale (1973), constructed to assess issues of competence and behavioral organization in the preterm and full-term organism.

2. HEURISTIC PRINCIPLES FOR A SYNACTIVE THEORY OF DEVELOPMENT

Many disciplines concentrate on the developmental process from many different perspectives and at different levels of functioning of the organism. All presumably are in some way reconcilable, since they derive their constructs from the study and observation of the same phenomenon; and all, no doubt, are important to consider in formulating a meaningful working framework for understanding the developing individual human organism. We would like to present several principles of development gathered from various disciplines concerned with development which have been helpful to our current thinking and have shaped our empirical investigation (Als, in press a,b).

2.1. Principle of Phylogenetic and Ontogenetic Adaptedness

The first principle concerns the individual organism as a member of a species in evolution. From the ethologist's construct of adaptedness, as applied to humans by Blurton Jones (1972, 1974, 1976), Hinde (1967, 1970), and others, we derive the principle that each organism at any stage of his development is evolved to implement not only a species-appropriate but also a species-parsimonious level of adaptedness to its particular adaptive niche. The process of selection takes place at the level of behavior, and over generations many essentials of a species repertoire become "hardwired," as experimental animal studies are showing. The classical single cell studies of Lettvin, Maturana, McCulloch, and Pitts (1959), for instance, have shown that a bug crossing a frog's visual field triggers a leaping and accurate catching response in the frog. More recently, Duffy, Mower, and Burchfiel (1978) have shown that cells in the cat's visual cortex will actively select normal environmental inputs and suppress firing to the same input if it is intermittently and randomly rotated via a prism. Along similar lines of investigation, Spinelli and Jensen (1979) augmented the normal visual input to developing kittens by specialized complex, unusual, and important (avoidance paradigms) inputs and were able to show that a significant number of visual cortex cells had become able to fire to these highly specialized, unusual inputs. Winter and Funkenstein (1973) extended this work to the auditory cortex and have shown that certain cells in the monkey's auditory cortex seek and need for proper firing the full monkey social call; if only partial calls are emitted, the cells remain silent. Freedman (1964, 1974, 1979) has approached the question from the behavioral side and has shown a functional preprogrammed selectiveness for human facial configurations in human newborn infants.

These studies suggest that in the course of a species' evolution a surprisingly specific and selectively operative organism–environment fit is continuously ensured at the level of the organism's central nervous system. The more primitive and simple the nervous system of the organism, the more behavioral configurations or sequences are likely to be hard-wired on a simpler level; the more complex the organism and the more complex and flexible the behavioral interconnections which are necessary to ensure species survival, the larger the association cortex is in comparison to primary sensory cortex and the more likely we are to find "soft wiring," that is, programmed but hugely complex propensities for actualization imbued with much energy reserve and resilience and backed up by a system of multiple checks and balances, such as buffering plasticity and organismic *Spielraum,* or idling space, augmented in a step wise fashion by the organismic availability of active coping strategies, by correction strategies, by holding patterns, by protective strategies, and eventually by a series of distortions in the service of cost-effective defense of the organism. Focusing on man at a species level makes apparent the extent of built-in relative flexibility.

The theory that human species-specific adaptation lies in the capacity to have material culture (Holloway, 1969; Vygotsky, 1978), that is, to see in a natural form an artificial form, transform it, and put it in the service of man, presupposes complexly integrated social, emotional, and cognitive functioning of its members and provides a compelling gating device to examine the adaptedness of human behavior from its very beginning.

Ethological studies of the human fetus and newborn in interaction with the caregiver have identified, from the prenatal period on, the complexity and subtlety of fine tuning of interactions. Humans are capable of more complex behavior than the ventro-ventral primate *Tragling* (being carried) configuration which Hassenstein (1973) describes; immediately after birth the connections between the newborn's attention and the parent's affection behavior demonstrate a homeostatic regulation not observed in other species (Als, 1975, 1977; Grossman, 1978; Minde, Morton, Manning, & Hines, 1980; Robson, 1967). These behaviors appear to function as mutual releasers launching both organisms on their path of complex affective and cognitive exchange and feedback, fueling mutual competence well beyond mere discomfort removal, caretaking, and feeding of the infant. The ethologist's principle of adaptedness thus sharpens our observations and tells us to take seriously all of the organism's behavior at each stage in the focus of his species's adaptation; and it further identifies the human newborn as a biologically social and active partner in a feedback system with the caregiver, eliciting and seeking that physiological, motoric, state, and attentional interactive organization from the environment that he himself needs in order to progress on his own course of self-actualization.

2.2. Principle of Continuous Organism–Environment Transaction

A second principle of development, from embryology, speaks of continuous organism–environment transaction. The key characteristic of the central nervous system, as Palay (1979) says, is to differentiate and develop by interaction with its environment. This is so from its unicellular stage on and continues to be its characteristic. The primitive neural streak differentiates into the neural groove, which by interaction with and incorporation of its immediate environment closes to become the neural tube, the most central part of the central nervous system. Not only the nerve cells but also the environment play an active role in this process. Just how complex this interaction may be is suggested by the work of Patterson, Potter, and Furshpan (1978), who have been able to show that young neural crest cells determine during their embryonic development what they will become, that is, whether they will secrete norephinephreine or acetylcholine, depending on the presence of the cells or medium near which they develop (see also Bunge, Johnson, & Ross, 1978; Landis, 1980; Patterson, 1979). The importance of organism–environment interaction has long been recognized by motor system physiologists such as Sherrington (1940), who speaks of the motor individual as driven from two sources, its own inner world and the world around it. The Bobaths (Bobath & Bobath, 1956; B. Bobath, 1967, 1971; K. Bobath, 1971), Milani-Comparetti and Gidoni (1967), Ayres (1973), and others have made this principle the basis for their clinical assessment and patterning schedules for the prevention and remediation of motor disorganization. Campbell and Wilhelm (in press) discuss various theories of motor neural recovery and conclude that neural mechanisms appear to operate differentially throughout the life span of the organism and in the presence of various environmental conditions. The developmental psychologist is most familiar with this basic principle of organism–environment transaction as it has been employed by Fowler and Swenson (1979), Hunt (1961), Piaget (1963), and others to specify the process of normal cognitive development. Sander has shown its applicability to various aspects of the full-term infant's functioning (Sander, 1975; Sander, Julia, Stechler, & Burns, 1972; Sander, Stechler, Burns, & Julia, 1970; Sander, Stechler, Burns, & Lee, 1979). A one-to-one caretaking situation during the first 30 days of life enhances the development of day–night cycles of sleep and activity and improves visual performance, in contrast to a regular nursery's practices. Horton, Lubchenco, and Gordon (1952) had found earlier that certain premature infants on a demand schedule with feeding intervals often of 5 hours and sometimes of 15 hours gained weight appropriately, in contrast to popular belief of the importance of routine feeding schedules for all premature infants. Korner's work has identified environmental ingredients of soothing and improved attention in full-term infants and has documented the fact that premature infants, if placed on oscillating waterbeds in their incubators, showed lowered rates of apneic spells and higher weight gain (Korner, 1979; Korner & Grobstein, 1966; Korner & Thoman, 1972; Korner, Kraemer, Haffner, & Cosper, 1975). Masi (1979) reviews several other studies suggesting various beneficial effects of positive input to the newborn organism. All this speaks to the continuous organism–environment transaction and its impact on the organism's optimal development.

2.3. Orthogenetic and Syncretic Principle

A third principle, the orthogenetic and syncretic principle, of organismic psychologists such as Werner (1948, 1957) extends the principles of cellular embryology. It has been applied more specifically to the embryology of behavior by, among others, Bruner (1965, 1968; Connolly & Bruner, 1974), Coghill (1929), Gesell (1945; Gesell & Armatruda, 1945), Hooker (1936, 1942), Humphrey (1968, 1970), and McGraw (1945). Wherever development occurs, it proceeds from a state of relative globality to a state of increasing differentiation, articulation, and hierarchic integration (Dabrowski & Piechowski, 1977; Dawkins, 1976). Qualities of functioning which are fused and apparently unifocal, as Herzog (1979) calls it, at one developmental level become discrete, multifocal, and differentiated at another.

Application of these principles to the study of the human newborn has led Sander (1962, 1964, 1970, 1976, 1980) to identify the interplay of various subsystems of functioning within the organism, such as the physiological system, motor activity, and state organization. Sander has identified the task of the newborn as synchronizing these three systems and has elucidated the impact of the caregiving environment in bringing about this synchronization. He describes the full-term infant (Sander, 1975, 1980; Sander, Stechler, Burns, & Lee, 1979) as a composite of semi-independent physiological subsystems. Each has its own periodicity, such as those of electric brain activity, heart beat, respiration activity, feeding, and sleeping. The adaptive tasks for the infant are to achieve phase synchrony between periodicities which characterize these different functions, as well as synchronization between internal events and events of the environment. He postulates that phase synchronization of internal subsystems is facilitated when the output from one subsystem occurs at the approximate time a shift in the cycle of another subsystem is occurring. Endogenous-exogenous phase synchronization or entrainment similarly, he postulates, is most effective when the exogenous event co-occurs with a shift in the endogenous cycle. This may apply at least partially; yet there may well be a weighting toward relative stability and integration, on the background of which shift becomes possible. This obviously remains to be explored more fully. Foremost, Sanders stresses the importance of the appropriateness of timing. Inputs that are poorly timed penetrate and disrupt all subsystems, while appropriately timed inputs maintain and enhance functional integration and support growth. The task, then, is the identification of synchronous and cohesive functioning and the identification of thresholds of disruption and opening of relative coherence which may be necessary in order for a shift to occur.

2.4. Principle of Dual Antagonist Integration

A principle of development from the neurophysiology of the motor system may be of help to us here. According to Denny-Brown (1962, 1966), the organism always strives for smoothness of integration. Underlying this striving is the tension between two basic antagonists of behavior as seen in movement disorders such as athetosis, two basic physiological types of response: the exploratory and the avoiding; the toward and the away; approaching, or reaching out, and withdrawing, or defending. The two at times are released together and in conflict with one another. If a threshold of organization-appropriate stimulation is passed, one may abruptly switch into the other.

How basic these two poles of motor behavior are to the organism's functioning was demonstrated by Duffy and Burchfiel (1971), who have identified single cells in the somatosensory cortex of the rhesus monkey programmed to produce upon stimulation total body-toward movements, while other single cells produce total body-avoidance movements.

The work of Twitchell (1963, 1965a,b,c; Twitchell & Ehrenreich, 1962) has shown the applicability of this dual antagonist integration theory to the study of the grasp and of reaching behavior in the human infant. Humphrey (1968, 1970) has carefully delineated aspects of the gradual differentiation of the excitation and inhibition components of face, trunk, and extremities to various tactile stimuli from the early fetal stage on. But not only is there a progression in differentiation, there is also an observable passing of the threshold from approach to avoidance at each stage. Schneirla and Rosenblatt (Schneirla, 1959, 1965; Schneirla & Rosenblatt, 1961, 1963; Rosenblatt, 1976) have discussed this principle on another level as operative in the process of the gradual specialization of central arousal processes leading to functionally adaptive action patterns such as suckling, nipple-grasping, huddling, and others, in altricial mammals. We feel, as their work also implies, that this principle of dual antagonist integration has applicability as a more general biological principle, and we think it is productive in investigating the behavioral patterns of the very young infant in order to assess the level of threshold from integration to stress. In the integrated performance, the two antagonists of toward and away modulate each other in bringing about an adaptive response. If an input is compelling to the organism and matches his interest and internal readiness, hitting on an open valence circuitry, as it were, the organism will approach the input, react and interact with it, seek it out, and become sensitized to it. If the input overloads the organism's circuitry, it will defend itself, actively avoid, and withdraw. Both responses are modulated by each other. For instance, a newborn is drawn to the animated face of the interacting caregiver. His attention intensifies, his eyes widen, eyebrows rise, and mouth shapes toward the interactor (Als, 1975, 1977). If the dampening mechanisms of this intensity are not established, as in the immature organism, the whole head may move forward, arms and legs may thrust toward the interactor, and fingers and toes will extend toward him. The response, which gradually is largely confined to the face, will early on involve the total body in an undifferentiated way (see Figure 9, p. 186 in Als, Lester, & Brazelton, 1979). The return of the excitation of the response to the baseline may be initiated by the organism through such homeostatic behavioral regulators as averting the eyes, yawning, sneezing, or hiccoughing, or it may be initiated by the caregiver through kissing, nuzzling, or moving the infant closer, thus resetting the cycle (Als, 1975, 1977). If neither of these regulation mechanisms is brought into play, or if the initial input is too strong, the organism may turn away, grimace, extend his arms to the side, arch his trunk, splay his fingers and toes, cry, spit up, have a bowel movement; in short, he will show active avoidance behavior on various levels of functioning.

We are attempting a synthesis of these various principles of development: the principle of continuous active organism intention to realize its species-specific adaptation at each stage of its development; the principle of continuous organism–environment interaction and setting up of feedback loops; the principle of increasing hierarchical differentiation of behavioral subsystems in their mutual interaction; and the principle of continuous balance of organism avoidance, defense, and inhibition with approach and activation. We have termed their interplay the principle of synaction (Als, in press a,b) and will apply this synactive perspective of development to the human newborn and explore our understanding of this functioning.

3. A SYNACTIVE MODEL OF NEWBORN BEHAVIORAL ORGANIZATION

On the basis of the above theoretical considerations and our work with the Brazelton scale (1973), we have developed the following conceptualization of stages of very early behavioral organization (Als, 1978, 1979, 1982, in press a,b; Als, Lester, & Brazelton, 1979; Brazelton & Als, 1979): Sequential developmental agenda are negotiated in the very young organism who finds himself prematurely *extra utero* in continuous interplay with the extrauterine environment.

1. Early on, the main issue with which the human prematurely born newborn is grappling is the stabilization and integration of physiological functions, such as respiration, heart rate, temperature control, digestive function, and elimination competence.

2. As the motor system becomes increasingly energized, movements and active postural adjustments may infringe on the balance of physiological systems. Similarly, tactile and vestibular manipulations by caregivers can also upset this balance, precipitating apneic episodes or other defensive strategies such as bowel movements or spitting up.

3. Gradually the full range of states from sleep to awake to aroused emerges, and the earlier diffuseness and indeterminateness of state referred to by Dreyfus-Brisac (1974), Prechtl (Prechtl, Fargel, Weinmann, & Bakker, 1979), and others disappear. Clear states emerge and increase in their flexibility, as Nystrom (1975, 1977; Nystrom, Bandmann, & Valentin, 1973) has identified for sleep states and Brazelton (1973) and Sander (e.g., 1980) have described for the full range of states. Increasing state differentiation initially can also impinge on motoric stability and possibly even on physiological stability.

4. Finally, the alert state becomes more flexible, robust, accessible, and well-differentiated from the other states, often initially disrupting motor control and physiological balance. Gorski, Davison, and Brazelton (1979) have based their clinical descriptions on this formulation and speak of the *in-turning* or physiological stage; the *coming out* or first active response to the environment stage; and *reciprocity,* the ultimate stage of environmental opportunity.

Thus, we are identifying a model of subsystem differentiation in the organism characterized by the *principle of synaction,* which lies in the simultaneity of all subsystems in negotiation with one another and with the environment. The process of development appears to be that of stabilization and integration of some subsystems, which allows the differentiation and emergence of others which then in turn feed back on the integrated system. In this process the whole system is reopened and transformed to a new level of more differentiated integration from which the next newly emerging subsystem can further differentiate and press to actualization and realization (Als, 1979, 1982, in press a,b). To paraphrase Erikson (1962), self-actualization is participation with the world and interaction with another with a "minimum of defensive maneuvers and a maximum of activation, a minimum of idiosyncratic distortion and a maximum of joint validation."

4. BEHAVIORAL ORGANIZATION OF THE PRETERM INFANT

Assessment of an individual preterm infant's functioning in this model, then, requires the identification of his current status on this dynamic filigree matrix of subsystem development embedded in transactions with the environment. First we shall offer a conceptualization of the preterm infant in the developmental framework outlined above. The consideration of the principle of species-specific adaptation makes it clear that the surviving human preterm infant is a product of modern medicine, that is, of cultural evolution.

There is no model from which to glean what is appropriate organismic adaptation of the preterm infant. The 32-week-old organism, for instance, is adapted to an intrauterine environment of a regulated temperature, contained movement pattern, suspension of gravity, muted and regular sensory inputs, and physiological supports which have evolved to ensure normal intrauterine development for a large percentage of fetuses. Should premature delivery ensue, one could predict that most fetuses would die, since their organismic adaptations do not fit the environment in which they find themselves. Modern technology and medicine have changed this, but we are still searching for how best to provide for such organisms after birth, given the incongruence of the situation. Artificial re-creation of the intrauterine environment for the preterm infant is inappropriate because the transitions at birth automatically trigger independent functioning or organ systems necessary for extrauterine survival, such as the respiratory, cardiac, and digestive systems. Not only is the preterm infant an organism living in an environment for which he has not yet evolved, but he is an organism whose biological program is called upon prematurely, so that the normal sequence of subsystem differentiation and integration generally found by term has not yet been accomplished. These premature infants face survival difficulties because some subsystems have already been activated and are functioning efficiently *in utero,* while other necessary subsystems have not matured and are not yet ready to function. For instance, the cardiac system has been functioning and has most likely developed well enough to handle stimulation from light, noise, movement, endocrine changes, and various other stresses. One might even expect that if the cardiac system's development has been normal *in utero,* it may be able to adapt with comparatively little difficulty. The respiratory system, on the other hand, may be activated at a point in time when the organ differentiation is not sufficient for activation. Lungs mature substantially during the last 3 months of gestation. The demand of autonomous breathing of ambient air is relatively impossible for the immature lung. One option that has recently become available to neonatologists is to accelerate lung maturation *in utero* by steroid treatments if the threat of prematurity is known (Liggins & Howe, 1974; Taeusch, Keitner, & Avery, 1972; Taeusch, Wang, Baden, Bauer, & Stern, 1973). Of course, the unanswered question is how such treatment focused on only one organ system influences the auto-regulation of the total organism. Another possible option is the introduction of synthetic surfactant (Fujiwara, Chida, Watabe, Maeta, Morita, & Abe, 1980). Again, the effects on other systems of introducing such substances is not known. One of the temporary options in enabling survival is to provide artificial respiratory aids such as mechanical ventilation. The obvious questions are how this influences the normal maturational course of the lung and how it affects the integration and functioning of the respiratory system with other subsystems within the total organism.

Similar questions are appropriate with respect to the motor system. The effects of gravity and of an unrestricted environment on a motor system which is adapted to several more weeks of maintenance in a fluid environment must be considered. The comparatively flexed position of head, shoulders, arms, and legs *in utero* during the last 3 months of gestation presumably aids the execution of modulated and adaptive movement of the fetus. *In utero,* most fetuses practice sucking their thumb or fingers, and most are born in left occipital anterior vertex presentation, indicating a complexly integrated movement and postural system and program during action and unfolding. The internal motoric drive of the fetus for activation is no doubt modulated by the continuous cutaneous input of the surrounding amniotic fluid and the comparatively plastic, yet finitely reactive, intrauterine enclosure. This adaptive system has to be considered contrasted to the flat surface of the incubator bedding, the lack of cutaneous input other than on the back, and often the impossibility of attaining a flexed trunk, head, and extremity posture because of necessary life support system applications.

State regulation is similarly disrupted. It is probably intimately tied to and interactive with that of various sensory processes, and the motor system and physiological regulation play an important role in its organization. *In utero,* the sleep-awake cycles of the mother influence the state regulation of the fetus (Sterman, 1967). The gradual differentiation of states *in utero* begins to include the capacity for higher states of quiet and active alertness and more aroused states of heightened motoric arousal and crying. Such state differentiation matures toward the last months of gestation *in utero.* How state differentiation is accomplished, and how states are regulated at the least cost for the organism in an environment as different from the uterus as is the premature nursery, with its visual, auditory, olfactory, and kinesthetic inputs, remains to be examined.

In addition to investigating the effect of differential discrepancies of subsystem readiness, we must discover how the infant eventually coordinates these various subsystems and how he achieves the continuous feedback between these subsystems and the input from the environment necessary to his survival and future development. When one realizes the current organizational issues for the preterm infant, one becomes aware of the flaws and possible dangers of intervention programs which consider premature infants to be deficient full-term infants and which, therefore, are intended to "train" infants in behavior appropriate for full-term babies. Our effort, then, must be directed at identifying the current organization of an individual preterm infant and tracking his development so that sensitive appropriate intervention may be planned.

5. ASSESSMENT OF THE PRETERM INFANT

In order to describe a process of examining the current organization of premature infants, let us take the case from our records of a preterm infant who was delivered at 32 weeks, appropriate weight for gestational age, and who is now 2 weeks of age (34 weeks post-conception). After several days of ventilatory assistance, the infant has been moved to room air, is maintained in a warming isolette, and is fed by bottle every 2 to 3 hours (Als, Lester, & Brazelton, 1979).

When first observed, the infant is lying on his stomach, legs pulled underneath him, buttocks raised, one hand on top of his face, which is turned to one side, the other arm extended along the side of his body, palm open. The infant is breathing regularly; his color is adequate; he is in deep sleep. Because of his modulated body position, good color, and effective respirations, we consider him to be well-organized during his sleep. We observe him for 2 minutes and notice one discrete whole-body startle which subsides with some facial grimacing. There is no perpetuation of activity. In order to test the stability of the inhibitory mechanisms in sleep, we first present a visual stimulus in the form of a flashlight beam moved before the infant's closed eyes. The first flashlight beam triggers an intense startle, followed by some squirming and a shift of the face to the other side. The color around his mouth changes to gray. The response subsides after approximately 4 seconds. Normal color returns. A second flashlight beam triggers a somewhat delayed motor response, no startle but active squirming, mild color change, head movement, arm movements, and stretching out of his legs. The response subsides, yet the tension in his arms and buttocks remains. A second cycle of activity starts, spreading from the arms and trunk to the legs, subsides, and starts again a third time, not quite as vigorous as in the second cycle; the fourth cycle is very mild, and the fifth ends in relaxed quietness. A third flashlight beam produces a diffuse, low-level motor response with only one recycling of activity. A fourth flashlight beam produces a smile only, and a fifth and sixth beam produce no response. Thus the infant, despite some difficulty in inhibiting the motor arousal which follows a stimulus, as indicated by the recycling of one response after another, finally is capable of maintaining his sleep state and of shutting out repeated light stimuli.

Using a rattle as the first auditory stimulus, we find that the response sequence is similar: The first several responses show some repetition of movement, then the perseveration subsides and eventually the response is successfully inhibited. The second auditory stimulus (a bell) produces no response at all, nor do repeated presentations arouse him. The implication is that the shutdown of responses from the first auditory stimulus has generalized.

The next maneuver is an attempt to place the infant on his back. As the infant is touched, his motoric arousal is intense. There is considerable color change spreading from the perioral region to cheeks and forehead; his hands become blue and his trunk mottled in appearance. The motor arousal builds up once the infant is actually put on his back because he is now lacking necessary frontal inhibition. Undifferentiated arm and leg movements in all directions ensue. The infant awakens with them, squirming in continuous motion. The examiner's hand placed at the soles of the infant's feet allows him to slow his leg movements and soften his arm movements. The infant then widens his eyes and visually locks onto the examiner's face intermittently as she bends over him and talks softly. Thus the impression is that it is possible to help the infant inhibit his motor responses using appropriate tactile, visual, and auditory stimulation.

The next maneuver, touching the sole of the infant's foot with a plastic probe, however, arouses his motor system again and results in cycles of undifferentiated, uncontrolled activity of arms, legs, and trunk, with each cycle punctuated by a prolonged tremor initially and ultimately. Accompanying this uncontrolled motor arousal are respiratory disruption and intense color change so that the infant appears very pale. No fussing or crying is evident. The impression is that tactile stimulation, especially while frontal inhibition is lacking, arouses the motor system beyond the infant's capacity to regain any control on his own. As the infant is gently placed back on his stomach, he squirms and moves for 1 minute in an attempt to regain regulation of this aroused system. Eventually, by pressing his feet against the bottom wall of the isolette and his back against a side wall, and by placing his hand across his face, he is able to inhibit this motor movement. His color and respirations are back under control.

As the examiner picks up the infant in order to assess his social skills, she swaddles him tightly to assist him with his motor inhibition. The infant's eyes are open, and he locks onto a bright visual stimulus to the side.

As the examiner attempts to engage the infant in eye-to-eye contact, the infant actively avoids her, closing his eyes and then moving his head in the opposite direction. There is mild color change periorally. The infant's arm movements during this sequence, however, are smooth, and there is some flexion at the elbow and wrist. With increased input from the examiner, the infant moves into a tonic neck reflex and fixates on an environmental stimulus, still avoiding the examiner's face. When the examiner speaks softly but makes her face unavailable, there is a noticeable relaxation of the infant's tone, his hands open and close rhythmically, his face softens, his mouth begins to purse, and eventually he turns his head smoothly in the direction of the sounds. As the infant's alertness and social interaction intensify to a high point, his arms and hands and his head and mouth lift toward the examiner in a generalized approach configuration. At the height of this approach, he sneezes, then yawns and returns to a resting state.

When the infant is then presented with an inanimate sound (rattle), his response is quite different. He turns with a large startle in the opposite direction, flinging his right hand over his face. He frowns and pales. Softening the sound results in his relaxing and now turning to the sound. With great effort, he locks onto it. The response is costly for the infant, as his color change and facial expression indicate. His movements, however, remain smooth.

This performance indicates that this infant's ability to shut out, actively to avoid stimulation that is too taxing for him, is quite sophisticated. Clearly, he processes information and can communicate whether he is able to deal with it. Once the examiner has helped him by swaddling or by placing him on her lap, holding his shoulders in flexion on a soft surface, he is capable of responding selectively to social and inanimate stimuli, organizing around those which are appropriate, such as animate auditory stimuli, and actively avoiding, with some cost, those which are too demanding for him. At the end of the examination he can make use of the examiner's help with motor inhibition to activate his own arousal regulation maneuvers, such as pressing his feet against the isolette wall, which then allows him to exert control over his movements such that he can suck on his own hand and return to a relaxed sleep state.

The overriding impression we have gained from this assessment is that this infant is comparatively well organized. He is capable of communicating behaviorally at what level he can function smoothly without undue cost to his system's balance. His sleep is quite stable, and the degree of motoric arousal due to distal visual and auditory stimuli is manageable. Tactile stimulation poses a more serious demand for his current regulation, especially when paired with ventral openness. The result is that both the state and the motor system balance become severely disrupted. The disruption, in turn, taxes the autonomic system, resulting in irregular respirations, color change, tremors, and uncontrolled motor discharge. However, environmental assistance in the form of ventral and general body surface inhibition, such as swaddling, produces motor regulation. The highest level of motor differentiation in this infant is seen during his calm alert state when smooth rhythmic flexion movements emerge. The maneuvers available to regulate input and avoid overwhelming stimulation and the degree of specificity of attending to appropriate inputs are indices of an intact, complex, and actively controlling nervous system which selects and rejects stimulation on the basis of self-regulation. Given a nurturant and supportive environment, one can predict several developmental achievements for this infant during the next 6 to 8 weeks: First, his great sensitivity to stimulation will diminish somewhat and his physiological regulation will mature to a degree so that more robust alertness will ensue. This speculation is based on the infant's current differential handling of soft and loud auditory stimuli, on his organizing around a face with a soft voice, and on relaxing through visual stimuli. A capacity for alertness and social interaction is already available to him; yet, he still shows a low threshold for stimulation. However, the basis for good prognosis is the infant's demonstration of a high degree of modulation over his systems when the input is appropriate.

A second expected developmental achievement is that the motoric arousal to tactile stimulation and to ventral disinhibition will diminish. This expectation is based on the infant's present ability to modulate his arousal when inhibitory aid is provided by the examiner, such as her hands placed on his feet, swaddling him, or allowing him to press his feet against a stationary surface. His seeking out of external inhibitory aids, such as pressing his hands and feet against the isolette wall, supports this impression. A third index to support this notion is the considerable smoothness and modulation of his movements when his motoric arousal is diminished. The ability to place his hands over his face, to bring his fingers to his mouth and suck on them, to fold his hands under his chin and keep them tucked together rather than throwing off jerky, startle-like movements, are all signs of increasing motor organization.

The third developmental anticipation is that higher states of arousal (vigorous crying) will become available. This prediction is based on the observation that the transitions between the states available now are well modulated and do not unduly tax the infant's physiological and autonomic systems. Since this infant exhibits considerable regulation of his sleep and quiet awake states, one expects the emergence of the crying state fairly soon.

Not all premature infants will show such stable organization and balanced progression of systems' differentiation and sequential integration as the infant described here. At one extreme of the continuum of preterm functioning is the hyperreactive infant who continuously is at the mercy of environmental and internal stimuli. This infant reacts to such stimuli without being able to shut them out in order to protect his own regulation. His continuous reactions to the environment are very expensive for him because they delay him in consolidating the regulation of his motor and state organization. He remains highly reactive and taxes his autonomic system severely. This vicious cycle may eventually prevent him from growing normally and may possibly even result in producing a hypersensitive infant who fails to thrive.

At the other extreme is the lethargic, depressed premature infant who will not respond to any stimulation, thereby preserving his fragile autonomic regulation. He cannot afford to activate new pathways by incorporating new information; this failure impedes him in developing a more complex repertoire. Such an infant may ultimately appear unreachably depressed, uncommunicative, and restricted, both emotionally and cognitively.

6. TOWARD A RESEARCH INSTRUMENT

Our goal has been to systematize our descriptions of the preterm infant's behavioral organization so that we may begin to study such clinical questions as, for instance: Are specific behavioral organizational issues associated with definable medical conditions such as bronchopulmonary dysplasia, respiratory distress syndrome, failure to gain weight, and others? If so, can we institute preventive or at least supportive measures to reduce developmental cost to the organism of such disorders? How effective are our individualized care practices? How do we time and institute environmental changes appropriate to the infant, for example, when might an infant be ready to be changed to demand feeding; when and for whom is reduced lighting important; what postural supports should be instituted, when, and for whom? All these questions require a way to assess systematically those behaviors we think reflect the preterm infant's psychobiological well-being. In this framework, an assessment tool requires a systematic situation or sequence of manipulations which clearly brings out the following aspects:

6.1. Current Level of Balance and Smooth Integrated Functioning

The assessment must show the infant's *current level of balance and smooth integrated functioning,* that is, in what situation, if there is one, with what facilitative inputs from the environment, does this infant appear to be functioning smoothly, relaxed, and comfortable. For example, when he is tucked together on the mother's or nurse's shoulder and softly rocked in a semi-dark section of the nursery, he is breathing regularly, his color looks good and his muscle tonus is even; yet when placed on his back with his arms and legs tied down, he can regain his physiological stability only after several minutes of frantic struggling and gasping, after which he exhaustedly lets his head fall to one side and appears to give up the fight, with obligatory breathing and tensed muscles, raw and on edge even in sleep, as evidenced by his startles to even mild events occurring around him.

6.2. Threshold of Disorganization

An assessment must bring out the *threshold of disorganization* indicated in the behaviors of defense and avoidance. Stress can serve as a sign of the next step in the sequence. The degree and kind of stress and the intensity of frustration or defense can serve as indicators of the degree of energy available to be brought to bear on the emergence of the next developmental agendum; for example, the infant's struggling and disorganization when placed on his back are an index of the emerging motor system energy and his beginning efforts at modulation. Prior to this stage, he would not struggle and actively attempt to regain an easier or more effective position but instead would limply react to passive displacements of his limbs. The threshold of disorganization can be assessed by testing whether some help at movement inhibition—for example, a hand on his feet, some support with flexing his shoulders, a finger placed in his palm—will suffice to initiate his own active body flexion, which in turn will allow his agitated movements to subside. Such an infant would indicate a threshold different from that which would be the case if relaxation could come about only with prolonged whole-body inhibition.

6.3. Degree of Differentiation

An assessment must bring out the *degree of differentiation within the level of integrated functioning:* that is, if the infant is quite well integrated if flexed and maintained quietly, how much leeway is there on either side of this maintenance; once relaxed, can he be moved off the shoulder softly and still retain his relaxed motor integration, or does every slightest shift in position throw him into disorganization? Can he be moved from shoulder to the lap and then to his isolette and positioned in such a way that he can maintain himself with only minor adjustments, or is disorganization unavoidable despite sensitive handling?

6.4. Degree of Modulation and Regulation

An assessment must bring out the *degree of modulation and regulation in the actualization of the differentiation.* For instance, one infant will be able to handle the shift from shoulder to isolette without any disruption of either the physiological, the motor, or the state organization. Another infant will be able to handle the shift yet will show changes in color, some movement disorganization, and grimacing, after which he will return to an integrated baseline; his modulation is not yet so well developed.

6.5. Strategies and Effectiveness of Self-Regulation

An assessment must bring out the *strategies and the effectiveness* of self-regulation of which an infant is capable. What strategies does the infant employ to return himself to an integrated balanced level? For example, when on his back does he flail and cycle his arms and legs until he has scooted towards the side of the bassinette, against which he then presses the soles of his feet, tucks in his arms, and relaxes; or is his arm and leg cycling so exhausting to him that he goes into apnea after he has exerted himself for a brief period? Both infants employ self-regulation strategies, but the second one is still much more at the mercy of environmental help.

6.6. Environmental Structuring Necessary

An assessment must bring out the *degree and kind of environmental structuring, support, and facilitation necessary,* not only to reestablish smooth functioning but also to facilitate optimal functioning. It also must identify what facilitative structuring is necessary to allow the infant to surmount the demanding defense against overloading inputs. This will then afford the infant the realization of the next emerging agendum; for example, the 34-week-old infant discussed earlier can eventually come to bright, focused alertness with the examiner when held wrapped and talked to very softly and at a moderate distance. To bring about this experience is, in our model of organism–environment interaction, important for the organism's realization of these pathways. It allows him to call on this realization again and gradually to solidify this competence, then differentiate and increasingly modulate it. This realization of the newly emerging capacity appears critical for development, especially when distortions and discrepancies must be conquered (Als, 1979, 1982, in press a,b; Als, Tronick, & Brazelton, 1980; Brazelton & Als, 1979). This realization is also important for the social interactors with the organism, the parents and caregivers, who in turn are biologically structured with expectations for positive social elicitation and feedback as is forthcoming in focused, bright eye opening on the background of relative motoric relaxation and physiological stability. To be able to free these small strands of the next level sets the path of development in a positive direction and avoids unwitting reinforcement of only the disturbing, negative, and seemingly distorted behavior, concomitants of the discrepant organism–environment fit with which we are confronted and which only too easily lead to a vicious cycle of increasing distortion (Herzog, 1979).

These considerations of integrative level, threshold, and cost have guided our construction over the years (Als, Tronick, & Brazelton, 1975; Als, Lester, & Brazelton, 1978, 1979) of a systematic behavioral assessment procedure for the preterm infant, referred to as the Assessment of Preterm Infants' Behavior, or APIB. (See Appendix for Manual and Scoresheets of APIB.)

7. DESCRIPTION OF INSTRUMENT

In the APIB the maneuvers of the Brazelton Neonatal Behavioral Assessment Scale (BNBAS) are seen as a graded sequence of increasingly demanding environmental inputs or packages, moving from distal stimulation presented during sleep to mild tactile stimulation, to medium tactile stimulation paired with vestibular stimulation, to more massive tactile stimulation paired with vestibular stimulation. The social interactive attentional package is administered whenever in the course of the examination the infant's behavioral organization indicates his availability for this sequence. It receives high priority in the examiner's attempts to facilitate the infant's organization (Als, 1978).

During the administration of each package, we are monitoring the infant's reactions and behaviors along five systems of functioning: his physiological system, his motor organizational system, his state organizational system, his attentional/interactive system, and his self-regulatory system. We are also monitoring what kind of graded examiner facilitation is necessary to bring the infant to optimal performance and/or to help him return to an integrated, balanced state. These system parameters are scaled from 9 (disorganized performance) to 1 (well-organized performance) and can also be graphed on a summary grid (see score sheet and systems grid).

The use of the assessment in this graded fashion with a consistent focus on the five subsystems outlined allows us to identify the threshold of integration and stable organization of the infant and the areas of currently poor functioning and regulation, as well as the areas of beginning modulation and differentiation.

Aside from the subsystem organization descriptors, the preterm scale furthermore allows one to document discrete behavioral dimensions as captured in the Brazelton full-term scale with a range of behaviors appropriate for the preterm infant. The manual and score sheet are organized so that the full-term BNBAS scores are distinctly identified in the APIB scoring system and are easily read off where appropriate. This makes the APIB useful for comparative work of preterm and full-term subgroups. Score Sheet III documents summary behavioral parameters. The scores from Sheets II and III can be examined as individual items or grouped into dimensions or clusters, as has been done successfully with the 26 BNBAS behavioral items (Adamson, Als, Tronick, & Brazelton, 1977; Lester, Als, & Brazelton, 1982; see also a review by St. Clair, 1978). The APIB scoring manual is the result of several revisions and expansions of earlier versions developed and refined in the course of our intensive work with preterm infants over the last eight years (Als, Tronick, & Brazelton, 1975; Als, Lester, & Brazelton, 1978, revised 1979). The administration of the examination itself usually takes at least 30 minutes. The examination should begin with the infant asleep about midway between two feedings (Brazelton, 1973); for the preterm infant it is usually better to come closer to the next feeding. This appears to facilitate state manipulations from the outside.

The in-hospital preterm examinations should be performed in a room where it is quiet and warm and where lighting is indirect and can be adjusted. The scoring of the APIB takes less than 1 hour, including the graphic display of the systems scores. The APIB reliability criterion is modeled after that for the BNBAS: for all the 9-point scales, including the system sheet, agreement within $\pm$ 1 point has to be at a rate of 90% or better. Training in the *scoring* procedure of the APIB manual is facilitated by reliability in the administration and scoring of the BNBAS and requires additional extensive training in behavioral observation of the additional parameters to be scored. Training in the *administration* of the APIB examination to preterm infants requires prior extensive training in the care and handling of preterm infants in the special care nursery. For the nonmedically trained researcher, this presupposes the establishment of an interdisciplinary work environment supervised and fully backed by the neonatology and nursing personnel. Behavioral observations and manipulations otherwise may add to the already massive barrage of inappropriate stimulation brought to bear on the infant and will potentially inflict more harm than good. Once the behavioral researcher has a good knowledge and understanding of preterm infants' medical and physiological issues and is competent in performing routine caregiving, such as changing diapers, feeding, cleaning, and bathing, behavioral observation training may begin.

The most important aspect of this training is the continuous monitoring of the infant's subtle signs of stress and availability, so that decisions of modification of behavioral manipulation, increase of support, or cessation of stimulation can be made reliably and appropriately. The most counterproductive procedure would be to force an infant through a whole examination for the sake of scoring all items when, in fact, his behavioral displays indicated clearly that he had reached his threshold of integrated organization and needed to be left alone. This in and of itself is much more important information than a sheet full of item scores and is recordable on the system sheet. The most important goal of training is to become an experienced, accurate, and confident reader of the varied cues of the infant and a sensitive interactor who is aware at all times of the impact of his own actions, his movements, and his own state on the infant, and who knows how to modify, pace, and structure his own behavior resourcefully in accordance with the infant's needs and cues in order to optimize and facilitate the infant's organization.

The examination as proposed here is appropriate for the infant who is in an open isolette or crib, in room temperature and room air. This is also usually the stage at which parents become more actively involved in caregiving with the infant and is therefore well suited in capturing their energies for the infant by means of the examination. As long as the infant must be maintained in oxygen or with other life support lines, behavioral manipulations are often inappropriate and additionally stressful. In these circumstances more can probably be learned from close behavioral observation of the infant in the course of routine care, and inferences as to the appropriateness of environmental changes can be drawn from such observations. A systematic approach to accomplish this is being developed.

8. IMPLICATIONS FOR MANAGEMENT, CARE, AND INTERVENTION

We have outlined a theoretical framework of organism development stressing the principles of species-specific adaptation at each stage of development, of continuous organism–environment interaction, and of continuous intraorganism subsystem interaction, leading to hierarchically organized transformations of functioning. We have discussed the idea of using relative differentiation and modulation at each stage as indices of an individual's competence. This framework is the backdrop for the APIB, the research instrument for the behavioral assessment of preterm infants which we have presented. The framework influences the instrument, and the instrument is worthwhile if it can help to uncover and articulate connections not immediately obvious before. Exploratory analyses performed on two case comparisons (Als & Brazelton, 1981) and on a sample of 10 preterm and term infants studied over time (Als, in press a) are very promising. At this point the conclusions we draw for the examination of the preterm infant are necessarily at a general level.

8.1. Infant Behavior: The Route of Communication

The behavior of the infant is his route to communication. It is possible and important to become aware and to document systematically the behavioral repertoire of the premature infant and to take seriously signs of a lowered and more vulnerable threshold, of being overtaxed, of relative inability to attend selectively, of efforts to reorganize motor activity which is stripped of inhibitory processes, and of poorly organized sleep states in which inhibitory mechanisms are not well organized as they are seen in the full-term infant. The premature infant may have difficulty coming into and maintaining states of alertness, especially in a distorted environment; yet it may be possible for the examiner to identify what environmental modification would facilitate the emergence of a more well modulated alert state.

The preterm infant's physiological mechanisms appear to be easily overtaxed and throw such an infant into a relatively exhausted state very quickly, and the autonomic nervous system exhaustion produces cyanotic color changes or periodic cessation of respirations when the cardiorespiratory systems are overtaxed.

A motor system with poor balance of flexors and extensors produces jerky, overshooting responses to most tactile manipulation. Premature infants are often unsuccessful in inhibiting these twitching, jerky responses, and this failure can result in the spreading of this response to the entire motor system from one stimulus. This also taxes the infant, particularly when the responses are circular and repeat themselves as cycles over and over until true exhaustion ensues. On the other hand, assessing reflex responses may be difficult because the musculature is so weak and energy stores so inadequate that a motor response may be delayed or inadequate in an intact premature infant.

A premature infant's intake of stimuli may appear to be delayed or inadequate as well. On occasion several stimuli may have to be presented before a single initial stimulus–response is elicited. Whether this is due to a high threshold established to protect the immature organism or whether it is a result of poor conduction at an intake level is yet to be determined.

Thus, a measure of the premature infant's capacity to organize himself in order to attend to or shut out stimuli may be a promising assessment of his CNS and autonomic functions. His ability to control and organize his awake and sleep states, his capacity to respond to a stimulus without undue delay and to inhibit repeated stimuli without exhaustion, the degree of self-organization as well as his capacity to incorporate and utilize environmental control and input—all may become predictors of optimal functioning in a well-organized premature. This behavioral repertoire becomes the basis for identifying and articulating an individual infant's developmental curriculum. His behavioral repertoire identifies those capacities of which he is easily capable, those which he can manage with aid from the environment, and those which tax him severely.

8.2. Observations of Change over Time

Observing changes in premature babies over time can possibly help us derive stepwise goals which are designed to maximize positive organization and regulation. Knowledge of the infant's current organization becomes the articulated base for his management. For instance, the 34-week-old infant described earlier enjoys auditory inputs *when* motorically organized, seeks visual stationary inputs in order to stabilize himself, has great difficulty with integration of tactile input, and does best with ventral inhibitory support. A concrete detailed chart of his behavioral strengths and problems, a specification of the input and handling procedures which are appropriate and which positively foster his organization, becomes possible. Detailing behavioral changes over limited time periods, such as a week, becomes helpful in measuring his progress and charting the course of the individual infant's development. With such a progress chart sensitively geared to each individual infant, it may become possible to avoid the danger of a routinized and potentially overloading program which might as easily retard his development and the organization of his central and autonomic nervous systems.

8.3. Implications for Parents

Implications from this model for working with parents of preterm infants are manifold. Parents appear to be biologically programmed to expect full-term normal newborn behavior. Not only are parents of preterm infants deprived of the realization of this expectation by having a preterm infant, but they are at a premature stage of development themselves, deprived of the last weeks and months of readying themselves for interaction with their infant. In addition, they are suffering through an inevitable grieving process in the early postnatal period which is likely to retard them in making adjustments in their behavior necessary for facilitating a premature with his fragile organization. Thus, we are dealing with two premature subsystems of an interactive feedback system in which both subsystems may be showing distorted behavioral patterns. Mutually beneficial adaptation can be facilitated by acknowledging the often extraordinary task at hand and the severity of the mutual deprivations. This will help free the parents from some of the guilt feelings of not being immediately drawn to the premature infant and legitimize, in fact, their feelings of helplessness, incompetence, anger, and deficiency. The next level of support to the parent then can come from giving the parents the license to read the infant's behaviors as meaningful communication (Als, Lester, & Brazelton, 1979; Brazelton & Als, 1979). For example, when the infant trembles and pales and extends his arms and legs, he is telling the parent that he is having difficulty keeping himself organized, but when the parent then puts his finger in the infant's palm or swaddles him, he becomes relaxed, thereby telling the parent that he also has ways to overcome this disorganization and that the parent is important to him.

The great sensitivity of preterm infants to environmental inputs is often frightening to parents. The immediate reaction is to let the professionals on the scene take care of the infant. Yet the more the parents can be helped to be concretely attuned to their infant's actions and reactions, the better able they will be to generate appropriate and development-promoting circumstances for this infant. In doing so, they will enhance the infant's early recognition of his own competence and facilitate his ever increasing autonomy. Clinically, we have used videotape successfully to help identify the ingredients of behavioral organization of preterm infants for ourselves and the parents. We have found this technique quite profitable because it allows the parents of a premature infant to stand outside the direct interaction with their infant and to watch how his and our behavior affects changes in the infant's behavior, before they are expected to take over care themselves. We are encouraged by the eagerness and creativity with which parents utilize the observations of their infants' behavior to modify and enhance their interactions with these infants. It is our hope that the framework and research tool presented here will be a step in the direction of better understanding and documenting of the developmental process of the preterm infant and therewith will increase our skills in supporting each infant's individuality and ego strength.

Acknowledgments

We would like to thank particularly F. H. Duffy, M.D., for his continuing and critical discussions from a neurologist's and neurophysiologist's point of view, especially concerning the motor system. We are grateful to the parents of all the infants we have observed, who so generously extended us the privilege to learn from them and their infants; to the infants themselves; and to the staff of the Special Care Nursery at the Boston Hospital for Women, who have made our observations possible over the last nine years. Pediatric Fellows involved in our work at earlier stages include Barry Zuckerman, M.D., Martha Davison, M.D., and Peter A. Gorski, M.D. Kevin Nugent, Ph.D., has been of continuing help with critical and encouraging feedback from his extensive experience as trainer and teacher of Brazelton scale examiners. Investigators from other centers have provided invaluable critical feedback and served as a testing ground for our ideas. They include Elsa Sell, M.D., and her staff at the University of Arizona Medical Center, Tucson, Arizona; Kathryn Barnard, R.N., Ph.D., and her staff at the University of Washington in Seattle; Francis Horowitz, Ph.D., and Patricia Linn, Ph.D., University of Kansas at Lawrence; Arthur H. Parmelee, M.D., University of California at Los Angeles; Gretty Mirdal, Ph.D., University of Copenhagen, Denmark; and Britt Venner, Ph.D., University of Oslo, Norway. We would like to thank Michael Yogman, M.D., who is coinvestigator on our current Prospective Follow-up Study of Preterm and Fullterm Infants and whose pediatric clinical and research expertise is critical to our ongoing work. Other staff involved in our projects to whom we are grateful for their unflagging dedication are Ann Berger, Emily Burrows, Joan Gilman, Carey Halsey, Sylvia Howe, Cassie Landers, and Carol Sepkoski. Special thanks go furthermore to Drs. Nancy Kozak, Margarite Moore, and Anna Lombroso for their support of our work. Preparation of the manuscript was carried out with expert knowledge, care, and great patience by Kate Neff.

9. REFERENCES

Adamson, L., Als, H., Tronick, E., & Brazelton, T. B. A priori profiles for the Brazelton neonatal assessment. In H. Als, E. Tronick, B. M. Lester, & T. B. Brazelton, The Brazelton Neonatal Behavioral Assessment Scale. *Journal of Abnormal Child Psychology,* 1977, *5*, 3–10.

Als, H. *The human newborn and his mother: An ethological study of their interaction.* Doctoral dissertation, University of Pennsylvania, 1975.

Als, H. The newborn communicates. *Journal of Communication,* 1977, *27*, 66–73.

Als, H. Assessing an assessment: Conceptual considerations, methodological issues, and a perspective on the future of the Neonatal Behavioral Assessment Scale. In A. J. Sameroff (Ed.), *Organization and stability of newborn behavior: A commentary on the Brazelton Neonatal Behavioral Assessment Scale.* Monographs of the Society for Research in Child Development, 1978, *43*(5–6), 14–29.

Als, H. Social interaction: Dynamic matrix for developing behavioral organization. In I. C. Uzgiris (Ed.), *New directions for child development* (Vol. 4). San Francisco: Jossey Bass, 1979, pp. 21–41.

Als, H. The unfolding of behavioral organization in the face of a biological violation. In E. Z. Tronick (Ed.), *Human communication and the joint regulation of behavior.* Baltimore: University Park Press, 1982.

Als, H. Infant individuality: Assessing patterns of very early development. In J. Call & E. Galenson (Eds.), *Frontiers in Infant Psychiatry.* New York: Basic Books, in press. (a)

Als, H. Towards a synactive theory of development: Promise for the assessment of infant individuality. *Infant Mental Health Journal,* in press. (b)

Als, H., & Brazelton, T. B. A new model of assessing the behavioral organization in preterm and fullterm infants. *Journal of the American Academy of Child Psychiatry,* 1981, *20*, 239–263.

Als, H., Tronick, E., & Brazelton, T. B. *Manual for the behavioral assessment of preterm and high risk infants: An extension of the Brazelton scale.* Unpublished manuscript, Child Development Unit, Children's Hospital Medical Center, Boston, 1975.

Als, H., Lester, B. M., & Brazelton, T. B. *Manual for the assessment of preterm infants' behavior (APIB).* Unpublished manuscript, Child Development Unit, Children's Hospital Medical Center, Boston, 1978. (Revised February 1979 and August 1979).

Als, H., Lester, B. M., & Brazelton, T. B. Dynamics of the behavioral organization of the premature infant: A theoretical perspective. In T. M. Field, A. M. Sostek, S. Goldberg, & H. H. Shuman (Eds.), *Infants born at risk.* New York: Spectrum, 1979, pp. 173–193.

Als, H., Tronick, E., & Brazelton, T. B. Affective reciprocity and the development of autonomy. The study of a blind infant. *Journal of the American Academy of Child Psychiatry,* 1980, *19,* 22–40.

Arend, R., Gove, F. L., & Sroufe, L. A. Continuity of individual adaptation from infancy to kindergarten: A predictive study of ego-resiliency and curiosity in preschoolers. *Child Development,* 1979, *50,* 950–959.

Ayres, A. J. *Sensory integration and learning disorders.* Los Angeles: Western Psychological Services, 1973.

Block, J., & Block, J. H. *The developmental continuity of ego control and ego resiliency: Some accomplishments and some problems.* Paper presented at the meeting of the Society for Research in Child Development, New Orleans, March 1977.

Blurton Jones, N. Characteristics of ethological studies of human behavior. In N. Blurton Jones (Ed.), *Ethological studies of child behavior.* Cambridge: Cambridge University Press, 1972, pp. 3–37.

Blurton Jones, N. Ethology and early socialization. In M. P. M. Richards (Ed.), *The integration of a child into a social world.* Cambridge: Cambridge University Press, 1974, pp. 263–295.

Blurton Jones, N. Growing points in human ethology: Another link between ethology and the social sciences? In G. Bateson & R. A. Hinde (Eds.), *Growing points in ethology.* Cambridge: Cambridge University Press, 1976, pp. 427–451.

Bobath, B. The very early treatment of cerebral palsy. *Developmental Medicine and Child Neurology,* 1967, *9,* 373–390.

Bobath, B. *Abnormal postural reflex activity caused by brain lesions.* London: William Heinemann, 1971.

Bobath, K. The normal postural reflex mechanism and its deviation in children with cerebral palsy. *Physiotherapy,* 1971, *57,* 1–11.

Bobath, K., & Bobath, B. The diagnosis of cerebral palsy in infancy. *Archives of Diseases of Children,* 1956, *31,* 408–414.

Brazelton, T. B. *The Neonatal Behavioral Assessment Scale.* Clinics in Developmental Medicine, No. 50. London: William Heinemann, 1973.

Brazelton, T. B., & Als, H. Four early stages in the development of mother–infant interaction. *The Psychoanalytic Study of the Child,* 1979, *34,* 349–369.

Bruner, J. S. *On knowing: Essays for the left hand.* New York: Atheneum, 1965.

Bruner, J. S. Processes of cognitive growth: *Infancy.* Worcester, Mass.: Clark University Press, 1968.

Bunge, R., Johnson, M., & Ross, C. D. Nature and nurture in development of the autonomic neuron. *Science,* 1978, *199,* 1409–1416.

Campbell, S. K., & Wilhelm, I. J. Developmental sequences in infants at high risk for central nervous system dysfunction: The recovery process in the first year of life. In J. M. Stack (Ed.), *An interdisciplinary approach to the optimal development of infants: The special child.* New York: Human Sciences Press, in press.

Coghill, G. E. *Anatomy and the problem of behavior.* New York: Macmillan, 1929.
Connolly, K., & Bruner, J. (Eds.). *The growth of competence.* New York: Academic Press, 1974.
Dabrowski, K., & Piechowski, M. M. *Theory of levels of emotional development: Multilevelness and positive disintegration* (Vol. 1). Oceanside, N.Y.: Dabor Science Publications, 1977.
Dawkins, R. Hierarchical organization. In P. P. G. Bateson & R. A. Hinde (Eds.), *Growing points in ethology.* Cambridge: Cambridge University Press, 1976, pp. 10–54.
Denny-Brown, D. *The basal ganglia and their relation to disorders of movement.* Oxford: Oxford University Press, 1962.
Denny-Brown, D. *The cerebral control of movement.* Springfield, Ill.: Charles C Thomas, 1966.
Dreyfus-Brisac, C. Organization of sleep in prematures: Implications for caretaking. In M. Lewis & L. A. Rosenblum (Eds.), *The effect of the infant on its caregiver.* New York: Wiley, 1974, pp. 123–141.
Duffy, F. H., & Burchfiel, J. Somato-sensory systems: Organizational hierarchy from single units in monkey area 5. *Science,* 1971, *172*, 273–275.
Duffy, F. H., Mower, G. D., & Burchfiel, J. L. Experimental amblyopia—Production by random monocular shifts of visual input. *Society for Neuroscience Abstracts,* 1978, *4*, 625.
Erikson, E. H. Reality and actuality. *Journal of the American Psychoanalytic Association,* 1962, *10*, 451–475.
Fowler, W., & Swenson, A. The influence of early language stimulation on development: Four studies. *Genetic Psychology Monographs,* 1979, *100*, pp. 73–109.
Freedman, D. G. Smiling in blind infants and the issue of innate versus acquired. *Journal of Psychology and Psychiatry,* 1964, *5*, 171–184.
Freedman, D. G. *Human infancy: An evolutionary perspective.* New York: Wiley, 1974.
Freedman, D. G. *Human sociobiology: A holistic approach.* New York: Macmillan, 1979.
Fujiwara, T., Chida, S., Watabe, Y., Maeta, H., Morita, T., & Abe, T. Artificial surfactant therapy in hyaline membrane disease. *Lancet,* 1980, *1*, 55–59.
Gesell, A. The ontogenesis of infant behavior. In L. Carmichael (Ed.), *Manual of Child Psychology.* New York: Wiley, 1946, pp. 295–331.
Gesell, A., & Armatruda, C. *The embryology of behavior.* Westport, Conn.: Greenwood Press, 1945.
Gorski, P. A., Davison, M. F., & Brazelton, T. B. Stages of behavioral organization in the high-risk neonate: Theoretical and clinical considerations. *Seminars in Perinatology,* 1979, *3*, 61–72.
Grossman, K. Die Wirkung des Augenöffnens von Neugeborenen auf das Verhalten ihrer Mütter. *Geburtshilfe und Frauenheilkunde,* 1978, *38*, 629–635.
Harmon, R. J. *Play, mastery and attachment in term and preterm infants.* Paper presented at the Annual Meeting of the American Academy of Child Psychiatry, Atlanta, 1979 (University of Colorado Medical Center, Dept. of Psychiatry, Denver, Colo. 80262).
Hassenstein, B. *Verhaltungsbiologie des Kindes.* Munich: Piper Verlag, 1973.
Herzog, J. M. *Attachment, attunement and abuse: An occurence in certain premature infant–parent dyads and triads.* Paper presented at the meetings of the American Academy of Child Psychiatry, Atlanta, 1979.
Hinde, R. A. *Animal behavior* (2nd ed.). London and New York: McGraw-Hill, 1970.
Hinde, R. A., & Spencer-Booth, Y. The behavior of socially living rhesus monkeys in their first two and a half years. *Animal Behavior,* 1967, *15*, 169–196.
Holloway, R. Culture, the human dimension. *Current Anthropologist,* 1969, *10*, 395–412.
Honzik, M. P. Value and limitations of infant tests: An overview. In M. Lewis (Ed.), *Origins of intelligence: Infancy and early childhood.* New York: Plenum Press, 1976, pp. 59–95.
Hooker, D. Early fetal activity in mammals. *Yale Journal of Biology and Medicine,* 1936, *8*(6), 579–602.

Hooker, D. Fetal reflexes and instinctual processes. *Psychosomatic Medicine,* 1942, *4*(2), 199–205.

Horton, F. H., Lubchenco, L. O., & Gordon, H. H. Self-regulatory feeding in a premature nursery. *Yale Journal of Biology and Medicine,* 1952, *24*, 263–272.

Humphrey, T. The development of mouth opening and related reflexes involving the oral area of human features. *Alabama Journal of Medical Sciences,* 1968, *5*, 126–157.

Humphrey, T. Reflex activity in the oral and facial area of the human fetus. In J. F. Bosma (Ed.), *Oral sensation and perception.* Springfield, Ill.: Charles C Thomas, 1970, pp. 195–233.

Hunt, J. McV. *Intelligence and experience.* New York: Ronald Press, 1961.

Hunt, J. McV. Motivation inherent in information processing and action. In O. J. Harvey (Ed.), *Motivation and social interaction: The cognitive determinants.* New York: Ronald Press, 1963, pp. 35–94.

Jennings, K. D., Harmon, R. J., Morgan, G. A., Gaiter, J. L., & Yarrow, L. J. Exploratory play as an index of mastery motivation: Relationships to persistence, cognitive functioning and environmental measures. *Developmental Psychology,* 1979, *15*, 386–394.

Korner, A. F. Maternal rhythms and waterbeds: A form of intervention with premature infants. In E. B. Thomas (Ed.), *Origins of the infant's social responsiveness.* Hillsdale, N.J.: Lawrence Erlbaum, 1979, pp. 95–125.

Korner, A. F., & Grobstein, R. Visual alertness as related to soothing in neonates: Implications for maternal stimulation and early deprivation. *Child Development,* 1966, *37*, 867–876.

Korner, A. F., & Thoman, E. B. The relative efficacy of contact and vestibular stimulation in soothing neonates. *Child Development,* 1972, *43*, 443–453.

Korner, A. F., Kraemer, H. C., Haffner, M. E., & Cosper, L. Effects of waterbed flotation on premature infants: A pilot study. *Pediatrics,* 1975, *56*, 361–367.

Landis, S. C. Developmental changes in the neurotransmitter properties of dissociated sympathetic neurons: A cytochemical study of the effects of medium. *Developmental Biology,* 1980, *77*, 349–361.

Lester, B. M., Als, H., & Brazelton, T. B. Scoring criteria for seven clusters of the Brazelton Scale. In Regional obstetric anesthesia and newborn behavior: A reanalysis toward synergistic effects. *Child Development,* 1982.

Lettvin, J. Y., Maturana, H., McCulloch, W., & Pitts, W. What the frog's eye tells the frog's brain. *Proceedings of the Institute of Radio Engineers,* 1959, *47*, 1940–1951.

Lewis, M. Infant intelligence tests: Their use and misuse. *Human Development,* 1973, *16*, 108–118.

Liggins, G. C., & Howie, R. N. The prevention of RDS by maternal steroid therapy. In L. Gluck (Ed.), *Modern perinatal medicine.* Chicago: Yearbook Medical Publishers, 1974, pp. 415–424.

Masi, W. Supplemental stimulation of the premature infant. In T. M. Field, A. M. Sostek, S. Goldberg, & H. H. Shuman (Eds.), *Infants born at risk.* New York: Spectrum, 1979, pp. 367–389.

McCall, R. B. Toward an epigenetic conception of mental development in the first three years of life. In M. Lewis (Ed.), *Origins of intelligence: Infancy and early childhood.* New York: Plenum Press, 1976, pp. 97–122.

McGraw, M. B. *The neuromuscular maturation of the human infant.* New York: Hafner, 1945.

Milani-Comparetti, A., & Gidoni, E. A. Pattern analysis of motor development and its disorders. *Developmental Medicine and Child Neurology,* 1967, *9*, 625–630.

Minde, K. K., Morton, P., Manning, D., & Hines, B. Some determinants of mother–infant interaction in the premature nursery. *Journal of the American Academy of Child Psychiatry,* 1980, *19*, 1–21.

Morgan, G. A., Harmon, R. J., Gaiter, J. L., Jennings, K. D., Gist, N. F., & Yarrow, L. J. A method of assessing mastery in one-year-old infants. *J.S.A.S. Catalog of Selected Documents in Psychology,* 1977, *7*, 68 (Ms. #1517).

Nystrom, M. Neonatal facial-postural patterning during sleep: II. Activity and states. *Psychological Research Bulletin,* Lund University, Sweden, 1975, *15*, 1–11.

Nystrom, M. Neonatal facial-postural patterning during sleep: V. Ethological models and temporal organization. *Psychological Research Bulletin,* Lund University, Sweden, 1977, *17*, 1–10.

Nystrom, M., Bandmann, Y., & Valentin, A. Neonatal facial-postural patterning during sleep. *Psychological Research Bulletin,* Lund University, Sweden, 1977, *17*, 1–10.

Palay, S. L. *Introduction to the nervous system: Basic neuroanatomy.* Lecture delivered at Harvard Medical School, 1979.

Patterson, P. H. Epigenetic influences in neuronal development. In S. O. Schmitt & F. G. Worden (Eds.), *Neuro Sciences Fourth Study Program.* Cambridge, Mass.: MIT Press, 1979, pp. 929–936.

Patterson, P. H., Potter, D. D., & Furshpan, E. J. The chemical differentiation of the nerve cells. *Scientific American,* 1978, *239*, 50–59.

Piaget, J. *The origins of intelligence in children.* New York: Norton, 1963.

Prechtl, H. F. R., Fargel, J. W., Weinmann, H. M., & Bakker, H. H. P. Postures, motility and respiration of low risk, preterm infants. *Developmental Medicine and Child Neurology,* 1979, *21*, 3–27.

Robson, K. The role of eye-to-eye contact in maternal–infant attachment. *Journal of Child Psychology and Psychiatry,* 1967, *8*, 13–25.

Rosenblatt, J. S. Stages in the early behavioral development of altricial young of selected species of non-primate mammals. In P. P. G. Bateson & R. A. Hinde (Eds.), *Growing points in ethology.* Cambridge: Cambridge University Press, 1976, pp. 345–385.

St. Clair, K. L. Neonatal assessment procedures: A historical review. *Child Development,* 1978, *49*, 280–292.

Sander, L. W. Issues in early mother–child interaction. *Journal of the American Academy of Child Psychiatry,* 1962, *1*, 141–166.

Sander, L. W. Adaptive relationships in early mother–child interaction. *Journal of the American Academy of Child Psychiatry,* 1964, *3*, 232–264.

Sander, L. W. Regulation and organization in the early infant–caretaker system. In R. Robinson (Ed.), *Brain and early behavior.* London: Academic Press, 1970, pp. 311–333.

Sander, L. W. Infant and caretaking environment: Investigation and conceptualization of adaptive behavior in a system of increasing complexity. In E. J. Anthony (Ed.), *Explorations in child psychiatry.* New York: Plenum Press, 1975, pp. 129–166.

Sander, L. W. Primary prevention and some aspects of temporal organization in early infant–caretaker interaction. In E. Rexford, L. W. Sander, & T. Shapiro (Eds.), *Infant psychiatry: A new synthesis.* New Haven: Yale, 1976, pp. 187–204.

Sander, L. W. Investigation of the infant and its environment as a biological system. In S. I. Greenspan & G. H. Pollock (Eds.), *The course of life: Psychoanalytic contribution toward understanding personality development.* Washington, D.C.: National Institute of Mental Health, 1980, pp. 177–201.

Sander, L. W., Stechler, G., Burns, P., & Julia, H. Early mother–infant interaction and 24-hour patterns of activity and sleep. *Journal of the American Academy of Child Psychiatry,* 1970, *9*, 103–123.

Sander, L. W., Julia, H. L., Stechler, G., & Burns, P. Continuous 24-hour interactional monitoring in infants reared in two caretaking environments. *Psychosomatic Medicine,* 1972, *34*, 270–282.

Sander, L. W., Stechler, G., Burns, P., and Lee, A. Change in infant and caregiver variables over the first two months of life: Integration of action in early development. In E. B. Thomas (Ed.), *Origins of the infant's social responsiveness.* Hillsdale, N.J.: Lawrence Erlbaum, 1979, pp. 349–409.

Schneirla, T. C. An evolutionary and developmental theory of biphasic processes underlying approach and withdrawal. In M. R. Jones (Ed.), *Nebraska Symposium on Motivation*. Lincoln: University of Nebraska Press, 1959, pp. 1–42.

Schneirla, T. C. Aspects of stimulation and organization in approach/withdrawal processes underlying vertebrate behavioral development. *Advances in the Study of Behavior,* 1965, *1*, 1–74.

Schneirla, T. C., & Rosenblatt, J. S. Behavioral organization and genesis of the social bond in insects and mammals. *American Journal of Orthopsychiatry,* 1961, *31*, 223–253.

Schneirla, T. C., & Rosenblatt, J. S. "Critical periods" in the development of behavior. *Science,* 1963, *139*, 1110–1115.

Sherrington, C. S. *Man on his nature.* Cambridge: Cambridge University Press, 1940.

Sigman, M., & Parmelee, A. H. Longitudinal evaluation of the preterm infant. In T. M. Field, A. M. Sostek, S. Goldberg, & H. H. Shuman (Eds.), *Infants born at risk.* New York: Spectrum, 1979, pp. 193–219.

Spinelli, D. N., & Jensen, E. E. Plasticity: The mirror of experience. *Science,* 1979, *203*, 75–78.

Sroufe, L. A. The coherence of individual development. *American Psychologist,* 1979, *34*, 834–841.

Sterman, M. B. Relationship of intrauterine fetal activity to maternal sleep stage. *Experimental Neurology,* 1967, Supplement 4, 98–106.

Taeusch, H. W., Keitner, M., & Avery, M. E. Accelerated lung maturation and increased survival in premature rabbits treated with hydrocortisone. *American Review of Respiratory Disease,* 1972, *105*, 971–973.

Taeusch, H. W., Jr., Wang, N. S., Baden, M., Bauer, C. R., & Stern, L. A controlled trial of hydrocortisone therapy in infants with respiratory distress syndrome: II. Pathology. *Pediatrics,* 1973, *52*, 850–854.

Thoman, E. G., & Becker, P. T. Issues in assessment and prediction for the infant born at risk. In T. M. Field, A. M. Sostek, S. Goldberg, & H. H. Shuman (Eds.), *Infants born at risk.* New York: Spectrum, 1979, pp. 461–485.

Twitchell, T. E. The neurological examination in infantile cerebral palsy. *Developmental Medicine and Child Neurology,* 1963, *5*, 271–278.

Twitchell, T. E. Normal motor development. *Physical Therapy,* 1965, *45*, 419–423. (a)

Twitchell, T. E. Attitudinal reflexes. *Physical Therapy,* 1965, *45*, 411–418. (b)

Twitchell, T. E. Variations and abnormalities of motor development. *Physical Therapy,* 1965, *45*, 424–430. (c)

Twitchell, T. E., & Ehrenreich, D. L. The plantar response in infantile cerebral palsy. *Developmental Medicine and Child Neurology,* 1962, *4*, 602–611.

Vygotsky, L. S. *Mind in society.* M. Cole, V. John-Steiner, S. Scribner, & E. Souberman (Eds.). Cambridge, Mass.: Harvard, 1978.

Werner, H. *Comparative psychology of mental development.* New York: International Universities Press, 1948.

Werner, H. The concept of development from a comparative and organismic point of view. In D. B. Harris (Ed.), *The concept of development.* Minneapolis: University of Minnesota Press, 1957, pp. 125–148.

Winter, P., & Funkenstein, H. H. The effect of species-specific vocalization on the discharge of auditory cortical cells in the awake squirrel monkey (Saimiri sciureus). *Experimental Brain Research,* 1973, *18*, 489–504.

APPENDIX

Manual for the Assessment of Preterm Infants' Behavior (APIB)

HEIDELISE ALS, BARRY M. LESTER,
EDWARD Z. TRONICK, AND T. BERRY BRAZELTON

INTRODUCTION

The following manual is an attempt to outline a strategy to systematically document behavioral ingredients of the prematurely born infant, from the stage when he can first be handled in room temperature and room air, without medical, technological aids, to the stage when the infant's attentional system is relatively independent from the other subsystems and when he can use it freely to regulate and control his inspection of the environment, i.e., by approximately one month post term for the healthy, fullterm infant.

The goal of this manual is to provide an instrument for the documentation of patterns of developing behavioral organization. The strategy of examination is broadly derived and adapted from the Brazelton Neonatal Behavioral Assessment Scale (Brazelton, 1973). The items of the scale are used as graded maneuvers in order to "test" the current status and organization of the infant's subsystems of functioning and their interplay.

The maneuvers are grouped into six larger packages, each of which places a specific demand on the infant and is intended to bring out the functioning of his various subsystems and their integration in the face of this demand.

This work is supported by grant #3122 from the Grant Foundation, grant #HD10899 from NICHD, and contract #278-78-0054 from NIMH. February 1979, revised August 1979 and April 1981.

ASSESSMENT OF PRETERM INFANT BEHAVIOR (APIB)

INFANT'S NAME | MED. REC. NO. | DATE OF BIRTH | AGE (Post-conception)

TIME – LAST FEEDING | TYPE OF FEEDING | CURRENT INTERVAL BETWEEN FEEDS

INITIAL CIRCUMSTANCES OF INFANT

POSITION: ☐ SUPINE ☐ PRONE ☐ SIDE
HEAD: ☐ RIGHT ☐ LEFT ☐ MIDLINE
COVERING: ☐ DIAPER ☐ SHIRT ☐ CLOTHES ☐ BLANKET(S)

INFANT'S INITIAL STATE | INFANT'S PREDOMINANT STATE

WEIGHT ____ LBS ____ OZS ____ GMS | HEIGHT ____ INCHES ____ CM | HEAD CIRCUMFERENCE ____ INCHES ____ CM | PONDERAL INDEX

DATE OF EXAM | TIME OF EXAM | PLACE OF EXAM | PERSONS PRESENT ☐MOTHER ☐FATHER ☐SIBLING(S) ☐OTHER ____

INTERFERING VARIABLES | EXAMINER | VIDEO | DURATION OF EXAM

SCORE SHEET I – SYSTEMS

LEGEND: B = Baseline R = Reaction P = Post-package Status

	ORDER OF PKG.	PHYSIOLOGY			MOTOR			STATE			ATTN/INTERACT			REGULATORY			EXAM FACIL
		B	R	P	B	R	P	B	R	P	B	R	P	B	R	P	
PACKAGE I SLEEP/DISTAL																	
PACKAGE II UNCOVER/SUPINE																	
PACKAGE III LOW TACTILE																	
PACKAGE IV MEDIUM TACTILE/VESTIBULAR																	
PACKAGE V HIGH TACTILE/VESTIBULAR																	
PACKAGE VI ATTENTION/INTERACTION																	

COMMENTS:

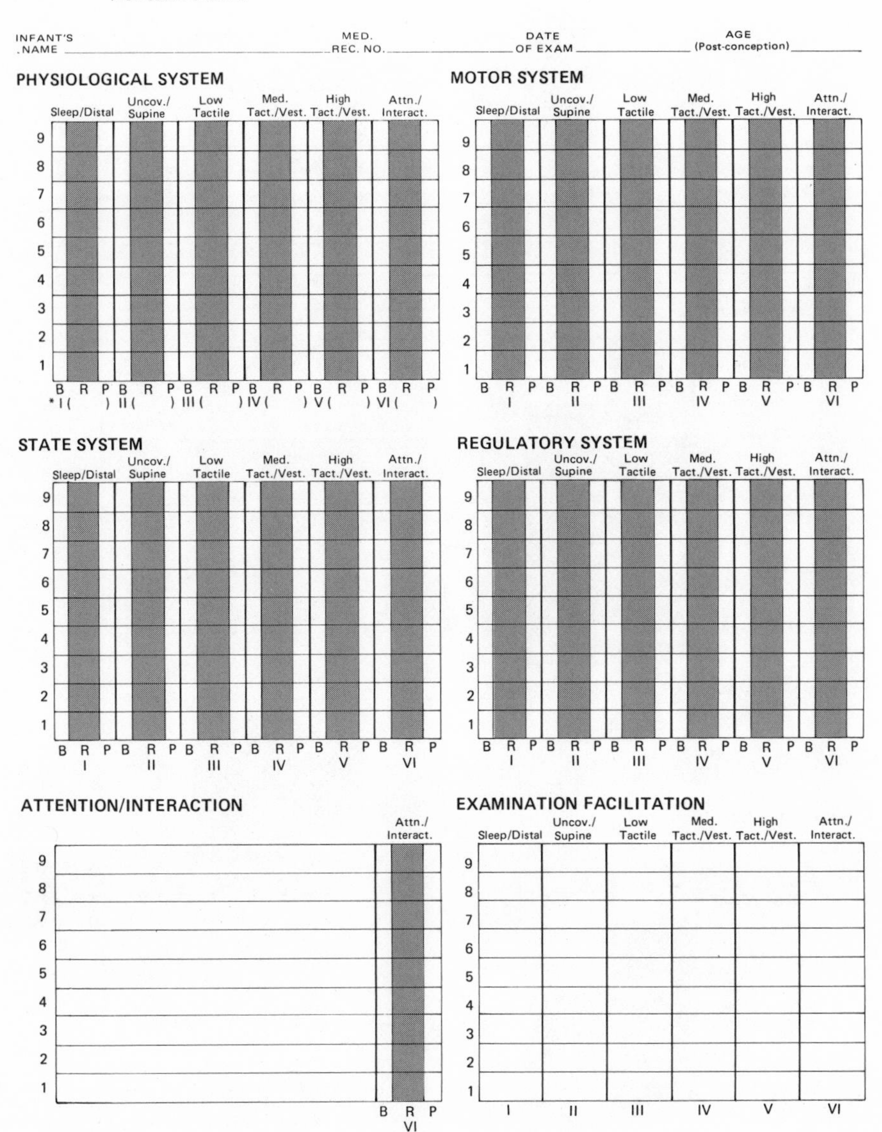

* Fill in order of administration

SCORE SHEET II – PACKAGES AND MANEUVERS

								ORDER
I: SLEEP/DISTAL								
LIGHT	Decrement	BNBAS	Ease of Elicitation	Timing	Recycling	Dis-organization	Discharge	
RATTLE	Decrement	BNBAS	Ease of Elicitation	Timing	Recycling	Dis-organization	Discharge	
BELL	Decrement	BNBAS	Ease of Elicitation	Timing	Recycling	Dis-organization	Discharge	
II: SLEEP PRONE/SUPINE								
UNCOVER	Capacity to deal with	→						
PRONE TO SUPINE	Capacity to deal with	→						
III: LOW TACTILE								
FREE FEET/HANDS	Capacity to deal with	→						
HEEL TOUCH	BNBAS	Ease of Elicitation	Timing	Recycling	Dis-organization	Discharge	→	
PLANTAR GRASP	BNBAS/R	BNBAS/L	→					
FOOT SOLE STROKE (Babinski)	BNBAS/R	BNBAS/L	→					
CLONUS	BNBAS/R	BNBAS/L	→					
PALMAR GRASP	BNBAS/R	BNBAS/L	→					
PALMAR MENTAL GRASP	APIB	→						
PASSIVE MOVEMENT ARMS	Resistance R	Resistance L	Recoil R	Recoil L	BNBAS/R	BNBAS/L	→	
PASSIVE MOVEMENT LEGS	Resistance R	Resistance L	Recoil R	Recoil L	BNBAS/R	BNBAS/L	→	
ARM/LEG DIFFERENTIATION	APIB	→						
GLABELLA	BNBAS	→						
ROOTING	BNBAS/R	BNBAS/L	→					
SUCKING	BNBAS	→						

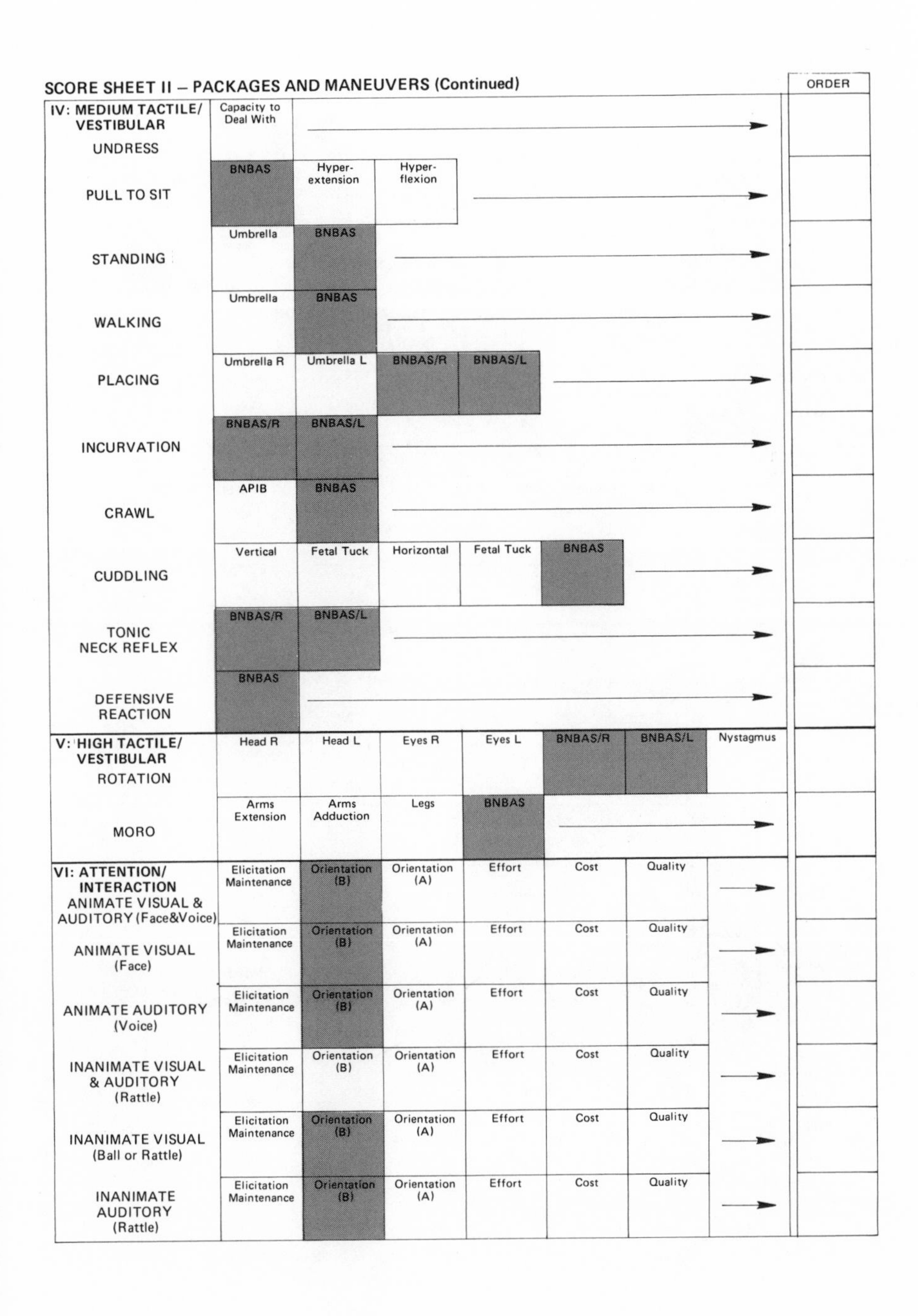

SCORE SHEET II – PACKAGES AND MANEUVERS (Continued)

								ORDER
IV: MEDIUM TACTILE/ VESTIBULAR UNDRESS	Capacity to Deal With	→						
PULL TO SIT	BNBAS	Hyper-extension	Hyper-flexion	→				
STANDING	Umbrella	BNBAS	→					
WALKING	Umbrella	BNBAS	→					
PLACING	Umbrella R	Umbrella L	BNBAS/R	BNBAS/L	→			
INCURVATION	BNBAS/R	BNBAS/L	→					
CRAWL	APIB	BNBAS	→					
CUDDLING	Vertical	Fetal Tuck	Horizontal	Fetal Tuck	BNBAS	→		
TONIC NECK REFLEX	BNBAS/R	BNBAS/L	→					
DEFENSIVE REACTION	BNBAS	→						
V: HIGH TACTILE/ VESTIBULAR ROTATION	Head R	Head L	Eyes R	Eyes L	BNBAS/R	BNBAS/L	Nystagmus	
MORO	Arms Extension	Arms Adduction	Legs	BNBAS	→			
VI: ATTENTION/ INTERACTION ANIMATE VISUAL & AUDITORY (Face&Voice)	Elicitation Maintenance	Orientation (B)	Orientation (A)	Effort	Cost	Quality	→	
ANIMATE VISUAL (Face)	Elicitation Maintenance	Orientation (B)	Orientation (A)	Effort	Cost	Quality	→	
ANIMATE AUDITORY (Voice)	Elicitation Maintenance	Orientation (B)	Orientation (A)	Effort	Cost	Quality	→	
INANIMATE VISUAL & AUDITORY (Rattle)	Elicitation Maintenance	Orientation (B)	Orientation (A)	Effort	Cost	Quality	→	
INANIMATE VISUAL (Ball or Rattle)	Elicitation Maintenance	Orientation (B)	Orientation (A)	Effort	Cost	Quality	→	
INANIMATE AUDITORY (Rattle)	Elicitation Maintenance	Orientation (B)	Orientation (A)	Effort	Cost	Quality	→	

SCORE SHEET III – BEHAVIORAL SUMMARY SCALES

PHYSIOLOGICAL PARAMETERS

TREMULOUSNESS	BNBAS			
STARTLES	BNBAS			
SKIN COLOR	Lability of Good Color	Lability of Comp. Color	Threshold	Jaundice
SMILES	APIB	BNBAS		

MOTOR PARAMETERS

TONUS	BNBAS	Balance		
MOTOR MATURITY	BNBAS	Threshold	Postural Control	Symmetry
ACTIVITY	Spontaneous Activity	Elicited Activity	BNBAS	
HAND-TO-MOUTH FACILITY	BNBAS			

STATE PARAMETERS

ALERTNESS	Degree (B)	Degree (A)	Quality	Am't. Manipulation
STATE REGULATION	Lability	Range and Flexibility	BNBAS	

SELF REGULATION PARAMETERS

WITHDRAWAL OR AVOIDANCE BEHAVIOR	CATALOG OF REGULATION MANEUVERS			
	Spit-ups	Gags	Hiccoughs	Bowel Mvt
	Grimace	Arching	Finger Splay	Airplane
	Salute	Sitting on Air		
	Sneezing	Yawning	Sighing	Coughing
	Averting	Frowning		

APPROACH OR GROPING BEHAVIOR	CATALOG OF REGULATION MANEUVERS			
	Tongue Extension	Hand on Face	Sounds	
	Hand Clasp	Foot Clasp	Fingerfold	Tuck
	Body Movement	Hand to Mouth	Grasping	Leg/Foot Bracing
	Mouthing	Suck Search	Sucking	Hand Hold
	Ooh Face	Locking	Cooing	

QUIETING	Self-quiet 6 ↑	Self-quiet Motor ↓	Consolability 6 ↑	Consolability Motor ↓
PEAK OF EXCITEMENT	BNBAS			
RAPIDITY OF BUILD-UP	Rapidity ↑ 6	Rapidity ↑ Motor		
IRRITABILITY	Irritability (B)	Irritability (A)		
ROBUSTNESS	Robustness			
CONTROL OVER INPUT	Control Over Input			
FACILITATION STIMULATION	Facilitation Stimulation			

SUMMARY
ATTRACTIVENESS

SUPPLEMENTAL LIST OF ASYMMETRIES

Check, rate degree and describe asymmetries noted; rate degree of asymmetry on a 0 – 3 continuum.

0 = no asymmetry noted (the item was not checked)
1 = subtly & mildly present and/or very transient
2 = moderately pronounced and/or intermittent
3 = pronounced, strong

Asymmetries	Check	Degree	Side	Description
1. Arm	______	______	________	______________________________
2. Hand	______	______	________	______________________________
3. Fingers	______	______	________	______________________________
4. Leg	______	______	________	______________________________
5. Foot/Toes	______	______	________	______________________________
6. Trunkal Posture	______	______	________	______________________________
7. Head Positioning	______	______	________	______________________________
8. Face	______	______	________	______________________________
9. Eyes	______	______	________	______________________________

Comments:

PART I. SYSTEMS ASSESSMENT

Each of the five systems—the physiological system, the motor system, the state system, the attentional-interactive system, and the regulatory system—is assessed and scored for each of the six packages. Three system scores, ranging from 1 to 9 each, are arrived at: a baseline score (B), a reaction score (R), and a post-package status score (P). A separate score (E) is derived for the degree and kind of facilitation of regulation necessary from the examiner to set the stage for a package or maneuver, to bring out the infant's best performance and to facilitate his return to a baseline for the next package.

Baseline (B)

In assessing the baseline of a system the infant is observed prior to the administration of the maneuvers making up a package, and an estimate of the baseline organization of the system is made. It may be necessary occasionally to give the infant up to a minute of time out between packages to have enough evidence to make a baseline assessment.

Reaction (R)

In assessing the reaction of a system the infant is observed throughout the administration of the maneuvers making up a package. Then the performance of the systems in the course of the administration of the maneuver of the package is assessed and scored. The degree of reactivity and relative disorganization of a system is captured.

Post Package Status (P)

This is the state of the system after the infant has been brought through the maneuvers of a package and reflects an aspect of the infant's own regulatory ability. This score captures the level of organization to which the system can return without aid from the examiner.

Examiner Facilitation (E)

Various maneuvers of the examiner may be necessary to maintain an infant's self-regulation or to bring the infant back to a stabler baseline from which the next package can be administered. The maneuvers are seen as graded and cumulative interventions designed to optimize the infant's performance and his balance. The score reflects the degree of aid necessary.

Order of Package Administration

In the first column, the order in which the packages were administered is indicated. This will allow the examiner to reconstruct the flow of the examination.

Systems

I. PHYSIOLOGICAL SYSTEM. Behavioral indices of the physiological system assessed by observation are the infant's respiratory pattern, occasionally his heart rate, his skin color, autonomically mediated movements such as tremors, startles, autonomic eye movements such as eye floating and eye rolling, sounds such as sighs and whimpers, and behavioral indices of visceral control such as hiccoughing, spitting up, gagging, bowel movement straining, and grunting. The intensity or severity of the signs present, not their absolute number, is decisive in the assignment of a particular score.

SCORING

1. Smooth regular respirations and heart rate (if assessed), good healthy color, no spontaneous tremors or startles, no facial twitches or autonomic movements, no signs of upset visceral response.

2. Very mild respiratory irregularity, very mild cyanosis, paleness, webbing or flushing, mild occasional tremors or startles, occasional mild facial twitches, sighs or whimpers, possibly a hiccough or mild gag.
3. Mild respiratory unevenness, some cyanosis, paleness, or webbing or flushing, some tremors or startles, some facial twitches, some floating eye movements, some mild gagging, whimpering, or occasional hiccough.
4. Mild to moderate respiratory unevenness, mild to moderate cyanosis, paleness, webbing or flushing, repeated tremors or startles, repeated facial twitches, occasional gagging or mild eye movements or hiccoughing or bowel movement straining or whimpering or mild floating eye movements.
5. Moderate respiratory unevenness, i.e., several periods of very mild apnea, or a period of moderate tachypnea, moderate cyanosis, paleness, webbing or flushing, moderate degree of tremors and startles, and/or moderate facial twitches, whimpers, hiccoughing, gagging, spitting up, eye floating, autonomic eye movement, or bowel movement straining.
6. Moderate to considerable respiratory unevenness, moderate to considerable tachypnea or mild apnea, moderate to considerable cyanosis, paleness, webbing, or flushing, quite considerable degree of tremors and/or startles, quite considerable facial twitches and autonomic eye movements, whimpers, hiccoughing, gagging, spitting up, or bowel movement straining.
7. Considerable respiratory unevenness, considerable tachypnea and/or mild to moderate apnea, considerable cyanosis, paleness, webbing or flushing, considerable degree of tremors and/or startles, considerable facial twitches, autonomic eye movements, hiccoughing, whimpers, gagging, spitting up or bowel movement straining.
8. Quite severe respiratory unevenness, apneaic episodes, retraction or nasal flaring, or quite severe tachypnea; quite severe cyanosis, paleness, webbing or flushing, quite severe degree of facial twitches, autonomic eye movements, whimpers, hiccoughing, gagging, spitting up and/or bowel movement straining.
9. Severe, definite apnea episodes with retraction or considerable nasal flaring, or considerable tachypnea accompanied by dusky color, or very pale or very "webbed" or flushed red color, tremors and/or severe facial twitches, and/or severe eye rolling.

II. MOTOR SYSTEM. Behavioral indices of the motor system assessed by observation are reflected in the infant's posture, movements, tonus, and amount and degree of differentiation of activity.

SCORING

1. Consistently smooth, well modulated controlled posture. If movement, smooth postural changes and adjustments, well modulated smooth movements of the extremities and head, good consistent tone throughout the body, moderate amount of smooth activity with well differentiated hand, arm, and leg movements.
2. Largely well modulated and controlled posture. If movement, almost consistently modulated tone; no sudden arm or leg extensions; activity is predominantly modulated, and hand, arm, and leg movements are moderately well differentiated.
3. Somewhat flaccid, flexed or extended posture. If movement, considerable periods of modulated controlled posture, only occasional fluctuations to hyperextension or hyperflexion, only an occasional arm or leg extension (salute) into midair, tone fairly consistent with mild hypertonic or mild hypotonic episodes, only brief periods of mildly frantic or diffuse activity, only occasional whole arm and leg movements.
4. Mildly to moderately flaccid, flexed or extended posture. If movement, moderate periods of modulated controlled posture alternating with brief episodes of hyperextension and/or hyperflexion, mild or infrequent arm and leg extensions (salute) into midair; tone only occasionally fluctuating between hypertonic or hypotonic, mainly consistently moderately one or

the other, infrequent periods of frantic or diffuse activity, infrequent sudden arm or leg movements.

5. Moderately flaccid, flexed or extended posture. If movement, occasional periods of modulated flexed posture alternating with episodes of hyperextension and hyperflexion in fluctuation, occasional moderate arm and leg extensions into midair (salute), tone largely synchronous, either hypertonic or flaccid with brief fluctuations, yet occasional periods of frantic diffuse movement, occasional undifferentiated whole arm and leg movements.
6. Moderately to considerably flaccid, flexed or extended posture. If movement, some fluctuation of hyperextended posture and hyperflexed posture, very brief periods of modulated flexion, occasional sudden, abrupt changes and adjustments. Some amount of jerky movements with several dramatic arm and leg extensions into midair (salute), tone somewhat variable between hypertonic and flaccid, moderate fluctuation, some periods of frantic, diffuse activity, some undifferentiated whole arm and leg movements.
7. Considerably hyperextended or hyperflexed or flaccid posture. If movement, moderate fluctuation of hyperextended posture and hyperflexed posture with sudden abrupt changes and adjustments; jerky movements with a considerable amount of dramatic arm and leg extensions into midair (salute), tone variable much of the time between flaccid and hypertonic with considerable fluctuation, moderately frantic activity or several bouts of frantic activity alternating with no or very mild activity, predominantly undifferentiated whole arm and leg movements.
8. Strongly hyperextended, hyperflexed, or almost completely flaccid posture. If movement, hyperextension alternating with sudden, abrupt changes and jerky adjustments; very jerky movements with dramatic arm and leg extensions into midair (salute), completely flaccid tone alternating with hypertonicity in dramatic fluctuation, excessive, frantic or diffuse activity alternating with no activity at all; undifferentiated whole arm and leg movements.
9. Completely flaccid posture essentially without active adjustments, tonus, or activity.

III. STATE SYSTEM. Various configurations of behaviors encompassing eye movements, eye opening and facial expressions, gross body movements, respirations, and tonus aspects are used in specific temporal relationships to one another to determine at what level of consciousness an infant is at a particular time. Although Prechtl *et al.* (1979) state that only by 36 weeks can states be identified, we believe that it is possible to make meaningful, systematic distinctions between dynamic transformations of various behavioral configurations which appear to correspond to varying states of availability and conscious responsiveness. We suggest the following spectrum of observable states. States labeled as A states are "noisy," unclean, and diffuse (premie states); states labeled as B states are clean, well defined states.

Sleep States

State 1A: Infant in deep sleep with momentary regular breathing, eyes closed, no eye movements under closed lids; relaxed facial expression; no spontaneous activity, oscillating fairly rapidly with isolated startles, jerky movements or tremors and other behavior characteristic of State 2 (light sleep).

State 1B: Infant in deep sleep with predominantly regular breathing, eyes closed, no eye movements under closed lids, relaxed facial expression; no spontaneous activity except isolated startles.

State 2A: Light sleep with eyes closed; rapid eye movements can be observed under closed lids; low activity level with diffuse or disorganized movements; respirations are irregular and there are many sucking and mouthing movements, whimpers, facial twitchings, much grimacing; the impression of a "noisy" state is given.

If the infant moves from states 1A or 2A to a more stressed state, becoming diffusely unreachable due to respiratory pauses, this is marked 1AA or 2AA, depending on the state in which this severe diffuseness is embedded. The AA notation may occasionally be necessary in 3A or 5A. The examination may have to be shortened in such cases.

State 2B: Light sleep with eyes closed; rapid eye movements can be observed under closed lids; low activity level with movements and dampened startles; movements are likely to be of lower amplitude and more monitored than in state 1; infant responds to various internal stimuli with dampened startle. Respirations are more irregular, mild sucking and mouthing movements can occur off and on; one or two whimpers may be observed, as well as an isolated sigh or smile.

Transitional States

State 3A: Drowsy
Drowsy or semi-dozing; eyes may be open or closed, eyelids fluttering or exaggerated blinking; if eyes open, glassy veiled look; activity level variable, with or without interspersed, mild startles from time to time; diffuse movement; fussing and/or much discharge of vocalization, whimpers, facial grimacing, etc.

State 3B: Drowsy, same as above but with less discharge of vocalization, whimpers, facial grimacing, etc.

Awake States

State 4: Alert

4AL: Awake and quiet, minimal motor activity, eyes half open or open but with glazed or dull look, giving impression or little involvement and distance; or focused, yet seems to look through, rather than at, object or examiner; or the infant is clearly awake and reactive but has his eyes open intermittently.

4AH: Awake and quiet, minimal motor activity, eyes wide open, "hyperalert" or giving the impression of panic or fear; may appear to be hooked by the stimulus, seems to be unable to modulate or break the intensity of the fixation.

4B: Alert with bright shiny look; seems to focus attention on source of stimulation and appears to process information actively and with modulation; motor activity is at a minimum.

State 5: Active

5A: Eyes may or may not be open, but infant is clearly awake and aroused, as indicated by his motor arousal, his tonus, and his mildly distressed facial expression, grimacing, or other signs of discomfort; fussing is diffuse.

5B: Eyes may or may not be open, but infant is clearly awake and aroused, with considerable, well defined motor activity. Infant is also clearly fussing but not crying.

State 6: Crying

6A: Intense crying, as indicated by intense grimace and cry face, yet cry sound may be very strained or very weak or even absent.

6B: Rhythmic, intense crying which is robust, vigorous, and strong in sound.

Behavioral indices of the state system assessed by observation are the range of states the infant has available and the degree and flexibility of modulation the infant has in moving from state to state and in maintaining a quiet, alert state. These are scaled from 1–9 and are recorded in the top portion of the boxes for B, R, and P for each package. In the lower portion of the boxes the predominant or typical state (1A, 1B, 2A, 2B, 3A, 3B, 4AL, 4AH, 4B, 5A, 5B, 6A, 6B) is recorded for B, R, and P of each package. For each box, up to 3 or more states can be recorded. This will preserve the overall fluctuation and range of states available and is merely an aid for the memory of the examiner.

Scoring

1. The infant comes to a well-defined state 4B and may move only briefly to 5B, 6B or 3B; or he maintains a solid state 1B or 2B in package I or II, or the baseline for package III.
2. The infant comes into a state 4B for brief periods; the states he comes from or goes to are well-defined states 5B, 6B, or 3B; or the infant maintains a less well-defined state 1A or 2A during package I or II, or the baseline for package III.
3. The infant may come into a state 4AH or 4AL, although there are also brief periods of 4B with oscillations to state 5B or 6B, except during package I or II, or the baseline for package III, when he stays in a poorly defined state IA, 2A, or 3A.
4. The infant comes into a state 4AL or 4AH and 4B minimally, but there are oscillations to 3A with B, 5A and/or 6A with B. One or the other state bordering on the 4 states is a clear B state, at least at times.
5. The infant comes into a state 4AH or 4AL, with oscillations to state 3A with B, 5A or 6A with B; or he is in 5A and B or 6A and B exclusively and 4A or B cannot be achieved.
6. The infant has clean states 1B, 2B, or 3B available and may come into 5B or 6B briefly; or he repeatedly oscillates between 3B and 5B or 6B.
7. The infant has periodic, sudden brief shifts from states 1A or B, 2A or B, or 3A or B to state 5A and 6A; or he is in 5A or 6A more or less continuously.
8. The infant has periodic, sudden brief shifts from states 1A with B or 2A with B to state 3A with B, or he maintains himself essentially mainly in state 3A and 3B. In package I, II, or the baseline for package III, he briefly moves to 1AA, 2AA, or 3AA from other sleep states: he recovers spontaneously.
9. The infant moves only between states 1A and 2A and at the most 3A. In package I, II, or the baseline for package III, he moves to AA states and needs facilitation to recover.

IV. ATTENTIONAL/INTERACTIVE SYSTEM. Behavioral indices of the attentional/interactive system are the quality of the infant's alert state, its robustness and its availability, the duration of the infant's responsivity to animate and inanimate stimuli, and the modulation and differentiation with which the infant utilizes his alertness to attend to and interact with various social stimuli and inanimate objects. Some infants cannot be brought to an alert state by manipulations, depending on the point of their internal state rhythmicity. They may come to an alert state spontaneously at a later time. Nevertheless, it is meaningful information to know that the infant could not be brought to alertness by manipulations at a given time, since this reflects the relative degree of flexibility and differentiation the infant has available to modify his ongoing state regulation by responding to external events. The timing of the examination in the infant's sleep/wake/feed cycle and the infant's initial state need to be recorded and taken into account. Very young infants tend to be more easily arousable closer to their next awakening or rousing (feeding) than at midpoint between 3-hour feeds. Timing of the examination should be controlled for as much as possible in respect to the infant's current endogenous, sleep/wake or rouse cycle. For packages I through V this score is optional.

Scoring

1. Attention and responsivity are robust, of long duration; the infant actively selects his target of attention and shifts smoothly from one target to another at his own initiative, beginning

free inspection of environment. His face reflects bright-eyed, focused interest, his mouth and eyebrows may be involved, there may be an occasional coo or smile and/or cyclical facial adjustments of heightened and relaxed attention in alternation.

2. The infant's attention is of considerable duration, predominantly bright, focused, robust, and well-modulated, and there may be an occasional episode of active environment inspection or target selection.
3. The infant's attention is of moderate duration and robustness, predominantly bright, focused, and well-modulated, directed and maintained by the stimulus.
4. The infant's attention is of moderate duration and consistency, at times bright, focused and well-modulated, at other times it is either low-keyed or slightly hyperreactive.
5. The infant's attention is consistently low-keyed, moderately well-organized, and of relatively short duration.
6. The infant's attention is moderately variable, may have periods of hyperalertness or diffuseness which can get intermittently modulated to more focused attention. There are some brief episodes of modulated processing embedded in more diffuse attention.
7. The infant's attention is highly variable, shifting from hyperalert or diffusely alert, possibly intermittent panicked facial expression or overly wide eyes to unavailable with floating or rolling eyes and/or eye close in rapid shifts. The duration is short, and there appears to be little modulation or control on the infant's part.
8. The infant's attention or alertness is only barely available for fleeting periods. Then the quality of his attention is either diffuse or, although focused, very transient.
9. Alertness and attention cannot be achieved at all.

V. REGULATORY SYSTEM. Behavioral indices of the regulatory system assessed by observation are reflected in the infant's use of varying physiological, postural, and/or state strategies to maintain himself and to return to a balanced baseline consisting either of solid sleep states or calm alert states, the degree to which the infant is able to maintain himself, and the level at which he maintains himself.

SCORING

1. The infant maintains himself easily either at a well-modulated state 4 level or in a calm sleep state, without losing regulatory balance.
2. The infant maintains himself successfully most of the time and/or can return to balance fairly easily and consistently. Some of his regulation strategies are highly sophisticated and differentiated (sneezing, yawning, subtle attentional cycling, etc.).
3. The infant makes consistent efforts at maintaining himself and at returning to balance. He is generally able to do so, although at times with some difficulty.
4. The infant makes repeated, prolonged, and differentiated efforts at maintaining himself, and at returning to balance. He repeatedly is able to maintain himself for moderate periods or to return to balance occasionally.
5. The infant makes repeated efforts at maintaining himself and at returning to balance, some of his efforts are quite differentiated and some are successful.
6. The infant makes several noticeable efforts at maintaining himself and at returning to balance; they may be gross and/or minimally and transiently effective.
7. The infant makes efforts at maintaining himself in balance and at returning to balance, yet they are unsuccessful.
8. The infant cannot regulate himself at all. He responds to maneuvers and then is completely at the mercy of the manipulations. He cannot regain even partial balance.
9. No effort at behavioral self-regulation is noticeable; the infant is essentially not responding to manipulations, and self-regulation is not an issue.

VI. EXAMINER FACILITATION. Behavioral indices of the examiner's facilitation are found in the degree and amount of facilitation necessary from the examiner to set the stage for a package or a maneuver within a package to bring out the infant's optimal performance and to help him return to a baseline for the next package or maneuver. This examiner facilitation is scored in a summary fashion for each package. It requires the examiner's sensitive awareness of the infant's own regulatory capacities so that he times and gauges his help appropriately. It furthermore requires substantial experience in handling infants in order to determine and implement appropriate facilitation.

SCORING

1. The infant is able consistently to maintain himself and the examiner can proceed with ease; the infant may actively seek out the examiner for differentiated interaction (play dialogue).
2. The infant needs occasionally very mild aids such as occasional brief time out or adjustment of the interaction presentation; i.e., the examiner will present a maneuver somewhat more delicately or slowly.
3. The infant needs some facilitation from the examiner in order to maintain or regain his balance. This facilitation may be mild, such as movement inhibition by occasional handholding or by postural shifting, allowing the infant to brace his feet against the crib or the examiner, or some time out; or the infant sucks on the examiner's finger or his own pacifier but can maintain this on his own, essentially.
4. The infant needs a moderate degree of facilitation from the examiner in order to maintain or regain his balance. This help may consist of motoric inhibition by wrapping the infant in a blanket or holding the infant's hands or feet for moderate periods or by repeated time out.
5. The infant needs a considerable degree of facilitation from the examiner in order to maintain or regain his balance. This facilitation may consist in repeated moderately long times out, motoric inhibition by hand holding, postural facilitation, and wrapping.
6. The infant needs quite a substantial degree of facilitation from the examiner in order to maintain or regain his balance. He may need repeated periods of considerable motoric inhibition and repeated time out, or he may be maintained with examiner induced and facilitated sucking. If these are provided he maintains himself quite easily each time.
7. The infant needs a substantial degree of help from the examiner in order to maintain or regain his balance. He may need prolonged and repeated motoric inhibition and sucking or lengthy periods of time out, with or without sucking, which eventually free his self-balance and maintenance for brief periods.
8. The infant needs a very substantial degree of help from the examiner, either consisting of vertical rocking which brings him to balance fairly readily or consisting of complete motoric inhibition and sucking and long resting periods, which work only moderately well.
9. The infant needs continuous, very carefully instituted and very considerable amounts of regulatory aid from the examiner, such as large vertical rocking for prolonged periods, to bring him at least momentarily to examinable state, prolonged complete motoric inhibition with sucking and prolonged resting periods, which work only barely, or the infant needs to be completely left alone, and the examination is inappropriate.

PART II. PACKAGES AND MANEUVERS

The maneuvers used in the Assessment of Preterm Infant Behavior (APIB) are grouped into the six larger packages outlined in the introduction. All maneuvers are scored separately in order to retain as much detailed behavioral information as possible. There are some differences in the administration of some of the maneuvers and the overall flow of the examination between this examination and the BNBAS (Brazelton, 1973). These are necessary because of the nature of the premature infant's organization and because of the continuous focus on the degree of organization and differentiation of each of the infant's subsystems of functioning when the various maneuvers are presented.

The "reflexes," for instance, are not primarily used in the traditional neurological sense to assess neurological intactness but provide Systematic Elicitations of Specific Movements (SEM), manipulations to bring out the range of various movements, postural and tonus capacities of the infant, and document their developmental course, while simultaneously watching for the effect of such manipulations on the autonomic, the motor, the state, and the regulatory systems. The administration of each of the Systematic Elicitations of Movement (SEM) is consistent with and derived from descriptions of Prechtl and Beintema (1964) where appropriate. Within each package these maneuvers are administered in as smooth a fashion as possible in order to control for idiosyncratic additional postural changes and manipulations between SEMs which would influence unsystematically the infant's performance and organization. The order in which SEMs are administered is recorded in the far right column of the score sheet. This will permit the reconstruction of the flow of the assessment in detail, which carries much implicit information about the infant's level of organization. If items had to be deleted, this is indicated by *A,* meaning the infant was too aroused, too weak, or otherwise too stressed to administer the item meaningfully; *X* indicates the examiner forgot to administer an item. *C* means the item was deleted because the circumstances were inappropriate, e.g., the infant was already undressed, etc.; *N* indicates an item is not scorable because another score makes it logically inappropriate to assign simultaneously this score.

In scoring each maneuver within a package, an attempt has been made to either retain the scoring of the BNBAS where appropriate or to provide a place and method for parallel scoring so that where applicable the BNBAS score for an item can be preserved. This will allow the user to compare data collected with the new manual to previously collected bodies of data on healthy term infants with the BNBAS. The aim of the new manual is usefulness for the assessment of preterm and term infants.

The behavior scales for all maneuvers and items are scaled from 1 to 9, with 1 meaning "little or none of a behavior" and 9 meaning "a lot of a behavior," rather than being scaled from good to bad or vice versa. This is in keeping with the BNBAS and protects against routinized generalization from one scale to another. Each scale has to be considered separately.

Administratively, a few items have been changed, which should not influence the overall flow of the examination considerably:

1. The use of the bell in the sleep/distal package is optional: If the infant stays well organized during the light and rattle items, the bell is used; if the infant is stressed during light or rattle, the bell is deleted.
2. Pinprick has been replaced by a touch with a dull-pointed plastic stick (orange stick).
3. An inanimate visual *and* auditory stimulus has been introduced, consisting of the red plastic rattle.

The flow of the examination is somewhat more systematized by the notion of the packages and by the examiner's attention to the infant's self-regulation and his own facilitation necessary to bring the infant repeatedly to a balanced state. The infant's organization always leads the examiner, yet he has the observation of clear organizational issues in mind at all times and interacts with the infant in such a way as to make these observations maximally possible.

The second part of this manual is structured to give a brief description of the goal of each package, the administration of each maneuver or SEM within each package, the scoring procedures for each, and the method to assess parallel scores where applicable. The following codes are used where numerical scores are not appropriate for the reasons specified:

A =	n/a	not administerable; infant too stressed
C =	n.a.c.	not appropriate circumstances, e.g., hands already free
R =	n.r.	no response
N =	n/N	not needed because item does not logically apply
X =		inadvertently omitted

Package I: Sleep/Distal

(Maneuvers: Repeated stimulation; light, rattle, bell)

One of the adaptive mechanisms developing in the newborn organism is his capacity to maintain a sleep state and decrease responses to repeated distal stimuli. In package I, an attempt is made to measure the infant's ability to deal with at least two repeated distal stimuli: a flashlight beam across his closed eyes repeatedly, and a soft rattle, rattled near one of his ears repeatedly. The bell may be deleted for the assessment of the premature infant since it may prolong this package inordinately and frequently becomes too taxing. Should an infant show good response decrement to light and rattle, the bell may be introduced. Some infants will not stay in a sleep state in the course of the first distal maneuver, the light. This is taken as a sign of the quality of their sleep organization and is assessed as such. Should this occur, the second response decrement item will not be administered. If there is no initial response, a second and, if necessary, a third stimulus is presented in 5-second intervals. If there is still no response, this will be indicated in the *Difficulty in Elicitation of Initial Response Scale.* The examiner should go on to the second response decrement item.

If the infant stays asleep, up to 10 stimuli are used to assess his ability to maintain a sleep state. An 11th stimulus is presented if the infant shows shut-down on trial 10. The passing of two consecutive trials without response is taken as the criterion for assessing "shut down."

Stimuli should be presented approximately 5 seconds after the end of the previous response. This implies the observer's ability to judge the end of each reaction. The test should be carried out using a standard 8-inch flashlight in good working condition. A semi-lighted quiet environment is desirable for this assessment. The sound stimulus used is a small red plexicontainer one-third filled with corn kernels. It is shaken in three consecutive shakes of approximately half a meter above or to the side of one of the infant's exposed ears.

LIGHT

Degree of Decrement. If an infant achieves a score of 9 on the APIB scale, then he should be scored on the BNBAS scale, and a score between 6 and 9 can be identified. BNBAS scale equivalents to the other scores are given in parentheses. If there is no response in 3 trials, give R (no response).

BNBAS Equivalent	Scoring
(1)	1. Increase of response over course of presentation and sleep state disrupted.
(1)	2. Increase of response over course of presentation, yet sleep state generally maintained.
(2)	3. Startles still present at end of 10 trials
(3)	4. Large body movement still present at end of 10 trials
(3)	5. Medium body movement and/or head movement still present at end of 10 trials
(3)	6. Facial grimaces and/or small sounds and/or minute body movements (finger, foot) still present at end of 10 trials
(4)	7. Blinks still present at end of 10 trials
(4)	8. Moderate respiration response still present at end of 10 trials
(6–9)	9. Shut-down achieved within 10 trials

Degree of Decrement (BNBAS)

1. No diminution of high responses over the 10 stimuli
2. Delayed startles and rest of responses are still present, i.e., body movement, eye blinks, respiratory changes continue over 10 trials
3. Startles no longer present but rest are still present, including body movement, in 10 trials

4. No startles, body movement delayed, respiration and blinks same in 10 trials
5. Shutdown of body movements, some diminution in blinks and respiratory changes in 9–10 stimuli
6. —in 7–8 stimuli
7. —in 5–6 stimuli
8. —in 3–4 stimuli
9. —in 1–2 stimuli

NA No response, hence no decrement.

Ease of Elicitation of Initial Response

Some premature infants are difficult to reach with environmental stimuli. This is not always a function of the level of sleep but may have to do with their generally less mature state control. Therefore, it becomes important to record the effort necessary in eliciting a response from which decrement can be observed.

Scoring

1. Initial response cannot be observed despite three stimulus presentations
2. Two or three stimuli needed before full initial response consisting of blinks and/or noticeable respiratory changes
3. Two or three stimuli needed before full initial response consisting of facial movement or sound
4. Two or three stimuli needed before full initial response consisting of startle or body movement
5. No difficulty, full initial response consisting of respiratory changes
6. No difficulty, full initial response consisting of eye blink
7. No difficulty, full initial response consisting of facial movement and sound
8. No difficulty, full initial response consisting of mild body movement or head move
9. No difficulty, good full initial response consisting of startle or body movement

Timing of Response

Some premature infants respond with excessive hyper-reactivity to any stimulation. Gradually inhibitory mechanisms mature. Before there is a modulated balance between inhibition and excitation, a period of varying degrees of delay to stimuli is observed. The degree of modulation in timing may reflect the capacity of the central nervous system to modulate its responsiveness. Little or no delay is expected in the healthy term infant, without excessively quick responsiveness.

Scoring

1. All responses instantaneous with *onset* of stimulus
2. Some responses instantaneous with onset of stimulus, some showing beginning inhibition
3. Considerable delay of most or all responses
4. Considerable delay of some responses
5. Moderate delay of most or all responses
6. Moderate delay of some responses
7. Minimal delay of most or all responses
8. Minimal delay of some responses
9. Responses in modulated interval from onset of stimulus

Recycling of Response

The quality of responses is variable in the premature newborn. A commonly observed feature is the lack of inhibition of an ongoing response leading to recycling of the initial response. The degree and amount of recycling can be used as a measure of developing inhibition.

SCORING

1. No recycling noted on any trials presented
2. Response shows mild recycling on one or a few trials only
3. Response shows moderate recycling on one or a few trials
4. Response shows moderate recycling on several trials
5. Response shows prolonged recycling on one or a few trials
6. Response shows prolonged recycling on several trials
7. Response shows prolonged recycling on most or all trials
8. Response shows a considerable degree of recycling on most or all trials
9. Response shows a considerable degree of recycling with increasing magnitude, which makes the continuation of the item impossible

Degree of Disorganization of Response

Some premature and at risk infants are unable to shut down to redundant stimuli and with each successive stimulus react with increasing behavioral disorganization. Other infants attempt to shut down but are unable to maintain shutdown criteria. Fluctuation of high and low level responses characterize these infants. For example, we may see delayed responses, recycling, diffuse body movements, facial expressions, color changes, hiccoughing, and other such signs of disorganization. Still other infants have variable responses but eventually achieve criteria for shutdown in 10 trials. And, finally, some infants are able to achieve decrement without variable responses.

SCORING

1. Gradual diminution of high level to low level responses or low level response maintenance without fluctuation; shutdown achieved
2. Gradual diminution of high level to low level response, or low level response maintenance without fluctuation, and without shutdown; or high level responses without sleep state disruption
3. Periods of sustained low level responses with occasional return of a high level response, followed by shutdown
4. Periods of sustained low level responses with occasional return of a high level response, without achievement of shutdown
5. Variable high and low level responses followed by shutdown
6. Signs of disorganization may be present repeatedly between trials, yet shutdown is achieved
7. Repeated fluctuation between high and low level responses with some disorganization and without meeting shutdown
8. Signs of disorganization may be present repeatedly between trials and shutdown is not achieved
9. Attains increasing disorganization with succeeding stimulus presentations. Unable to proceed with decrement item(s)

Degree of Discharge after Termination of Stimulus Sequence

Some premature infants may be able to shut down body movement, facial movement, and even respiratory reactivity to repeated stimuli presented in sleep, but once the stimulation is terminated they show a physiological imbalance, motoric discharge, and state disorganization, indicating the degree of imbalance between inhibition and excitation.

SCORING

1. There is no discharge noted after termination of stimulus sequence.
2. There is a mild respiratory response, possibly the sigh response, after termination of stimulus sequence.
3. There is some facial movement after termination of the stimulus sequence.
4. There is mild motor discharge after termination of stimulus sequence.
5. There is moderate motor discharge after termination of stimulation sequence.
6. There is prolonged motor discharge after termination of stimulus sequence.
7. There is noticeable tachypnea or a brief episode of apnea, nasal flaring, gasping, etc., after termination of stimulus sequence.
8. There is prolonged tachypnea or a moderate episode of apnea after termination of stimulus sequence.
9. There is prolonged discharge of motoric or physiological nature leading to state change after termination of stimulus sequence.

RATTLE

Degree of Decrement. Scoring 1 through 9.
Degree of Decrement (BNBAS). Scoring 1 through 9.
Ease of Elicitation of Initial Response. Scoring 1 through 9.
Timing of Response. Scoring 1 through 9.
Recycling of Response. Scoring 1 through 9.
Degree of Disorganization of Response. Scoring 1 through 9.
Degree of Discharge after Termination of Stimulus Sequence. Scoring 1 through 9.

BELL (OPTIONAL)

Degree of Decrement. Scoring 1 through 9.
Degree of Decrement (BNBAS). Scoring 1 through 9.
Ease of Elicitation of Initial Response. Scoring 1 through 9.
Timing of Response. Scoring 1 through 9.
Recycling of Response. Scoring 1 through 9.
Degree of Disorganization of Response. Scoring 1 through 9.
Degree of Discharge after Termination of Stimulus Sequence. Scoring 1 through 9.

Package II: Sleep: Uncover and Prone to Supine

(Maneuvers: Uncovering and placing into supine)

The infant's ability to deal with and adjust to being uncovered and then placed into supine position is assessed. There will be differences in the way the baby is found at the beginning of the examination. Some infants will be wrapped tightly in several blankets and placed in prone, possibly with blanket rolls against their backs, sides, or legs; others may be covered only lightly or not at all; some may already be in supine. The assessment of the degree of autonomous regulation demanded and available to the infant throughout by these two procedures is the goal of these two scales. In certain circumstances this cannot be assessed. It should then be scored C. On the basis of the infant's behavior during the response decrement items during the initial observation period, a judgment is arrived at as to how carefully the infant needs to be uncovered and placed into supine. Once these maneuvers are initiated, the infant is observed closely, and the degree of care in handling can be adjusted.

CAPACITY TO DEAL WITH UNCOVERING

Scoring

1. The infant is uncovered or unwrapped very gradually and very gently. The infant goes into hyperextension or flexion and/or changes out of sleep state. The disorganization is severe and does not abate.
2. The infant is uncovered with great care; there is moderate disorganization physiologically and motorically. The infant attempts to regain control by postural adjustment but fails to do so.
3. The infant is uncovered with great care; there is moderate to mild disorganization physiologically and motorically. The infant successfully regains adjustment eventually.
4. The infant is uncovered with great care; there is no, minimal, or very brief disorganization physiologically and motorically, and the infant regains adjustment gradually.
5. The infant can be uncovered with moderate care, and there is moderate disorganization, which abates.
6. The infant can be uncovered with moderate care, and there is minimal, brief, or no disorganization.
7. The infant can be uncovered without special adjustments and care, and there is moderate disorganization which abates.
8. The infant can be uncovered without special adjustments and care, and there is minimal disorganization.
9. The infant is already uncovered and shows no disorganization, or the infant can be uncovered without special adjustments and care and shows no disorganization.

CAPACITY TO DEAL WITH BEING PLACED IN SUPINE

Scoring

1. The infant is placed in supine very gradually, carefully and gently; the infant goes into hyperextension and/or hyperflexion and/or changes out of sleep state. The disorganization is severe and does not abate.
2. The infant is placed in supine very carefully and gently; there is moderate disorganization physiologically and motorically. The infant attempts to regain control but fails to do so.
3. The infant is placed in supine very carefully and gently; there is moderate to mild disorganization physiologically and motorically. The infant successfully regains adjustment eventually.
4. The infant is placed in supine very carefully and gently; there is minimal and brief disorganization motorically and physiologically, and the infant regains adjustment gradually.
5. The infant can be placed in supine with moderate care; there is only moderate disorganization which abates.
6. The infant can be placed in supine with moderate care, and there is only minimal brief disorganization.
7. The infant can be placed in supine without special adjustments and care, and there is only moderate disorganization which abates.
8. The infant can be placed in supine without special adjustments and care, and there is only minimal disorganization.
9. The infant is already in supine and shows no disorganization; or the infant can be placed in supine without special adjustments and care and shows no disorganization.

Package III: Low-Grade Localized Tactile Input to Extremities and Face/Supine

(Maneuvers: Freeing of the feet and hands; foot:SEM; hand:SEM; passive movements of arms and legs; heeltouch; glabella; root and suck)

The maneuvers in this package all provide opportunities to observe the infant's capacity to deal with delimited tactile inputs to the extremities or to the face. Passive movement of arms and legs is the most massive maneuver included. Each specific elicited movement (SEM) is scored for its execution for left and right side separately where appropriate on a range from 0 (not elicitable) to 3, very strong hyperreactive or obligatory response, unless otherwise indicated.

FREEING OF FEET AND HANDS

The infant's ability to deal with having his feet and hands uncovered is an opportunity to observe the infant's organizational stability. Some infants find this manipulation very taxing and disturbing, while others are only minimally bothered by it or not bothered at all. Of course, not all infants will have their hands and feet covered. In such a case this item should be scored C.

Scoring

1. The infant's feet and/or hands are uncovered very gradually, carefully and gently; the infant goes into hyperextension and/or hyperflexion and/or changes to a disorganized state. The disorganization is severe and does not abate.
2. The infant's feet and/or hands are uncovered very carefully and gently; there is moderate disorganization physiologically and motorically. The infant attempts to regain control but fails to do so.
3. The infant's feet and/or hands are uncovered very carefully and gently; there is moderate to mild disorganization physiologically and motorically. The infant successfully regains adjustment eventually.
4. The infant's feet and/or hands are uncovered very carefully and gently; there is minimal and brief disorganization motorically and physiologically, and the infant regains adjustment gradually.
5. The infant's feet and/or hands are uncovered with moderate care; there is only moderate disorganization, which abates.
6. The infant's feet and/or hands are uncovered with moderate care, and there is only minimal brief disorganization.
7. The infant's feet and/or hands are uncovered without special adjustments and care, and there is only moderate disorganization which abates.
8. The infant's feet and/or hands are uncovered without special adjustments and care, and there is only minimal disorganization.
9. The infant's feet and/or hands are uncovered without special adjustments and care, and there is no disorganization.

HEEL TOUCH

As a test of response decrement to repeated localized tactile stimulation, a plastic stick with a dull point (such as the orange stick included with the scrub brush in the nursery) may be used to touch the heel of the infant's foot when he is asleep. If the infant, after foot freeing, is no longer asleep, this item is not administered and is scored C. If the infant is asleep, this touch may be repeated several times. This is preferred over the pin of the BNBAS since the premature infant is often tactually very sensitive and gets too taxed with the pin. The examiner watches for how totally and how rapidly the whole body responds to this touch. The degree, rapidity, and repetition of the "spread" of stimulus to the rest of the body is measured. The other aspect is the infant's capacity to shut down this spread of generalized response.

The foot should be touched up to six times. If no response decrement occurs, the stimulation should be stopped after the fourth touch. The APIB and BNBAS scoring of the degree of decrement is the same.

Degree of Decrement

SCORING

1. Response generalized to whole body and increases over trials.
2. Both feet withdraw together; no decrement of response.
3. Variable response to stimulus. Response decrement but return of response.
4. Response decrement after five trials; localized to stimulated leg. No change to alert state.
5. Response decrement after five trials; localized to stimulated foot.
6. Response limited to stimulated foot after 3–4 trials. No change to alert state.
7. Response limited to stimulated foot after 1–2 trials. No change to alert state.
8. Response localized and minimal.
9. Complete response decrement.

Ease of Elicitation of Initial Response

SCORING

1. Initial response cannot be observed despite three stimulus presentations.
2. Two or three stimuli needed before full initial response consisting of blinks and/or noticeable respiratory changes.
3. Two or three stimuli needed before full initial response consisting of facial movement or sound.
4. Two or three stimuli needed before full initial response consisting of startle or body movement.
5. No difficulty, full initial response consisting of respiratory changes.
6. No difficulty, full initial response consisting of eye blink.
7. No difficulty, full initial response consisting of facial movement and sound.
8. No difficulty, full initial response consisting of mild body movement or head move.
9. No difficulty, good full initial response consisting of startle or body movement.

Timing of Response

SCORING

1. All responses instantaneous with onset of stimulus.
2. Some responses instantaneous with onset of stimulus, some showing beginning inhibition.
3. Considerable delay of most or all responses.
4. Considerable delay of some responses.
5. Moderate delay of most or all responses.
6. Moderate delay of some responses.
7. Minimal delay of most or all responses.
8. Minimal delay of some responses.
9. Responses in modulated interval from onset of stimulus.

Recycling of Response

SCORING

1. No recycling noted on any trials presented.
2. Response shows mild recycling, on a number of trials only.
3. Response shows moderate recycling on a few trials.
4. Response shows moderate recycling on several trials.
5. Response shows prolonged recycling on a few trials.

6. Response shows prolonged recycling on several trials.
7. Response shows some degree of recycling on most or all trials.
8. Response shows a considerable degree of recycling on most or all trials.
9. Response shows a considerable degree of recycling with increasing magnitude, which makes the continuation of the item impossible.

Degree of Organization of Response

SCORING

1. Gradual diminution of high level to low level responses or low level response maintenance without fluctuation. Shutdown achieved.
2. Gradual diminution of high level to low level response, or low or high level response maintenance without fluctuation, and without shutdown.
3. Periods of sustained low level responses with occasional return of a high level response, followed by shutdown.
4. Periods of sustained low level responses with occasional return of a high level response, without achievement of shutdown.
5. Variable high and low level responses followed by shutdown.
6. Signs of disorganization may be present repeatedly between trials, yet shutdown is achieved.
7. Repeated fluctuation between high and low level responses without achieving shutdown.
8. Signs of disorganization may be present repeatedly between trials and shutdown is not achieved.
9. Attains increasing disorganization with succeeding stimulus presentations. Unable to proceed with decrement item(s).

Degree of Discharge after Termination of Stimulus Sequence

SCORING

1. There is no discharge noted after termination of stimulus sequence.
2. There is a mild respiratory response, possibly the sigh response, after termination of stimulus sequence.
3. There is some facial movement after termination of stimulus sequence.
4. There is mild motor discharge after termination of stimulus sequence.
5. There is moderate motor discharge after termination of stimulation sequence.
6. There is prolonged motor discharge after termination of stimulus sequence.
7. There is noticeable tachypnea or a brief episode of apnea, nasal flaring, gasping, etc., after termination of stimulus sequence.
8. There is prolonged tachypnea or a moderate episode of apnea after termination of stimulus sequence.
9. There is prolonged discharge of motoric or physiological nature leading to state change after termination of stimulus sequence.

SYSTEMATICALLY ELICITED MOVEMENTS

Feet

Plantar Grasp (SEM scores equivalent BNBAS)

SCORING

(0) Not elicitable
(1) Weak, unsustained

(2) Good, sustained response
(3) Very strong obligatory response
(A) Infant too aroused or too stressed to administer or score item meaningfully
(X) Inadvertently omitted

Foot Sole Stroke (SEM scores equivalent BNBAS Babinski Response)

Scoring

(0) Not elicitable
(1) Minimal spread of toes
(2) Marked spread of toes
(3) Obligatory spread of toes
(A) Infant too aroused or too stressed to administer or score item meaningfully
(X) Inadvertently omitted

Clonus (SEM scores equivalent to BNBAS)

Scoring

(0) No clonus
(1) One beat only
(2) Two or more beats, up to 4 or 5 if gradual decrease in intensity
(3) More than 5 beats
(A) Too tight in ankle or too aroused or stressed to administer or score meaningfully
(X) Inadvertently omitted

Hands

Palmar Grasp (SEM scores equivalent to BNBAS)

Scoring

(0) No grasping movement at all
(1) Short, weak flexion
(2) Strong, sustained, modulated grasp
(3) Obligatory grasp with tips of baby's fingers or knuckles going white, difficult to terminate
(A) Too aroused or too stressed to administer or score meaningfully
(X) Inadvertently omitted

Palmar Mental Grasp (SEM scores only)

Scoring

(0) No grasping or mouth opening at all
(1) Brief grasping and minimal mouth opening
(2) Strong modulated grasping and mild yet recognizable mouth opening, possible mild head straining
(3) Obligatory excessive grasping with pronounced mouth opening and/or head lifting
(A) Too aroused or too stressed to administer or score meaningfully
(X) Inadvertently omitted

Passive Movements: Arms

Resistance to Extension

SCORING

(0) No resistance to extension
(1) Little resistance to extension
(2) Moderate and modulated resistance to extension
(3) Excessive resistance to extension; extension may not be possible
(A) Too aroused or too stressed to administer or score meaningfully
(X) Inadvertently omitted

Degree of recoil

SCORING

(0) No flexion at all, possibly hyperextension
(1) Minimal flexion
(2) Moderate and modulated flexion
(3) Excessive hyperflexion
(A) Too aroused or too stressed to administer or score meaningfully
(X) Inadvertently omitted

BNBAS Scores

If resistance and recoil scores are equivalent, they can be used as BNBAS scores.

Arms (BNBAS)

(0) No resistance to extension and no recoil
(1) Little resistance to extension and weak recoil
(2) Moderate and modulated resistance to extension and good or moderate recoil
(3) Hypertonic resistance to extension and obligatory recoil

Passive Movements: Legs

Resistance to extension

SCORING

(0) No resistance to extension
(1) Little resistance to extension
(2) Moderate and modulated resistance to extension
(3) Excessive resistance to extension; extension may not be completely possible
(A) Too aroused or too stressed to administer or score meaningfully
(X) Inadvertently omitted

Degree of recoil

SCORING

(0) No flexion at all, possibly hyperextension
(1) Minimal or very delayed flexion
(2) Moderate and modulated flexion
(3) Excessive hyperflexion

(A) Too aroused or too stressed to administer or score meaningfully
(X) Inadvertently omitted

BNBAS Scores

If resistance and recoil scores are equivalent, they can be used as BNBAS scores.

Legs (BNBAS)

(0) No resistance to extension or recoil
(1) Little resistance to extension and weak recoil
(2) Moderate and modulated resistance to extension and good or moderate recoil
(3) Hypertonic resistance to extension and obligatory recoil

Arm/Leg/Head/Trunk Differentiation during Passive Movements

SCORING

(0) No differentiation of arms, legs, head, and trunk, e.g., if arms are extended, legs and head come up; if legs are extended, arms come up, or trunk and head lift.
(1) Some differentiation of arms, legs, head, and trunk: Only occasionally is there overflow to other parts when arms or legs are manipulated.
(2) Good differentiation of arms, legs, head, and trunk; the movement of one part of the body does not elicit obligatory reactions in the other parts of the body.
(3) Excessive differentiation of arms, legs, head, and trunk; it appears that the movement of one part of the body occurs in complete isolation from all other parts of the body; there seems much disconnection of body parts.
(A) Too aroused or too stressed to administer or score meaningfully.
(X) Inadvertently omitted.

Glabella (SEM scores equivalent BNBAS)

SCORING

(0) No reaction
(1) Weak response, barely discernible
(2) Modulated response
(3) Overly brisk closure, and total facial grimace and/or startle
(A) Too aroused or too stressed to administer or score meaningfully
(X) Inadvertently omitted

Rooting (SEM scores equivalent BNBAS)

SCORING

(0) No lip or tongue movement
(1) Only a weak turn or lip movement and/or slight tongue protrusion
(2) Turn to stimulated side; mouth may open and grasp; tongue may move to stimulated side; lips may curl to stimulated side
(3) Obligatory unmodulated turn and grasping movement
(A) Too aroused or too stressed to administer and score meaningfully
(X) Inadvertently omitted

Sucking (SEM scores equivalent BNBAS)

Scoring

(0) No sucking movement at all; possibly expulsion or clamping
(1) Weak or barely discernible sucking and stripping action of tongue, and/or intermittent single sucks
(2) Modulated rhythmical suck
(3) Exaggerated obligatory suck
(A) Too aroused or too stressed to administer or score meaningfully
(X) Inadvertently omitted

Package IV: Medium Tactile Input Combined with Medium Vestibular Input

(Maneuvers: Undressing; Pull-to-Sit; Standing; Walking; Placing; Incurvation; Crawl; Cuddling; Tonic Neck Response; Defensive Reaction)

The various maneuvers in this package demand repeated whole body postural adjustments of the infant aside from the distinct responses expected to the specifically elicited movements. Again, the examiner aims for maximum smoothness in administering these maneuvers so as to keep low unnecessary extraneous manipulation beyond the maneuvers. The order of the maneuvers may vary somewhat depending on the infant's capacities, yet the examiner should keep in mind that certain postures and positions are natural transitions to the next maneuver, e.g., he should not take the infant from standing to placing and then back to walking; but from standing, through walking, to placing, or from placing to incurvation, to crawl, etc. Placing is most easily administered by leaning the infant's back against the examiner's chest, tucking one of the infant's legs up under the respective buttock, and then testing the free leg; then with minimal shifting of the infant's body the other leg is tucked and the released leg is tested. This procedure reduces gross body manipulation of the infant, which is not wanted at this point.

CAPACITY TO DEAL WITH UNDRESSING

The infant's ability to deal with the tactile and vestibular manipulations necessary to remove his shirt, gown, or other clothing (the diaper need not be removed), combined with the change in temperature, can be a good index of his general organization. The examiner will need to take the ambient temperature into account. A cool environment can be very taxing for the preterm infant and should be avoided. Some infants will only be wearing a diaper; then this item has to be scored C. The infant who has great difficulty dealing with undressing may have to be dressed or wrapped again, at least partially, in order to prevent undue stress and exhaustion. This would be recorded under (E) examiner faciliation.

Scoring

(1) The infant is undressed very gradually and very gently. He goes into hyperextension and/or hyperflexion. The physiological, motor, and state disorganization is severe and does not abate.
(2) The infant is undressed with great care; there is moderate disorganization physiologically and motorically. The infant attempts to regain control but fails to do so.
(3) The infant is undressed with great care; there is moderate to mild disorganization physiologically and motorically. The infant successfully regains adjustment eventually.

(4) The infant is undressed with great care; there is minimal and brief disorganization physiologically and motorically, and the infant regains adjustment gradually.
(5) The infant can be undressed with moderate care and there is moderate disorganization, which abates.
(6) The infant can be undressed with moderate care, and there is minimal brief disorganization.
(7) The infant can be undressed without special adjustments, and there is moderate disorganization, which abates.
(8) The infant can be undressed without special adjustments, and there is minimal disorganization.
(9) The infant can be undressed without special adjustments, and he shows no disorganization.

PULL TO SIT

The examiner places a forefinger in each of the infant's palms. With the arms extended, the infant's automatic grasp is used to pull him to sit. The shoulder girdle muscles respond with tone, and muscular resistance to stretching his neck and lower musculature as he is pulled into a sitting position. Usually he will also attempt to right his head into a position which is in the midline of his trunk and parallel to his body. Since his head is heavy and out of proportion to the rest of his body mass, this is not usually possible, and his head falls backward as he comes up. In a seated position, he may attempt to right his head, and it may fall forward. Several attempts to right it can be felt via the shoulder muscles as the examiner maintains his grasp on the infant's arms. A few infants make no attempt at all. The range of this performance is scored on the original BNBAS scale.

Some infants resist flexion and head-righting by arching backward. Their bodies may become rigid and hyperextended. This is presumably due to an imbalance of extensor and flexor tone. The degree to which this occurs is scored on the scale of Hyperextension of Head and Trunk. Other infants will come with handgrasp to an extreme flexed position in an exaggerated palmar-mental grasp response. Occasionally the infant will come to stand in this maneuver, since his legs hyperextend as his head and trunk go to hyperflexion. The degree to which this occurs is scored on the scale of Hyperflexion of Head and Trunk.

Scoring (BNBAS)

(1) Head flops completely in pull to sit, no attempts to right it in sitting
(2) Futile attempts to right head but some shoulder tone increase is felt
(3) Slight increase in shoulder tone, seating brings head up once but not maintained; no further efforts
(4) Shoulder and arm tone increase, seating brings head up, not maintained but there are further efforts to right it
(5) Head and shoulder tone increase as pulled to sit, brings head up once to midline by self as well, maintains it for 1–2 seconds
(6) Head brought up twice after seated, shoulder tone increase as comes to sit, and maintained for more than 2 seconds
(7) Shoulder tone increase but head not maintained until seated, then can keep it in position 10 seconds
(8) Excellent shoulder tone, head up while brought up but cannot maintain without falling, repeatedly rights it
(9) Head up during lift and maintained for one minute after seated, shoulder girdle and whole body tone increases as pulled to sit

Hyperextension of Head and Trunk

SCORING

(1) The infant shows no hyperextension of head and/or trunk and can be scored on the full-term scale.
(2) There is initially noticeable head and/or trunk extension which can be gradually overcome.
(3) There is moderate head and/or trunk extension which can be gradually overcome.
(4) There is considerable head and/or trunk extension which can, however, be gradually overcome.
(5) There is some mild head and trunk extension which cannot be overcome despite several efforts.
(6) There is moderate head and/or trunk extension which cannot be overcome.
(7) There is considerable head and/or trunk extension which cannot be overcome.
(8) There is consistent head *and* trunk extension which cannot be overcome.
(9) There is severe head and trunk extension which cannot be overcome.

Hyperflexion of Head and Trunk

SCORING

(1) The infant shows no hyperflexion of head and/or trunk and can be scored on the full-term scale.
(2) There is initially noticeable head and/or trunk flexion which can gradually be overcome.
(3) There is moderate head and/or trunk flexion which can be gradually overcome.
(4) There is considerable head and/or trunk flexion which can, however, be gradually overcome.
(5) There is some mild head and/or trunk flexion which cannot be overcome despite several efforts.
(6) There is moderate head and/or trunk flexion which cannot be overcome.
(7) There is considerable head and/or trunk flexion which cannot be overcome.
(8) There is consistent head *and* trunk flexion which cannot be overcome.
(9) There is severe head *and* trunk flexion which cannot be overcome.

SYSTEMATICALLY ELICITED MANEUVERS

Standing

Many premature infants cannot come to a stand in which both legs are next to one another but stand with a wide base. This is referred to as umbrella stand, since the shape of their legs during this maneuver resembles an open, curved umbrella roof. Varying degrees of support are assessable for umbrella stand and for the more modulated regular stand.

Umbrella Stand

SCORING

(0) No support
(1) Minimal response felt; brief transitory support
(2) Supports weight
(3) Obligatory hyperextension of legs and/or feet
(A) Infant is too aroused or too stressed to administer or score item meaningfully
(X) Inadvertently omitted

Standing (BNBAS)

SCORING

(0) No support
(1) Minimal response felt; brief transitory support
(2) Supports weight
(3) Obligatory hyperextension of legs and/or feet
(A) Infant is too aroused or too stressed to administer or score item meaningfully
(X) Inadvertently omitted

Walking

The umbrella position of the legs is also observed in walking. It should be assessed as to its differential degree of modulation. If an infant shows a mixture of umbrella walking and more narrow based walking, a judgment as to the more typical performance should be made.

Umbrella Walk

SCORING

(0) No hip or knee flexion at all
(1) Some indication of stepping action with slight hip or knee flexion
(2) Modulated, discernible steps with knee and hip flexion
(3) Obligatory hyperreactive response with hip and knee flexion and ankle extension
(A) Infant is too aroused or too stressed to administer or score item meaningfully
(X) Inadvertently omitted

Walking (BNBAS)

SCORING

(0) No hip or knee flexion at all
(1) Some indication of stepping action with slight hip or knee flexion
(2) Modulated discernible steps with knee and hip flexion
(3) Obligatory hyperreactive response with hip and knee flexion and ankle extension
(A) Infant is too aroused or too stressed to administer or score item meaningfully
(X) Inadvertently omitted

Placing

The equivalent in midair of the umbrella position in standing is often observed when the infant is held up to elicit placing. This response should be scored separately to document the poor leg posture which may be independent of the placing response as such.

Umbrella Placing

SCORING

(0) No flexion or extension of leg or foot and no fanning of toes
(1) Minimal flexion and extension of knee and hip and/or minimal ankle flexion and flaring of toes
(2) Modulated flexion of knee and hip and ankle flexion and foot extension with toe fanning
(3) Obligatory flexion of knee and hip and/or obligatory ankle flexion and extension of the foot

(A) Infant is too aroused or too stressed to administer or score item meaningfully
(X) Inadvertently omitted

Placing (BNBAS)

SCORING

(0) No flexion or extension of leg or foot and no fanning of toes
(1) Minimal flexion and/or extension of leg and/or foot and minimal fanning of toes
(2) Modulated flexion and/or extension of leg and/or foot with toe fanning
(3) Obligatory flexion and extension of leg and foot
(A) Infant is too aroused or too stressed to administer or score item meaningfully
(X) Inadvertently omitted

Incurvation (SEM scores equivalent BNBAS)

This is assessed as the infant is held across the examiner's hand in prone position.

SCORING

(0) No response
(1) Minimal incurvation on movement with minimal hipswing
(2) Good incurvation, modulated with moderate hipswing
(3) Exaggerated response with excessive hipswing and/or leg extension
(A) Infant is too aroused or too stressed to administer or score meaningfully
(X) Inadvertently omitted

Crawling (SEM)

The crawling maneuver assesses various components of the infant's posture and movement when he is placed prone on a smooth surface: the freeing of the face, flexion of legs and trunk, and the ability to relax and stop moving after some adjustment. BNBAS equivalents will be given in the margin.

BNBAS EQUIVALENT	SCORING	
(0)	(0)	No freeing of the face and no attempt at movement; legs flaccidly extended
(0)	(1)	No freeing of the face and no attempt at movement; legs flexed or in fetal tuck position, buttocks up
(0)	(2)	No freeing of the face but some attempt at movement; legs partially in fetal tuck with effort to get toes pushing off
(1)	(3)	Freeing of the face, but legs extended and no movement
(1)	(4)	Freeing of the face, legs flexed and no movement
(1)	(5)	Freeing of the face, legs flexed and some movement
(2)	(6)	Freeing of the face, legs flexed and coordinated, modulated movement which can be inhibited
(3)	(7)	Freeing of the face, legs flexed and coordinated, modulated movement which cannot be inhibited
(3)	(8)	Freeing of the face and/or lifting of the head; legs flexed and extended in alternation, some arching and movement, which can or cannot be inhibited

(3) (9) Freeing of the face and/or lifting of the head; legs extended, trunk arched at times, frantic movement which cannot be inhibited
(A) Infant too aroused or too stressed to administer or score the item meaningfully
(X) Inadvertently omitted

Cuddling

Cuddling is a two-part maneuver which mirrors the maneuvers crawling and supine but uses the examiner's body as surfaces. This can make a remarkable difference to the infant. First, the examiner places the infant ventrovertically against his own body and observes how the infant adjusts. If the infant cannot settle in, the examiner makes adjustments to facilitate the infant's cuddle response. Then the examiner moves the infant into the horizontal position and nestles him supine in his arm. Again, he first waits to observe how the infant adjusts. If the infant cannot settle in the examiner's arm, he makes adjustments to facilitate the infant's cuddle response. Some infants will assume in either the vertical or horizontal position or both a fetal tuck position (hyperflexion) and do not stretch out enough to be able to cuddle. Others will be completely flaccid or will hyperextend, an index of poor balance of flexor and extensor tone. The degree to which this occurs is assessed. Cuddling as such may not be applicable under these circumstances and may have to be scored N. If vertical and horizontal cuddling are discrepent by not more than one point, the infant can be scored on the original BNBAS scale.

Vertical Position

SCORING

(1) Actually resists being held, continuously pushing away, thrashing, or stiffening
(2) Resists being held most but not all of the time; or is quite floppy
(3) Resists being held some of the time; or is somewhat floppy
(4) Eventually molds into arms, but after a lot of nestling and cuddling by the examiner
(5) Usually molds and relaxes when first held, i.e., nestles head in crook of neck and of elbow of examiner. Turns toward body when held horizontally; on shoulder he seems to lean forward
(6) Always molds initially with above activity
(7) Always molds initially with nestling, and turning toward body, and leaning forward
(8) In addition to molding and relaxing, he nestles and turns head, leans forward on shoulder, fits feet into cavity of other arm; all of body participates
(9) All of the above, and baby grasps hold of the examiner to cling to him

Fetal Tuck (Hyperflexion) and Extension

SCORING

(1) The infant can be scored on the 9-point scale.
(2) The infant shows minimal fetal positioning but predominantly stretches out.
(3) The infant shows some fetal positioning but also some stretching out.
(4) The infant shows predominantly fetal positioning and rarely stretches out.
(5) The infant always shows fetal positioning and does not stretch out.
(6) The infant occasionally is completely limp, oscillates with fetal positioning and some hyperextension.
(7) The infant oscillates between limpness, some effort at fetal tuck but also shows moderate hyperextension.
(8) The infant oscillates between fetal tuck and hyperextension; hyperextension predominates; limpness may be observed occasionally.

(9) The infant is almost continuously in hyperextension or complete limpness; flexion can only be induced with difficulty or not at all.

Horizontal Position

SCORING

(1) Actually resists being held, continuously pushing away, thrashing or stiffening
(2) Resists being held most but not all of the time or is quite floppy; sucking may prove facilitative
(3) Resists being held some of the time; sucking may prove facilitative
(4) Eventually molds into arms, but after a lot of nestling and cuddling by examiner
(5) Usually molds and relaxes when first held, i.e., nestles head in crook of elbow of examiner, turns toward body
(6) Always molds initially with above activity
(7) Always molds initially with nestling, and turning toward body, and leaning inward
(8) In addition to molding and relaxing, he nestles and turns head, leans inward to body of examiner, fits feet into cavity of other arm, i.e., all of body participates
(9) All of the above, and baby grasps hold of examiner to cling

Fetal Tuck (Hyperflexion) and Extension

SCORING

(1) The infant can be scored on the 9-point scale.
(2) The infant shows minimal fetal positioning but predominantly stretches out.
(3) The infant shows some fetal positioning but also some stretching out.
(4) The infant shows predominantly fetal positioning and rarely stretches out.
(5) The infant always shows fetal positioning and does not stretch out.
(6) The infant occasionally is completely limp, oscillates with fetal positioning and some hyperextension.
(7) The infant oscillates between limpness, some effort at fetal tuck but also shows moderate hyperextension.
(8) The infant oscillates between fetal tuck and hyperextension; hyperextension predominates.
(9) The infant is almost continuously in hyperextension. Flexion can only be induced with difficulty.

Cuddliness (BNBAS)

(If vertical and horizontal scores are equivalent or within one point of each other)

SCORING

(1) Actually resists being held, continuously pushing away, thrashing, or stiffening
(2) Resists being held most but not all of the time
(3) Resists being held some of the time
(4) Eventually molds into arms but after a lot of nestling and cuddling by examiner
(5) Usually molds and relaxes when first held, i.e., nestles head in crook of neck and of elbow of examiner; turns toward body when held horizontally, on shoulder he seems to lean forward
(6) Always molds initially with above activity
(7) Always molds initially with nestling, and turning toward body, and leaning forward

(8) In addition to molding and relaxing, he nestles and turns head, leans forward on shoulder, fits feet into cavity of other arm, i.e., all of body participates
(9) All of the above, and baby grasps hold of examiner to cling

Tonic Neck Response (SEM score equivalent to BNBAS)

The infant's postural adjustment is assessed when he is placed in supine position on a smooth surface and his head is passively moved first to one side and then to the other. The examiner's one hand is placed on the infant's chest so that the infant's body does not turn onto its side. The examiner may hold the infant in this position up to 30 seconds or more. Premature infants are often slow in responding but eventually may show a modulated adjustment of arms and legs.

Scoring

(0) No adjustment of arms and legs
(1) Transient adjustment of arms and legs, not maintained
(2) Gradual modulated adjustment of arms and legs
(3) Obligatory response of arms and legs
(A) The infant is too aroused or too stressed to administer or score the item meaningfully
(X) Inadvertently omitted

Defensive Movements

A small cloth is placed with the examiner's fingers asserting light pressure over the upper part of the face without occluding the nose. It is kept in place for up to one minute, or until the infant responds with a series of responses graded as to their degree of differentiation. The infant's hands should not be under the cloth. The scoring is the same for the APIB as for the BNBAS. A may have to be scored if the infant is too aroused or too stressed.

Scoring

(1) No response
(2) General quieting
(3) Nonspecific activity increase with long latency
(4) Same with short latency
(5) Rooting and lateral head turning
(6) Neck stretching
(7) Nondirected swipes of arms
(8) Directed swipes of arms
(9) Successful removal of cloth with swipes

Package V: Massive Tactile Input Combined with Massive Vestibular Input

(Maneuvers: Rotation (SEM); Moro (SEM))

The two maneuvers in this package provide opportunities to observe the infant's capacities in the five subsystem dimensions outlined, when the infant's whole body is repeatedly moved through space in a brisk manner in upright position and from a horizontal position. Aside from the infant's system organization, they allow specifically for the observation of the degree and kind of balance and differentiation of relative extensor and flexor tone and posture.

ROTATION (SEM)

The infant is suspended vertically by holding him with two hands under his arms, stabilizing his head with one's thumbs under his chin. The infant is raised slightly above the examiner's eye level and facing the examiner; the infant's body is tilted forward about 30 degrees so that his face is closer to the examiner than his legs. Once the infant's head is in midline, the examiner briskly rotates him to one side at least 90 degrees. He stops the rotation having watched for the infant's head adjustment and eye adjustment. The examiner then brings the infant back to the original position and, when the infant's head is stabilized, rotates the infant now briskly to the other side, through an at least 90-degree excursion, again watching for head and eye adjustment during the rotation. Head and eye adjustments are scored separately, as is the degree of optokinetic nystagmus observed during the rotation. If head and eye adjustments are scored equivalently, the respective Tonic Deviation of Head and Eyes BNBAS score is appropriate. Nystagmus SEM scores are the same as on the BNBAS.

Head (SEM)

SCORING

(0) No head movement
(1) Weak head movement in the direction of the rotation
(2) Good modulated head turn in the direction of the rotation
(3) Immediate, obligatory head turn in the direction of the rotation
(A) Infant too aroused or too stressed to administer or score item meaningfully
(X) Inadvertently omitted

Eyes (SEM)

SCORING

(0) No eye movement
(1) Weak eye adjustment in the direction of the rotation
(2) Good modulated eye adjustment in the direction of the rotation
(3) Immediate obligatory eye adjustment in the direction of the rotation
(A) Eyes cannot be observed
(X) Inadvertently omitted

BNBAS Tonic Deviation of Head and Eyes

SCORING

(0) No head or eye movement
(1) Weak, response barely discernible
(2) Good modulated response
(3) Immediate, obligatory head and eye turn

Nystagmus (SEM scores equivalent to BNBAS)

SCORING

(0) No saccadic movement
(1) 1 or 2 saccades during rotation
(2) 3 or 4 saccades per rotation
(3) Many sustained saccades per rotation

MORO

The examiner holds the infant suspended in supine horizontally in midair by placing his arm and hand under the infant's trunk and the other hand under the infant's head. When the infant is stabilized in midline and his arms and legs are in symmetrical position, the examiner drops the hand supporting the head and observes the reaction of the infant's arms and legs. Arms and legs are scored separately and for the arms the extension and adduction component are scored separately. If the extension and adduction component are scored equivalently, the respective BNBAS score can also be given.

Arms (SEM)

Extension

SCORING

(0) No response
(1) Weak response, minimal extension
(2) Modulated arm extension
(3) Obligatory, excessive arm extension
(A) Infant too aroused or too stressed to administer or score item meaningfully
(X) Inadvertently omitted

Adduction

SCORING

(0) No response
(1) Weak response, minimal adduction
(2) Modulated, smooth adduction back to midline
(3) Excessive, exaggerated adduction
(A) Infant too aroused or too stressed to administer or score item meaningfully
(X) Inadvertently omitted

Legs (SEM)

Extension and Adduction

SCORING

(0) No response
(1) Minimal leg extension noticeable
(2) Modulated, moderate leg extension and relaxation
(3) Excessive, exaggerated leg extension followed by no relaxation or exaggerated flexion
(A) Infant too aroused or too stressed to administer or score item meaningfully
(X) Inadvertently omitted

BNBAS Equivalent Score

(0) No response
(1) Weak response with minimal adduction of shoulders and extension of elbow and wrist; followed by minimal or no adduction of shoulder and flexion of elbow and wrist; minimal extension of hip and knee
(2) Modulated, good adduction of shoulders and extension of elbow and knee; followed by modulated adduction of shoulders and flexion of elbow and wrist
(3) Obligatory, excessive adduction of shoulders and extension of elbow and wrist; obligatory brisk extension of hip and knee; followed by obligatory unmodulated adduction of shoulders and flexion of elbow and wrist

Package VI: Social Interaction and Inanimate Object Orientation

(Maneuvers: Attention to examiner's voice and face (animate visual and auditory); attention to examiner's voice alone (animate auditory alone); face alone (animate visual alone); attention to red rattle (inanimate visual and auditory); attention to rattle alone (inanimate auditory alone); attention to rattle or ball (inanimate visual alone).

The maneuvers in the orientation and interaction package provide opportunities to observe the infant's attentional capacities and his social interaction capacities when the infant has been brought to his optimally alert state. To bring the infant to an optimally alert state is the examiner's primary administrative goal, and he will aim for it and take advantage of it whenever the infant's behavioral cues indicate this is appropriate. In the very immature or the very poorly organized infant, this may be very early on in the assessment. Often behaviors during the sleep/distal package alert the examiner to the fragility of the infant's organization. He may decide to move immediately to Package VI. If he then finds that the infant is too sleepy but more robust than he thought, he may decide to continue with Package II and possibly III and come back to Package VI later. Some infants will indicate by their severe physiological and motoric responses that Package VI is inappropriate, in which case the examiner will not pursue these items. Some infants can be brought into a more alert and well organized state by items from Package VI. This should usually be attempted. In order to make appropriate decisions for the pursuit or the deletion of certain items, the examiner needs extensive training in the manipulation of the preterm infant and in the reading of his behavioral cues. He has to be a skilled observer and interactor simultaneously. This is important for the assessment of all infants but is particularly important for the preterm infant.

The order in which the various interaction and attentional items are administered may obviously vary. It is often easiest to organize the infant's alert state with one's own face and voice. The infant is gently taken onto the examiner's lap. He may need to be dressed and/or swaddled (see E). The examiner has to be in a comfortable sitting position, preferably with his feet on a footstool so that his lap provides an incline for the infant, propping him up.

The infant is highly sensitive to the examiner's own body tension and emotional and physiological state. The examiner needs to be aware of this at all times and attempts to be relaxed. Appropriate to training is an effort to control for examiner variability.

It is very important that there is no direct light shining on the infant's face. Direct light prevents many infants from opening their eyes. Optimally the examiner faces a window or other light source in midline, and the infant is faced away from it. The ambient lighting of the environment should be quite dim. Light sources to either side of the infant or from above can make for particular difficulty, even if they are fairly dim, and should be eliminated. It is recommended that the examiner arrange the circumstances as optimally as possible for the assessment prior to beginning the examination. This precaution will prevent extraneous manipulation and movement of the infant during the examination. Once the examiner is comfortably settled with the infant on his lap, he will need to be sensitive to the infant's cues and responses in order to time and space his administration optimally for the particular infant. He will first attempt to bring the infant to alert, then to have him follow the stimulus horizontally to assess the degree of lateral excursions possible, then vertically, and then in a circle. The infant's performance is the base for the decision to continue or end a particular sequence. Several parameters of attention are scored separately for each item.

1. The Degree to Which a Particular Stimulus Elicits and Maintains an Alert State and Responsivity: This is a measure of the relative flexibility and availability of the infant's attention. Some infants, once in an alert state, can perform quite well. Yet to achieve this state is a difficult task for them due to their low flexibility and autonomous control of the attentional state. Other infants, capable of coordinated attentional excursions, can maintain these only very briefly. Others may be attentive for longer periods, yet will not achieve the complexity of excursions. Maintenance of attention is a measure of the relative stability of a particular level of attentional flexibility.

2. The Infant's Orienting Capacity as Such: This is a measure of the degree of flexibility of the specific coordination of eyes and/or head the infant is capable of, once he is alert and when the various stimuli are presented. The original BNBAS scoring is retained for the infant who achieves an alert state of 4B and is assessed under Orientation B. The infant who comes predominantly to a state 4AL or 4AH is scored for this specific capacity in these states, and the respective score is noted under Orientation A. Orientation B is scored N in most of these cases unless there are usable periods in state 4B as well as in 4A. In such a case both scores may be given. If the infant cannot be brought to alertness, this is scored N.

Some mature infants are capable of actively selecting the targets of their attention. They may avoid the examiner's stimulus presentation; an initial stage setting for social play may be necessary to overcome their differential responsiveness.

3. Effort to Shut Out and Effort to Attend: The interplay of these two parameters is seen as an index of the developing balance of attentional excitatory and inhibitory capacities, mirroring the flexion–extension balance of the motor system. Some quite well organized infants make great efforts *not* to attend. They are actively attempting to shut out to preserve their current level of self-regulation. Once sufficiently stabilized, they may make efforts to varying degrees to attend. Other infants are overly drawn to each presented stimulus and the inhibitory balance is not yet achieved. This scale assesses the relative balance of these two poles of the attentional modulation process.

4. Cost of Attention: This parameter assesses the degree to which the other subsystems of organization are taxed or possibly enhanced by the activation of the attentional system. This provides an index of the degree and level of stable system integration. If no effort and no responsiveness are observable, this item is scored A.

5. Quality of Responsivity: This is a summary rating of the degree of attentional involvement as the infant's *face* communicates his attention to the examiner. It reflects the relative differentiation of his expressive interactive and attentional capacities. If an infant is robust and well organized in all subsystems but cannot be brought into alertness for whatever reason, this scale may not be applicable and should be scored N.

ATTENTION AND INTERACTION: FACE AND VOICE (ANIMATE VISUAL AND AUDITORY)

The examiner presents his face and voice to the infant, speaking softly and with animation. He attempts to cycle with the infant's attention. Once this synchrony is achieved, he draws the infant along, moving his face to one side, while continuing to talk to the infant. He brings the infant back into midline before drawing the infant to the other side. Then he attempts to draw the infant's attention vertically. If the infant can follow with him vertically, he moves his face in a 180-degree arc, the center of which is approximately at the infant's chest and the radius of which is the infant's saggital axis.

Elicitation and Maintenance of Attention

SCORING

(1) The infant is not in an alert state to start with, and alertness cannot be achieved by the use of this stimulus; or the infant is in an alert state, but the stimulus moves him out of this state immediately and consistently.

(2) The infant is not in an alert state to start with, and alertness can be achieved only momentarily and with difficulty by the use of this stimulus; or the infant is in an alert state and with this stimulus alertness can be maintained only minimally.

(3) The infant is not in an alert state and by the use of this stimulus he can be brought to alerting with moderate difficulty and for a very brief period; or the infant is in an alert state and, with this stimulus, can be maintained with difficulty for a brief period.

(4) The infant is not in an alert state and by the use of the stimulus can be brought to alerting fairly easily once or twice and maintained for a fleeting period; or the infant is in an alert state and can be maintained moderately easily by the stimulus for a brief period.
(5) The infant is not in an alert state and by the use of the stimulus can be brought to alerting once or twice easily and can be maintained for a brief period; or the infant is in an alert state and can be maintained easily for at least a brief period by the stimulus.
(6) The infant is not in an alert state and by the use of the stimulus can be brought to alerting repeatedly and can be maintained for a brief period each time; or the infant is in an alert state and can be maintained easily for a moderate period by the stimulus.
(7) The infant is not in an alert state and by the use of the stimulus can be brought to alerting readily and frequently and then can maintain alertness for a moderate period; or the infant is in an alert state and can be maintained quite easily for a considerable period by the stimulus.
(8) The infant is not in an alert state and by the use of the stimulus can be brought to alerting reliably, then can maintain alertness for considerable periods; or the infant is in an alert state and can be maintained quite easily for extended periods by the stimulus.
(9) The infant is not in an alert state and by the use of the stimulus he can be brought to alerting every time and then maintain alertness easily and for long periods; or the infant is in an alert state and the stimulus enhances and prolongs this state consistently.

Orienting Capacity (B) (BNBAS Equivalent)

SCORING

(1) Does not focus on or follow stimulus
(2) Stills with stimulus and brightens
(3) Stills, focuses on stimulus when presented, brief following
(4) Focuses on stimulus, follows for 30-degree arc, jerky movements
(5) Focuses and follows with eyes horizontally and/or vertically for at least a 30-degree arc. Smooth movement, loses stimulus but finds it again
(6) Follows for 30 degree arcs with eyes and head. Eye movements smooth
(7) Follows with eyes and head at least 60 degrees horizontally, maybe briefly vertically, partially continuous movement, loses stimulus occasionally, head turns to follow
(8) Follows with eyes and head 60 degrees horizontally and 30 degrees vertically
(9) Focuses on stimulus and follows with smooth, continuous head movement horizontally, vertically, and in a circle. Follows for at least 120-degree arc

Orienting Capacity (A)

SCORING

(1) Does not focus on or follow stimulus
(2) Stills with stimulus and brightens
(3) Stills, focuses on stimulus when presented, brief following
(4) Focuses on stimulus, follows for 30-degree arc, jerky movements
(5) Focuses and follows with eyes horizontally and/or vertically for at least a 30-degree arc. Smooth movement, loses stimulus but finds it again
(6) Follows for 30-degree arcs with eyes and head. Eye movement smooth
(7) Follows with eyes and head at least 60 degrees horizontally, maybe briefly vertically, partially continuous movement, loses stimulus occasionally, head turns to follow
(8) Follows with eyes and head 60 degrees horizontally and 30 degrees vertically
(9) Focuses on stimulus and follows with smooth, continuous head movement horizontally, vertically, and in a circle. Follows for at least 120-degree arc.

Effort to Shut Out and Effort to Attend

SCORING

(1) The infant shows no active effort in either direction. Attention is not an appropriate issue. He is unavailable.
(2) The infant makes initially a weak effort to attend. No further efforts are noted and he is unavailable.
(3) The infant makes initially repeated mild efforts to attend, then no further efforts are noted and he is unavailable.
(4) The infant makes some real efforts to attend; there may or may not be several efforts not to attend.
(5) The infant makes repeated strong efforts to attend. At other times he may make very strong efforts not to attend.
(6) The infant makes repeated very strong efforts to attend. At other times he may make very strong efforts not to attend. The amplitude of the vacillation is high.
(7) The infant is showing some degree of smoothness in regulating his efforts to attend.
(8) The infant is relatively smooth and well controlled in regulating his efforts to attend.
(9) The infant is very smooth and autonomous in the regulation of his efforts to attend.

Cost of Attention

SCORING

(1) There is no apparent cost to the other systems of behavioral organization. The infant's overall organization is enhanced by the interaction with the stimulus.
(2) There is very mild cost mainly in terms of brief state fluctuations up or downward. These may be brief eye closure, mild sneezing, or an occasional yawn. The motor system and physiological systems are not taxed at all, or are taxed only minimally.
(3) There is very mild cost involving very mild motor arousal and possibly a mild physiological reaction such as repeated sneezing, a mild hiccough, or a mild bowel strain, or mild color change.
(4) There is mild to moderate cost involving the state system, the motor system, and/or the physiological system. There may be some fluctuation of states up or downwards, some motor arousal accompanied by physiological responses such as mild to moderate color change, some hiccoughing, or mild to moderate respiratory irregularity.
(5) There is moderate cost involving the state fluctuations with gaze aversion, intermittent eye closure, brief occasional eye floating, and motor disorganization; or there is moderate motoric disorganization with motoric arousal and disorganized movements; or there is moderate physiological stress with hiccoughing, gagging or color change with moderate cyanosis, flushing, or paling.
(6) There is moderate to considerable cost involving the state system: the infant oscillates into very aroused states repeatedly and sharply; or the motor system: the infant gets aroused motorically to a moderate to considerable degree; or to the physiological system: the infant shows moderate to considerable color change, hiccoughing, and repeated eye aversions or spitting up or gagging, or moderate to considerable tremors.
(7) There is considerable cost involving the state system with repeated oscillations into aroused states accompanied by motor disorganization and/or physiological stress, such as intermittent eye floating, considerable paling, cyanosis, spitting up, apnea, possibly nasal flaring, bowel movement grunting and straining, and/or considerable tremors.
(8) There is high cost to the motor and/or the physiological systems with frantic or disorganized motor activity and/or arching, or with shifts to motoric tuning out and flaccidity,

with very considerable eye floating, considerable hiccoughing and bowel movement straining, very considerable paling, flushing, or cyanosis, and/or tremulousness.

(9) There is very high cost to the motor and/or the physiological systems with excessively frantic or disorganized motor activity and/or arching with or without repeated shifts to complete motoric tuning out and flaccidity. There may be much eye floating, paling, cyanosis, spitting up, gagging, apnea, bowel movement grunting and straining, and repeated prolonged tremors, or the infant may become completely diffusely unavailable.

Quality of Responsivity

SCORING

(1) Attentional responsivity is not an appropriate issue for this infant.
(2) The infant attempts to stay in lower states; he is only barely available for fleeting periods. His face may or may not appear pained and bothered.
(3) The infant attempts to rouse himself and is fleetingly successful.
(4) The infant comes to alertness with low intensity and flat or pained responsiveness, although he may show good following.
(5) The infant comes to alertness but then appears to be in overly heightened, strained, and/or almost panicked alertness. He may appear easily overloaded or at the mercy of the input. He may show good following.
(6) The infant shows brighteyed, modulated, focused alertness at times for brief periods; he may be flat or hyperalert at other times. When he is modulated, his face is softened and participates in the response; his mouth may open and round; he may raise his cheeks and eyebrows for brief periods.
(7) The infant's responsiveness is usually of the modulated quality described above. Only rarely is he flat or hyperalert.
(8) The infant's responsiveness is almost consistently modulated and differentiated; his face participates as described above.
(9) The infant's responsiveness is consistently well modulated, and there may also be active elicitation and initiation of interaction by the infant.

ATTENTION AND INTERACTION: FACE (ANIMATE VISUAL)

The same criteria for scoring are used as in the preceding item. The examiner places his face in the infant's line of vision, once the infant is alert, and then moves it slowly in horizontal, then vertical, then sagittal arcs, until the infant stops following.

Elicitation and Maintenance of Attention. Score 1 through 9.

Orienting Capacity (B) (BNBAS Equivalent)

SCORING

(1) Does not focus on or follow stimulus
(2) Stills with stimulus and brightens
(3) Stills, focuses on stimulus when presented, brief following
(4) Stills, focuses on stimulus, follows for 30-degree arc, jerky movements
(5) Focuses and follows with eyes horizontally and/or vertically for at least a 30-degree arc. Smooth movements, loses stimulus but finds it again
(6) Follows for 30-degree arcs, with eyes and head. Eye movements smooth
(7) Follows with eyes and head at least 60 degrees horizontally, maybe briefly vertically, partially continuous movement, loses stimulus occasionally, head turns to follow
(8) Follows with eyes and head 60 degrees horizontally and 30 degrees vertically

(9) Focuses on stimulus and follows with smooth, continuous head movement horizontally, vertically, and in a circle. Follows for at least 120-degree arc.

Orienting Capacity (A). Score 1 through 9.
Effort to Shut Out and Effort to Attend. Score 1 through 9.
Cost of Attention. Score 1 through 9.
Quality of Responsivity. Score 1 through 9.

ATTENTION AND INTERACTION: VOICE (ANIMATE AUDITORY)

The examiner removes his face from infant's line of sight and talks to him from one side (6 to 12 inches from ear). Continuous, soft and high-pitched speech is the best stimulus, e.g., infant's own name. The examiner has to take care not to move his own face to one side with the infant following him visually before he presents his voice. This can be a problem with very alert infants. Motorically poorly organized infants will need much containment of their own interfering movements by swaddling and hand holding. The examiner needs to be aware of the impact his shifting body has as he attempts to call to the infant from one side, while simultaneously providing the infant with optimal facilitation.

Elicitation and Maintenance of Attention. Score 1 through 9.

Orienting Capacity (B) (BNBAS Equivalent)

Scoring

(1) No reaction
(2) Respiratory change or blink only
(3) General quieting as well as blink and respiratory changes
(4) Stills, brightens, no attempt to locate source
(5) Shifting of eyes to sound, as well as stills and brightens
(6) Alerting and shifting of eyes and head turn to source
(7) Alerting, head turns to stimulus, and search with eyes
(8) Alerting prolonged, head and eyes turn to stimulus repeatedly
(9) Turning and alerting to stimulus presented on both sides on every presentation of stimulus

Orienting Capacity (A). Score 1 through 9.
Effort to Shut Out and Effort to Attend. Score 1 through 9.
Cost of Attention. Score 1 through 9.
Quality of Responsivity. Score 1 through 9.

ATTENTION TO INANIMATE SOUND AND SIGHT (INANIMATE VISUAL AND AUDITORY: RED RATTLE)

The infant's ability to alert to inanimate sights and sounds is assessed. The red plexibox rattle makes an attractive stimulus with its bright, shiny appearance and its soft sound. As with the social stimulus of face and voice, the infant's best performance should be brought out by holding the infant comfortably on the examiner's lap, slightly propped up. The examiner's face may be a competing stimulus and needs to be moved back once the infant's attention is available. The optimal distance for visual fixation varies considerably from infant to infant, and the examiner should be flexible. Some infants do better if the stimulus is further away, since it prevents them from getting overwhelmed. If the rattle is suddenly shaken harshly for a brief period, this stimulus can be used occasionally to break through the disorganization of hyperarousable infants; once their attention is caught, they can be modulated gradually by reducing the intensity of the stimulus as they are brought down. The rattle is moved gently to emit a soft continuous but varied sound, first in horizontal, then vertical, and then sagittal excursions, always starting in the infant's midline.

Elicitation and Maintenance of Attention. Score 1 through 9.

Orienting Capacity (B)

SCORING

(1) Does not focus on or follow stimulus
(2) Stills with stimulus and brightens
(3) Stills, focuses on stimulus when presented, brief following
(4) Stills, focuses on stimulus, follows for 30-degree arc, jerky movements
(5) Focuses and follows with eyes horizontally and/or vertically, for at least a 30-degree arc. Smooth movements, loses stimulus but finds it again
(6) Follows for 30-degree arcs, with eyes and head. Eye movements smooth
(7) Follows with eyes and head at least 60 degrees horizontally, maybe briefly vertically, partially continuous movement, loses stimulus occasionally, head turns to follow
(8) Follows with eyes and head 60 degrees horizontally and 30 degrees vertically
(9) Focuses on stimulus and follows with smooth, continuous head movement horizontally, vertically, and in a circle. Follows for at least 120 degree arc

Orienting Capacity (A). Score 1 through 9.
Effort to Shut Out and Effort to Attend. Score 1 through 9.
Cost of Attention. Score 1 through 9.
Quality of Responsivity. Score 1 through 9.

ATTENTION TO INANIMATE OBJECT (INANIMATE VISUAL: RATTLE WITHOUT SOUND OR RED BALL)

This item is scored and administered like the preceding one, but without the sound component. The infant's ability visually to attend to and track a bright silent stimulus is assessed. The rattle, if held quietly, or a bright red ball may be used.

Elicitation and Maintenance of Attention. Score 1 through 9.

Orienting Capacity (B) (BNBAS Equivalent)

SCORING

(1) Does not focus on or follow stimulus
(2) Stills with stimulus and brightens
(3) Stills, focuses on stimulus, little spontaneous interest, brief following
(4) Stills, focuses on stimulus, following for 30-degree arc, jerky movement
(5) Focuses and follows with eyes horizontally for at least a 30-degree arc. Smooth movement, loses stimulus but finds it again
(6) Follows for 30-degree arc, with eyes and head. Eye movements are smooth
(7) Follows with eyes and head at least 60 degrees horizontally, maybe briefly vertically, continuous movement, loses stimulus occasionally, head turns to follow
(8) Follows with eyes and head 60 degrees horizontally and 30 degrees vertically
(9) Focuses on stimulus and follows with smooth, continuous head movement horizontally, vertically, and in a circle. Follows for 120-degree arc.

Orienting Capacity (A). Score 1 through 9.
Effort to Shut Out and Effort to Attend. Score 1 through 9.
Cost of Attention. Score 1 through 9.
Quality of Responsivity. Score 1 through 9.

ATTENTION TO INANIMATE SOUND (INANIMATE AUDITORY: RATTLE)

This is a measure of the infant's response to the rattle or a soft ball as an inanimate stimulus. The auditory stimulus should be presented to each side and out of sight so that one can observe the infant's eyes and head as they respond to the lateralized stimulus. Alerting, eye shift, and head turning to the stimulus are scored. Brightening of face and eyes are evidence of the infant's attention to the stimulus. The examiner should present the sound stimulus very softly; many preterm infants are easily overloaded by sound and will avert or become disorganized. If the infant alerts to the soft sound but does not attempt to orient to it, the sound may be increased slowly, but not beyond the infant's threshold of tolerance. Again, occasionally the sound may be used to break through an infant's disorganization and draw him into calmer availability.

Elicitation and Maintenance of Attention. Score 1 through 9.

Orienting Capacity (B) (BNBAS Equivalent)

SCORING

(1) No reaction
(2) Respiratory change or blink only
(3) General quieting, as well as blink and respiratory changes
(4) Stills, brightens, no attempt to locate source
(5) Shifting of eyes to sound, as well as stills and brightens
(6) Alerting and shifting of eyes and head turns to source
(7) Alerting, head turns to stimulus, and search with eyes
(8) Alerting prolonged, head and eyes turn to stimulus repeatedly
(9) Turning and alerting to stimulus presented on both sides on every presentation of stimulus.

Orientation Capacity (A). Score 1 through 9.
Effort to Shut Out and Effort to Attend. Score 1 through 9.
Cost of Attention. Score 1 through 9.
Quality of Responsivity. Score 1 through 9.

PART III. BEHAVIORAL SUMMARY SCALES

Each of the summary scales documents the degree to which the infant has displayed specific organizational capacities or parameters of organizational functioning in the overall course of the examination.

These summary scales are grouped along the behavioral-organizational subsystems.

Specific Physiological Parameters

Tremulousness
Startles
Skin Color: Lability of Skin Color; Lability of Compromised Skin Color; Threshold of Color Change; Degree of Jaundice.
Smiles

Specific Motor Organization Parameters

Tonus: General Tone; Balance of Tone.
Motor Maturity: General Motor Maturity; Threshold of Motoric Imbalance; Control over Posture; Symmetry of Tonus, Posture and Movement.
Activity: Spontaneous Activity, Elicited Activity; General Activity.
Hand-to-Mouth Facility

Specific State Organization Parameters

Alertness: Degree of Responsivity B & A; Quality of Responsivity; Amount of Manipulation Necessary.
State Regulation: Lability of States A & B; Range and Flexibility of States.

Specific Self-Regulatory Parameters

Catalog of Regulation Maneuvers
Quieting Activity: Self-Quieting from Crying; Self-Quieting from Motor Arousal.
Consolability: Consolability with Intervention from Crying; Consolability with Intervention from Motor Arousal.
Peak of Excitement
Rapidity of Buildup: Rapidity of Buildup to Crying; Rapidity of Buildup to Motor Arousal.
Irritability: Irritability with Crying; Irritability with Motor Arousal.
Robustness and Endurance
Control over Input
Need for Facilitation and Use of Stimulation

One Overall Summary Parameter

Attractiveness

Many of these parameters are directly based on the BNBAS.

Where possible the original scoring is retained or equivalents are indicated. Additional scales were developed to capture the range of functioning along these parameters observable in preterm infants. Other scales are newly developed to document essential parameters of emerging organization.

Physiological Parameters

TREMULOUSNESS

In its severe form, tremulousness may be a measure of central nervous system irritation; it may occur for metabolic reasons, or may be a sign of motoric physiological imbalance. Milder forms of tremulousness are demonstrated at the end of a startle, and as a baby comes from sleeping to awake states. In light sleep or as he startles in deep sleep, tremors of the extremities are often noted. As he becomes alert and active, the tremulousness is often overcome with smoother behavior of the limbs. In some infants tremors may reflect the intensity of alerting. Tactile stimulation often is followed by tremulousness of the chin and extremities. Gradually tactile stimuli will no longer cause tremors. Tremulousness can be seen as one index of relative physiological lability.

BNBAS Equivalent	Scoring
(1)	(1) No tremors or tremulousness noted
(2)	(2) Tremors only during sleep
(3)	(3) Tremors only after the Moro or massive tactile stimulation
(4)	(4) Tremulousness seen 1 or 2 times in state 5 or 6, or with moderate tactile vestibular stimulation
(5)	(5) Tremulousness seen 3 or more times in state 5 or 6, or with moderate tactile and vestibular stimulation
(6)	(6) Tremulousness seen 1 or 2 times in state 4, or with very mild tactile stimulation
(7)	(7) Tremulousness seen 3 or more times in state 4, or with very mild tactile stimulation
(8)	(8) Tremulousness seen in several states and with most tactile or vestibular stimulation
(9)	(9) Tremulousness seen consistently in all states and with any kind of stimulation

STARTLES (BNBAS EQUIVALENT)

Both spontaneous startles and those which have been elicited in the course of the stimulation are included in this scale assuming the examiner manipulates the infant sensitively. Some infants never startle during an exam, except when a Moro is elicited. Highly sensitive infants react to many disturbing stimuli with a startle, and many have observable startles for no obviously observable reason. A startle is scored when there is sudden body jump or "jumping" of the extremities.

SCORING

(1) No startles noted
(2) Startle as a response to the examiner's attempts to set off a Moro reflex only
(3) Two startles, including Moro
(4) Three startles, including Moro
(5) Four startles, including Moro
(6) Five startles, including Moro
(7) Seven startles, including Moro
(8) Ten startles, including Moro
(9) Eleven or more startles, including Moro

SKIN COLOR

This measures the changes of color which take place during the period of exam, e.g., the acrocyanosis or peripheral mild cyanosis, the change from pink to pale or purple—mottling and a weblike appearance may occur occasionally in some infants, or paling. A normal full-term newborn is likely to demonstrate mild color changes several times in an exam. The changes are based on good color, and return to good color. The frequency and degree of change is scored. These scores are equivalent to the BNBAS; scorepoint one has been modified. No change in poor color may be the result of depressed or stressed autonomic system, as seen in pale or cyanotic infants. Marked changes are also frequently seen in preterm infants or in infants whose central and autonomic nervous systems are unable to master the changes during an exam. Some premature infants may start out with good color in sleep states before they are stimulated; in the course of the examination they change to varying degrees of paleness and/or webbing, color characteristics not typically observed in well-organized term infants. The recovery from these changes can be varied. Other infants may start out with poor color and throughout the examination may improve somewhat or may get even worse. These changes are scored under *Lability of Compromised Color.* The BNBAS score should be N in such a case.

The immediacy of color changes is an index of relative physiological stability and is scored under *Threshold of Color Change.*

Presence and degree of *jaundice* also is scored separately.

Lability of Skin Color

BNBAS EQUIVALENT	SCORING
	(1) Good color which is stable
(2)	(2) Good color which changes only minimally during exam
(3)	(3) Good color; no changes except change to slight blue around mouth or extremities when uncovered or to red when crying; recovery of original color is rapid
(4)	(4) Good color; mild cyanosis around mouth or extremities only when undressed; slight change in chest or abdomen but rapid recovery
(5)	(5) Good color, but changes color when uncovered or crying; face, lips, or extremities

may pale or redden; mottling may appear on face, chest, or limbs; original color returns quickly

(6) (6) Good color; change in color all over body during exam, but color returns with soothing or covering

(7) (7) Good color at outset, changes color to very red or blue when uncovered or crying; recovers slowly if covered or soothed

(8) (8) Good color which rapidly changes with uncovering; recovery is slow but does finally occur when infant is dressed

(9) (9) Good color with marked, rapid changes to very red or blue; good color does not return during rest of the exam

Lability of Compromised Skin Color

SCORING

(1) The infant has initially good color, but in the course of the examination he gets somewhat pale or webbed; he can recover quite well.
(2) The infant has initially good color, but in the course of the examination he gets moderately pale, or webbed and can recover the original color with soothing and relaxing.
(3) The infant is initially somewhat pale, flushed, or blue, but in the course of the examination he gets better, especially with containing or other facilitation.
(4) The infant is initially somewhat pale, flushed, or blue, gets more webbed, blue, or pale during examination, but does recover with resting or soothing.
(5) The infant is initially somewhat pale, flushed, or blue but in the course of the examination he gets very pale or flushed or webbed and cannot recover the original color.
(6) The infant is initially moderately pale or shows some blueness or flushedness; in the course of examination he gets somewhat worse, but recovers eventually.
(7) The infant is initially moderately pale or shows slight bluing or flushedness; he gets much worse in the course of the examination and recovers the initial color only barely.
(8) The infant is initially very pale, gray, flushed, or blue, gets somewhat worse during the examination and recovers to originally poor color.
(9) The infant is initially very pale, gray, flushed, or blue and gets worse in the course of the examination. He does not recover.

Threshold of Color Change

SCORING

(1) The infant's color is stable throughout.
(2) Color change ensues with massive tactile stimuli only.
(3) Color change ensues with medium tactile stimuli.
(4) Color change ensues in the course of low tactile stimuli.
(5) Color change ensues with early low tactile stimuli.
(6) Color change ensues during social interaction or object orientation.
(7) Color change ensues with being placed supine.
(8) Color change ensues with being uncovered.
(9) Color change ensues with distal stimulation during sleep or if infant is awake at start with first distal stimulation in awake state.

Degree of Jaundice

SCORING

(1) No jaundice is noticed.
(2) There is only very mild jaundice.

(3) There is mild to moderate jaundice.
(4) There is moderate jaundice.
(5) There is moderate to considerable jaundice.
(6) There is quite considerable jaundice.
(7) There is considerable jaundice.
(8) There is pronounced jaundice.
(9) There is very pronounced jaundice.

SMILES

Smiles are seen in the neonate in various circumstances. They can be fleeting to soft auditory and/or visual cues, in drowsy or in alert states. Occasionally, when the infant is handled and restrained in a cuddling position, a smile comes across his face as he relaxes. Prolonged or frequent undifferentiated smiles mainly involving the mouth region during sleep and, in an occasional infant, also during awake state 4AL give an eerie impression and may be discharge behaviors of a more primitive kind, reflecting mild facial avoidance. There is no apparent connection to external stimulation. State and nature of the smile not only influence interaction, but may well reflect physiological differentiation. If no smiles are observed, this should be scored N.

SCORING

(1) Frequent undifferentiated smiles which come on in any state and are apparently stimulus independent, and internally triggered
(2) Frequent undifferentiated smiles which come on in sleep and drowsy states only, with an occasional spill-over into higher states, internally triggered
(3) Frequent undifferentiated internally triggered smiles in sleep states only
(4) Some mild undifferentiated internally triggered smiles in sleep states only
(5) An occasional internally triggered smile in sleep state and some smiles to soft auditory stimuli (voice, rattle), or in the course of relaxing, or in sleep or drowsy states
(6) Several smiles to soft auditory stimuli or in the course of relaxing in sleep or drowsy states
(7) An occasional smile with eyes open to soft auditory stimuli or in the course of relaxing
(8) One or two eyes-open smiles with focused attention to social or inanimate objects
(9) Several eyes-open, differentiated smiles with focused attention to social or inanimate objects

Specific Motor Organization Parameters

TONUS

This assesses the characteristic motor tone the infant shows in the course of the assessment. The infant's posture reflects tonus to a large extent. In well-modulated tone there is a continuous smooth tonic balance between extensor and flexor tone, between avoidance and groping or approach. In the hypertonic infant, both flexor and extensor tone, avoidance posture and approach postures of trunk, head, and the extremities, are exaggerated, and the resistance to overcoming these postures is high. In the hypotonic infant, there is flaccidity and floppiness in these postures and movement; the infant shows limp extremities with no resistance to manipulations and his trunk and head have the behavior of a ragdoll.

Continuity of consistency of tone is a sign of increasing balance. Some infants show great fluctuation in tonicity, shifting suddenly and frequently from hypertonic to flaccid and back. Other infants show fluctuation in tone in different body parts, and floppier arms than legs is a common occurrence in young infants. The degree of these differences and the frequency of fluctuation can be seen as an index of relative maturity of the motor system.

The first tonus scale presupposes a fair degree of consistency in *overall tone;* it is equivalent to the BNBAS tonus scale. If this degree of consistency in tone cannot be observed, this scale is scored N. The second tonus scale grades the degree of *tonus balance.*

General Tone (BNBAS Equivalent)

SCORING

(1) Flaccid, limp like a ragdoll, no resistance when limbs are moved, complete head lag in pull-to-sit
(2) Little response felt as he is moved, but less than about 25% of the time
(3) Flaccid, limp most of the time, but is responsive 25% of the time with some tone
(4) Some tone half the time, responds to being handled with average tone less than half of the time
(5) Tone average when handled, lies in fairly flaccid state in between handling
(6) Variable tone in resting, responsive with good tone as he is handled approximately 75% of the time
(7) Is on the hypertonic side approximately 50% of the time
(8) When handled he is responsive with hypertonicity about 75% of the time
(9) Hypertonic at rest (in flexion) and hypertonic all the time (abnormal)

Balance of Tone

SCORING

(1) The infant has essentially no tone.
(2) Arms and legs and trunk are alternately and independently hypertonic and completely hypotonic in repeated sudden fluctuations in the course of the examination.
(3) Arms, legs, and trunk are quite hypertonic; there are some periods of sudden complete flaccidity.
(4) Arms, legs, and trunk may differentially alter tone between hypertonicity and hypotonia, yet with facilitation there is some balance.
(5) Arms are somewhat more flaccid than legs which are very hypertonic, yet this remains fairly consistent throughout the examination.
(6) Arms are relatively well-modulated and legs are predominantly hypertonic; there is some decrease in discrepancy with facilitation.
(7) Arms are relatively well-modulated and legs are somewhat hypertonic, yet this is mainly due to tactile manipulation.
(8) Arms are somewhat softer than legs, but this remains fairly constant during manipulation.
(9) Arms, legs and trunk are of consistent modulated tone during resting and during manipulation.

MOTOR MATURITY

Motor maturity is demonstrated by smooth movements of the extremities and a free, wide range of movements. The arm movements are the easiest to score. The assessment of smoothness versus jerkiness reflects the balanced flexor and extensor tone and its differentiation. The degree to which flexors and extensors are competing also comes out in freedom of arcs of movement (45–90°) versus restricted arcs (45° or less). The preterm infant may have apparent unlimited freedom of movement in lateral, sagittal, and cephalad areas, but the movements are jerky and uncontrolled, sudden extensions of arms and legs with no modulation from the flexors, i.e., the avoidance and approach components are executed independently without respective modulation on one another.

The mature infant has controlled freedom of movement in all directions associated with a smooth respective balance of avoidance and approach components, making for smooth, differentiated, controlled movements.

Motor maturity is scored with very minor changes on the original BNBAS scale. If there is not enough activity to judge motor maturity, N is given.

The threshold of movement imbalance in the course of the examination is an index of the stability of the balance and should be scored separately on *Threshold of Motoric Imbalance*. The degree of postural control is reflected in the occurrence of occasional or predominant characteristic postures which changes with maturation. *Control over Posture* is another reflection of relative motor maturity. The fetal "natural" resting position is complete trunkal flexion which the preterm attempts to maintain once his extrauterine tonicity emerges. Gradually this fetal posture is more and more freed up and differentiated. In the process of this differentiation the struggle of balance of groping and avoidance behavior, of flexion and extension components is apparent. It comes out in characteristic hand-on-face behavior, fetal tucking, midline hand-grasping, all indices of flexion and groping; and on the other side, "salute" position of arms and legs, i.e., sudden arm and leg extensions into midair, and trunkal inversion, i.e., U-shaped body posture with head and feet off surface in a total body arch, arms flailing in airplane position and finger splaying, all indices of avoidance. The facial accompaniment of the postures is, on the one hand, tongue extensions and lip-pursing (groping), on the other hand, lip retraction, cheek retraction, and grimacing (avoidance). On the physiological level the extensions are sighing, yawning, and sneezing, and at a more intense level, the visceral analogues of avoidance are gagging, spitting up, and bowel movements. The specific postures of groping and avoidance are catalogued individually and graded on a range from 0 to 3 in the Catalogue of Regulation Behaviors. This will permit the identification of the level of modulation, e.g., balance may be achieved at the visceral level, yet is still being negotiated at the motoric level; or balance may be achieved at the motoric level but is being negotiated at the attentional sensory level, exemplified in visual locking as groping and yawning or sneezing as avoidance. All these behaviors gradually come under mutual balance and mutual inhibition.

For control over posture the motoric components are considered. The infant who is not yet grappling with the postural differentiation is typically in a flat out, at times frog-like, posture. The next stage is the beginning emergence of the imbalance postures mentioned, which gradually come under more and more balanced control.

Symmetry of Tonus, Posture, and Movement is a further sign of differentiation and intactness of the motor system. Right-left symmetry along the sagittal body axis is considered here. Symmetry is conventionally assessed for each of the standard reflexes separately. Aside from such systematically elicited movements, symmetry of tonus, posture, and movement should be assessed on a continuous basis in the course of the examination by observing facial expression and the spontaneously assumed preferred total body or head postures of the infant, differential movement of the extremities, paying attention to arms, legs, hands, and feet, and fingers and toes specifically. Tonus asymmetries may be observed during spontaneous movement or resting postures or when the infant is moved about or is interacted with socially. The locus and degree of asymmetry are noted on the supplemental List of Asymmetries (see Score Sheet).

Motor Maturity (BNBAS Equivalent)

SCORING

(1) Overshooting of legs and arms in all directions
(2) Jerky movements and mild overshooting
(3) Jerky movements, no overshooting
(4) Only occasional jerky movements predominating arcs to 45°
(5) Smooth movements predominate, arcs predominantly 60° half the time

(6) Smooth movements, arcs predominantly 60°
(7) Smooth movements and arcs of 90° less than half of the time
(8) Smooth movements and unrestricted arms laterally 90° most of the time
(9) Smoothness, unrestricted (90°) all of the time

Threshold of Motoric Imbalance

SCORING

(1) The infant shows no motoric imbalance.
(2) The infant shows motoric imbalance only with medium or massive tactile and vestibular stimulation in crying states or with massive tactile and vestibular stimulation in any state.
(3) The infant shows motoric imbalance once medium tactile and vestibular stimulation is applied, when he is still in semi-alert and alert states.
(4) The infant shows motoric imbalance with low tactile stimulation in a semi-awake or awake state.
(5) The infant shows motoric imbalance when visual and auditory stimulation is presented, when he is in alert and semi-alert states.
(6) The infant shows motoric imbalance with low tactile stimulation when in sleep state.
(7) The infant shows motoric disorganization and imbalance as soon as he is uncovered and placed supine.
(8) The infant shows motoric imbalance even during distal stimuli in sleep state.
(9) The infant shows motoric imbalance even during the initial observation period in sleep state.

Control over Posture

SCORING

(1) Frog-like or flat posture without tonicity
(2) Some beginning fetal tuck, weak salutes of arms and legs, otherwise frog-like or complete flatness
(3) Strong fetal tuck, well maintained; occasional salutes of arms and legs; only occasional flatness
(4) Fetal tuck and/or flatness with strong and frequent salutes of arms and legs or rigid extension
(5) Moderately soft fetal tuck, moderate salutes, beginnings of differentiated flexion and extension balance
(6) Some fetal tuck, occasional salutes, but on the whole infrequent; at times modulated flexion and extension balance
(7) Only occasionally fetal tuck components, isolated arm or leg salutes or frantic movement; usually quite modulated flexion and extension balance
(8) Almost consistently well-modulated flexion and extension balance, with predominantly balanced arm, leg, and trunkal tone and movement
(9) Well-modulated flexion and extension balance, differentiated "free" posture with balanced arm, leg, and trunkal tone and movement

Symmetry of Tonus, Posture, and Movement

SCORING

(1) There is repeated, quite pronounced, consistent asymmetry in elicited and spontaneous movements and elicited and spontaneously assumed postures.

(2) There is definite, reproducible, consistent asymmetry of movement, posture, and tonus which is somewhat modifiable by relaxation and organization.
(3) There is intermittently fixed, reproducible asymmetry of movement and/or posture and/or tonus. This is not always consistent and is sometimes more pronounced in spontaneous movement, sometimes in elicited movement.
(4) There is transient, but fairly reproducible asymmetry of movement and/or posture and/or tonus. This is reliably observable, although it is not fixed.
(5) There is repeatedly definite but limited asymmetry of posture, tone, or movement. The occurrence may be either mild in intensity or quite limited in the extent of involvement, yet it is consistent.
(6) There is repeatedly mild asymmetry of posture, of tone, of several systematically elicited movements, or of spontaneous movement. This asymmetry is fluid and modifiable by increased attention or facilitation.
(7) There is an occasional transient asymmetry of posture, tonus, or of a systematically elicited movement. Two or three SEMs are more pronounced on the same side of the body, or a certain posture such as leg tucking or leg bracing is mildly but repeatedly more pronounced on one than the other side.
(8) There is an occasional, mild asymmetry in a spontaneous preferred posture or in systematically elicited movement. This is not easily reproducible.
(9) There is no asymmetry of tone, movement, or posture, neither spontaneously nor on elicited motor patterns.

Supplemental List of Asymmetries

Check, rate degree, and describe asymmetries noted; rate degree of asymmetry on a 0–3 continuum.

0 = no asymmetry noted; the item was not checked
1 = subtly and mildly present and/or very transient
2 = moderately pronounced and/or intermittent
3 = pronounced, strong

Asymmetries

1. Arm	4. Leg	7. Head positioning
2. Hand	5. Foot/toes	8. Face
3. Fingers	6. Trunkal posture	9. Eyes

ACTIVITY

This is a summary of the activity seen during the entire observation, especially during the alert states. The activity consists of two kinds—(1) spontaneous and (2) in response to the stimulation of handling and the stimuli used by the observer, which are scored separately. The discrepancy between spontaneous and elicited activity reflects the relative imbalance of the motor system. All or none activity is common in immature infants. If spontaneous and elicited activity are not more than a point apart, the infant can also be scored on the BNBAS scale.

A further dimension of activity is reflected in the inaccessibility of the activity, i.e., when the activity can be inhibited by the examiner's maneuvers. Amount of activity is graded. Frantic excessive activity is uncontrolled, intense activity; intense high activity is somewhat more controlled activity at a high level; much activity is controlled activity which builds up first, perpetuates itself for a period after activity is initiated, and then dies out. Average activity has no buildup, but at least 3 cycles of activity which are decreasing all the time; little activity has 2 or 3 cycles of activity which die out quickly.

Spontaneous activity is observed when the infant is on his own, having returned to baseline after a manipulation sequence, either in the crib or on the examiner's lap. Opportunities for observation are before uncovering, after supine, before the orientation sequence, and between packages II through V. Elicited activity is the degree of activity and motor arousal produced by the infant by any of the examiner's manipulations.

Spontaneous Activity

SCORING

(1) No activity at all
(2) Slight activity
(3) Moderate activity
(4) Much activity
(5) Continuous activity which is easily consolable
(6) Continuous activity which is increasingly difficult to control
(7) Continuous activity which is very difficult to control
(8) Continuous, intense and at times frantic activity which is, at times, very difficult to control
(9) Continuous intense and frantic activity which is not consolable

Elicited Activity

SCORING

(1) No activity
(2) Slight activity
(3) Moderate activity
(4) Much activity
(5) Continuous activity which is easily consolable
(6) Continuous activity which is increasingly difficult to control
(7) Continuous activity which is very difficult to control
(8) Continuous, intense, and at times frantic activity which is, at times, very difficult to control
(9) Continuous, intense, frantic activity which is not consolable

Activity (BNBAS Equivalent)

SCORING

Spontaneous and elicited activity is scored separately on a four point scale: 0 = none, 1 = slight, 2 = moderate, 3 = much. Then add up the two scores. If spontaneous and elicited activity have a difference of more than one point, this scale should not be used.

(1) = a total score of 0
(2) = a total score of 1
(3) = a total score of 2
(4) = a total score of 3
(5) = a total score of 4
(6) = a total score of 5
(7) = a total score of 6
(8) = continuous but consolable movement
(9) = continuous, unconsolable movement

HAND-TO-MOUTH FACILITY (BNBAS EQUIVALENT)

Hand-to-mouth coordination develops *in utero*. It is seen spontaneously as the infant attempts to control himself or comfort himself when aroused. It is a measure of differentiated motor coordination reflected in his ability to bring his hand to his mouth in supine as well as his success in insertion and maintaining it there. Some infants bring their hands to their mouths repeatedly, insert a part of the fist or fingers, and suck actively on the inserted part, which requires a fair degree of motoric balance, postural stability, and integration. The scoring is the same as on the BNBAS.

SCORING

(1) No attempt to bring hands to mouth
(2) Brief swipes at mouth area, no real contact
(3) Hand brought to mouth and contact, but no insertion, once only
(4) Hand brought next to mouth area twice, no insertion
(5) Hand brought next to mouth area at least 3 times, but no real insertion, abortive attempts to suck on fist
(6) One insertion which is brief, unable to be maintained
(7) Several actual insertions which are brief, not maintained, abortive sucking attempts, more than 3 times next to mouth
(8) Several brief insertions in rapid succession in an attempt to prolong sucking at this time
(9) Fist and/or fingers actually inserted and sucking on them for 5 seconds or more for several brief insertions

Specific State Organization Parameters

ALERTNESS

This assesses the *responsivity* shown in the course of the examination when the infant is in an alert state. This is best assessed during package VI when the examiner facilitates the infant's other systems to maximally free up and bring out his alertness and responsivity. Since very young infants are alert for only short periods if at all and are more responsive when they come to alertness spontaneously in the course of their endogenous circadian rhythm, the ability to come to alertness in the course of this examination is an index in itself of the infant's increasing overall state differentiation and state control. The degree of responsivity when alert becomes an index of the emerging and expanding differentiation of the alert state. Duration of responsivity and delay with which attention to stimuli can be brought about are indices of the degree of responsivity. The well-differentiated infant will no longer need the examiner's prompting to attend to stimuli, but will actively select and even initiate social interaction and visual exploration of stimuli.

The *Degree of Responsivity (B)* is scored essentially as on the BNBAS. This scale should be used only if the infant reaches state 4B during orientation. The score of 9 is changed to reflect active selection and exploration of the environment. *Degree of Responsivity (A)* is scored if the infant attains states 4AL or 4AH. Then degree of responsivity (B) should be scored N.

Aside from the degrees of responsivity, the *Quality of Responsivity* is scored on a facial animation continuum, again capturing increasing modulation and differentiation. Another component of responsivity assessed is the relative degree of autonomous stability of the level of responsivity achieved, measured by the *Degree of Manipulation Necessary* from the examiner.

Degree of Responsivity (B)

Alertness

BNBAS EQUIVALENT	SCORING
(1)	(1) Inattentive—rarely or never responsive to direct stimulation
(2)	(2) When alert, responsivity brief and generally quite delayed—alerting and orientation very brief and general. Not specific to stimuli
(3)	(3) When alert, responsivity brief and somewhat delayed—quality of alertness variable
(4)	(4) When alert, responsivity somewhat brief but not generally delayed, though variable
(5)	(5) When alert, responsivity of moderate duration and response generally not delayed and less variable
(6)	(6) When alert, responsivity moderately sustained and not delayed. May use stimulation to come to alert state
(7)	(7) When alert, episodes are of generally sustained duration, etc.
(8 & 9)	(8) Always has sustained periods of alertness in best periods. Alerting and orientation frequent and reliable. Stimulation brings infant to alert state consistently
	(9) Always alert in best periods. Actively selects stimuli and explores the inanimate environment visually or actively initiates social interaction

Degree of Responsivity (A)

(1)	(1) Inattentive—rarely or never responsive to direct stimulation
(2)	(2) When alert, responsivity brief and generally quite delayed—alerting and orientation very brief and general. Not specific to stimuli
(3)	(3) When alert, responsivity brief and somewhat delayed—quality of alertness variable
(4)	(4) When alert, responsivity somewhat brief but not generally delayed, though variable
(5)	(5) When alert, responsivity of moderate duration and response generally not delayed and less variable
(6)	(6) When alert, responsivity moderately sustained and not delayed. May use stimulation to come to alert state
(7)	(7) When alert, episodes are of generally sustained duration, etc.
(8 & 9)	(8) Always has sustained periods of alertness in best periods. Alerting and orientation frequent and reliable. Stimulation brings infant to alert state consistently
	(9) Always alert in best periods. Actively selects stimuli and explores the inanimate environment visually or actively initiates social interaction

Quality of Responsivity

SCORING

(1) Attentional responsivity is not an appropriate issue for this infant.
(2) The infant attempts to stay in lower states; he is only barely available for fleeting periods. His face may appear pained and bothered.
(3) The infant attempts to rouse himself, but is only fleetingly successful.
(4) The infant comes to alertness with low intensity and flat responsiveness, although he may show good following.
(5) The infant comes at times to alertness but then appears overly heightened, pained, or almost panicked. He may appear easily overloaded or at the mercy of the input. He may show good following.

(6) The infant shows bright-eyed, modulated, focused alertness at times for brief periods; he may be flat or hyperalert at other times; when he is modulated, his face is softened and participates in the response; his mouth may open and round; he may raise his cheeks and eyebrows for brief periods.
(7) The infant's responsiveness is usually of the modulated quality described above. Only rarely is he flat or hyperalert.
(8) The infant's responsiveness is almost consistently modulated and differentiated; his face participates as described above.
(9) The infant's responsiveness is consistently well-modulated, and there may also be active elicitation and initiation of interaction by the infant.

Amount of Manipulation Necessary

SCORING

(1) The baby was in an alert state to begin with or came to alert early with one attempt. From then on he stayed in alert state and responsive essentially throughout the examination.
(2) The baby is almost continuously in alert states, after easily being brought to it or being there spontaneously.
(3) The baby is easily brought into alertness and spontaneously prolongs alert periods.
(4) The baby is spontaneously alert at some time during the examination and can maintain alertness for moderate periods.
(5) The baby is spontaneously in an alert state some time during the examination, and can maintain this alertness for brief periods.
(6) The baby's states need to be manipulated by the examiner a few times to elicit alerting; then alerting is of moderate duration.
(7) The baby's states need to be manipulated most of the time to elicit or maintain alerting; alerting is brief.
(8) The baby's states need to be manipulated continuously to elicit or maintain some alerting.
(9) The baby cannot be manipulated by any means into responsiveness.

STATE REGULATION

The availability of certain states, the degree of fluctuation between them, the stability of the alert states and other well-defined states as sleep and robust crying, in contrast to more diffuse states as state 1A, 2A, 3A, 4AL, 4AH, 5A, 6A, which should gradually diminish, all are indices of expanding state organization.

Lability of states measures the infant's fluctuation of states by counting the number of state changes in the course of the examination. Each change is counted as soon as the state is recognizable which means operationally probably a period of 3 seconds or more.

Range and flexibility assess the degree of state expansion and respective robustness of the range available. Again the more diffuse states are expected to get solidified over time, and the range from sleep to solid alert to robust crying states is expected to unfold.

Lability of States

SCORING

The score corresponds to the frequency of swings.

(1) 1–2 swings over course of examination
(2) 3–5
(3) 6–8
(4) 9–10

(5) 11–13
(6) 14–15
(7) 16–18
(8) 19–22
(9) 23 on up

Range and Flexibility of States

SCORING

(1) The infant is only in low states (1A or B, 2A or B, or 3A) in the course of the examination.
(2) The infant is mainly in low states, but can come to state 3 with maybe one excursion into state 4AL or 5A.
(3) The infant is mainly in low states, including state 3, but has brief periods in state 5A.
(4) The infant is mainly in low states, including state 3, and has some state 4AL or may have 4AH available, with or without an occasional 5A or 6A.
(5) The infant is mainly in low states, has states 1, 2, and 3 available, no longer shows states 1A or 2A, also has state 5B available and may or may not briefly state 4B; or he is oscillating exclusively between states 3 and 5 and 6.
(6) The infant has states 1, 2, 3, and 5 available, shows the beginnings of state 4B, possibly embedded either in 4AL or 4AH; state 6A is occasionally observable, embedded in state 5A and 5B or 6B; or the infant is continuously in states 5 and 6.
(7) The infant has state 4B available and actively keeps himself there with minimal excursions to 5A or 5B indicating stress. The sleep states, if observed, are well organized; or the infant has state 6B available while state 4B is still embedded in 4AH or 4AL or is not very prolonged or stable, and the sleep states are well organized.
(8) The infant has state 6B available, as well as state 4B; the oscillations may still be abrupt and unmodulated.
(9) The infant has the full range of organized states available with only fairly brief periods in either state 3 or states 5 and 6, or he may actively control himself in state 4B without stress to any of the other systems.

Lability of States (BNBAS)

SCORING

The score corresponds to the frequency of swings.

(1) 1–2 swings over 30 minutes
(2) 3–5
(3) 6–8
(4) 9–10
(5) 11–13
(6) 14–15
(7) 16–18
(8) 19–22
(9) 23 on up

Specific Self-Regulatory Parameters

CATALOGUE OF REGULATION MANEUVERS

All behaviors in this catalogue are scored on a range from 0 to 3: 0 = not observed; 1 = only occasionally used; 2 = used moderately frequently; 3 = frequently used; appears a reliably discernible pattern.

As indicated on pages 42 and 114, these maneuvers are seen on a continuum of developmental differentiation, from involving the physiological system (level 1) to involving the motor system (level 2) to involving the attentional system (level 3). Within each of these levels, there are the *groping* maneuvers, maneuvers directed *toward* the stimulus, representing the approach component of behavior; and there are *withdrawing* maneuvers, maneuvers directed *away* from the stimulus, representing the avoidance component of behavior. The task of the organism's development is seen as a gradual smoothing of the mutual acting of these two components, approach and avoidance, on each other, until this regulation as such becomes imperceptible, i.e., they are, in fact, smoothly regulated, and their relative preponderance is no longer behaviorally apparent, i.e., the organism appears to execute an action smoothly and freely. For example, when a visual stimulus is presented to the very young organism, the organism may first be reacting by being drawn toward the stimulus in a generalized fashion: The mouth may open into a groping "ooh" configuration, the tongue may extend strongly and repeatedly toward the stimulus, undifferentiated whimper-like sounds may be emitted, the eyes may widen and lock onto the stimulus; arms, legs, and head may strain toward the stimulus; hands, fingers, and toes may open in a groping fashion; respirations and heartrate may possibly speed up, and the whole organism is engaged in a strong, unchecked approach configuration. The modulation, regulation, and dampening configuration is then imposed by the subsequently ensuing avoidance components of behavior: Now the face withdraws into a grimace, mouth narrows and retracts, grunting and bowel movement straining ensue, eyes narrow, the tongue draws in, the infant may sneeze or yawn, or intensely gag, or spit up; the arms draw back or extend sideways and backwards, the trunk arches, hands open and fingers splay. The organism has swung from unchecked approach to unchecked avoidance. Now, as the second cycle starts and the approach component comes into play again, it may be more dampened, due to the action of the avoidance component on its amplitude. If the organism does not endogenously possess such regulatory facility, it may nevertheless be modulatable by environmental input, e.g., the amplitude of arm retraction and finger splay can be reduced by placing the examiner's finger into the infant's hands; this may induce the groping component, the infant's hands may close around the examiner's finger, his arms in turn may begin to flex, and the groping component may spread to trunk and face; trunkal arching will soften and gradually go into trunkal flexion; the facial retraction and grimace will relax, bowel movements, gags, and spit-ups may be prevented, and the face may return to a neutral baseline or move gradually into a new cycle of groping. Thus, the continuous action of avoidance and approach components at the various levels of organization upon each other will result in increasing periods of well-regulated balance, freeing up the organism for the grappling of the next level of functional balance and competence.

The catalogue of behaviors first lists the avoidance behaviors, then the approach and groping behaviors. Within each category the behaviors are roughly grouped by level. Many of these behaviors are self-explanatory. A brief description of each is given.

Withdrawal or Avoidance Behaviors

1. ***Spit-Ups***
   The infant spits up; more than a passive drool is required, although the vomitus as such may be no more than a drool or quite minimal.
2. ***Gags***
   The infant appears to choke momentarily or gulp or gag; the swallowing and respiration patterns are out of synchrony. This is often, but not necessarily, accompanied by at least mild mouth-opening.
3. ***Hiccoughs***
   The infant hiccoughs.
4. ***Bowel Movement Grunting or Straining***
   The infant's face and body display the straining often associated with bowel movements and/or he emits the grunting sounds often associated with bowel movements.

5. ***Grimace, Lip Retraction***
   The infant's lips retract noticeably and/or his face is distorted in a retracting direction (eyebrow-knitting alone is not sufficient but is a likely part of this configuration).
6. ***Trunkal Arching***
   The infant's trunk arches and/or his head extends in an arching fashion. The upper extremities do not have to extend; often the legs may be extending.
7. ***Finger Splay***
   The infant's hands open strongly, and the fingers are extended and separated from each other.
8. ***Airplane***
   The infant's arms either are fully extended out to the side at approximately shoulder level or upper and lower arm are at an angle to each other but are extended out at the shoulder.
9. ***Salutes***
   The infant's arms are fully extended into midair, either singly or simultaneously.
10. ***Sitting on air***
    The infant's legs are extended into midair, either singly or simultaneously. This may occur when the infant is supine or upright.
11. ***Sneezing***
    The infant sneezes.
12. ***Yawning***
    The infant yawns.
13. ***Sighing***
    The infant sighs.
14. ***Coughing***
    The infant emits coughing sounds.
15. ***Averting***
    The infant actively averts his eyes. He may momentarily close them.
16. ***Frowning***
    The infant knits his brows or darkens his eyes by contracting his periocular musculature.

Approach or Groping Behaviors

1. ***Tongue Extension***
   The infant's tongue either is extended toward a stimulus, or it repeatedly extends and relaxes.
2. ***Hand on Face***
   The infant's hand or hands are placed onto his face or over his ears and are maintained there for brief period.
3. ***Sounds***
   The infant emits undifferentiated, at times whimper-like, sounds.
4. ***Hand Clasp***
   The infant grasps his own hands or clutches his hands to his own body; the hands each may be closed but they touch each other.
5. ***Foot Clasp***
   The infant positions his feet against each other, footsole to footsole, or folds his legs in a crossed position with his feet grasping his legs or resting on them.
6. ***Finger Fold***
   The infant interdigitates one or more fingers of each hand.
7. ***Tuck***
   The infant curls or tucks his trunk or shoulders, pulls up his legs, and tucks his arms, or uses the examiner's hands or body to attain tuck flexion.
8. ***Body Movement***
   The infant adjusts his body or extremities or head into a more flexed position, such as turning to the side, attempting to attain a tonic neck response, etc.

9. ***Hand to Mouth***

The infant attempts to bring his hand or fingers to his mouth. He does not have to be successful.

10. ***Grasping***

The infant makes grasping movements with his hands either directed to his own face, his body, or in midair, or to the examiner's hands or fingers or body, or toward the side of the bassinette, etc.

11. ***Leg/Foot Brace***

The infant extends his legs and/or feet toward the examiner's body, hands, the surface he is on, the sides of the bassinette, etc., in order to stabilize himself. Once touching, he may flex his legs or he may restart the bracing.

12. ***Mouthing***

The infant makes mouthing movements with his lips and/or jaws.

13. ***Suck Search***

The infant extends his lips forward or opens his mouth in a searching fashion, usually moving his head while doing so.

14. ***Sucking***

The infant sucks on his own hand or fingers, on clothing, the examiner's finger, a pacifier or other object that he has either obtained himself or that the examiner has inserted into his mouth.

15. ***Hand Holding***

The infant holds on to the examiner's hands or finger or arm, etc., with his own hands. He may have placed them there himself, or the examiner may have positioned them there; the infant then actively holds on.

16. ***"Ooh" Face***

The infant rounds his mouth and purses his lips or extends them in an "ooh" configuration; this may be with his eyes open or closed.

17. ***Locking Visually and/or Auditorially***

The infant locks onto the examiner's face or an object or sight in the environment, e.g., he may lock on above or to the side examiner's face but maintains his gaze in one direction for observable periods. The sound component of an environmental stimulus may contribute to his locking.

18. ***Cooing***

The infant emits pleasurable cooing sounds.

Maneuvers

Avoidance or Withdrawal Behaviors (Scoring 0–3)

(1) Spit up
(2) Gag
(3) Hiccough
(4) Bowel movement grunt and strain
(5) Grimaces, lip retraction
(6) Trunkal arching
(7) Finger splay
(8) Airplane
(9) Salutes
(10) Sitting on air
(11) Sneezing
(12) Yawning
(13) Sighing
(14) Coughing
(15) Averting
(16) Frowning

Approach or groping behaviors (Scoring 0–3)

(1) Tongue extension
(2) Hand on face
(3) Sounds
(4) Hand clasp

	(5) Foot clasp
	(6) Finger fold
	(7) Tuck
BNBAS equivalent	(8) Body movement
BNBAS equivalent	(9) Hand to mouth
	(10) Grasping
	(11) Leg/foot brace
BNBAS equivalent	(12) Mouthing
	(13) Suck search
BNBAS equivalent	(14) Sucking
	(15) Hand holding
	(16) "Ooh" face
BNBAS equivalent	(17) Locking visually and/or auditorily
	(18) Cooing

QUIETING

Two measures of the infant's quieting ability are considered: first, his ability for self-quieting, and second, his ability to use graded input from the examiner to become consoled, i.e., consolability with intervention.

Self-Quieting

This is a measure of the activity the infant initiates when in an aroused state in an observable effort to quiet or control himself. The activities which are observed in these efforts are scored in a separate catalogue of self-regulating maneuvers. Some of these are also scored on the BNBAS; this will be indicated. Their success can be of varying degrees of quieting and control which the infant can achieve on his own. The range is based directly on the BNBAS. The arousal level of the infant has to be kept in mind in assessing these effects. Infants who can achieve solid state 6 or 5 behavior with real crying are obviously different infants from those who come to either a state 6A or 5A or to motoric arousal in state 3. States 6A and 5A are characterized by motoric arousal without clear fussing; 5B always has clear fussing; state 3 may or may not have one or the other. Self-quieting from these two types of arousal is scored separately. If the infant does not achieve state 6B or 5B, the respective scale is scored N. The infant may use for his self-quieting efforts any behaviors listed in the catalogue, including use of the examiner's face and voice. Once the examiner has to intervene actively either by presenting his face or by talking to the infant or by the use of tactile manipulations to help the infant, this no longer qualifies as self-quieting.

Self-Quieting from State 6B or 5B (BNBAS Equivalent)

SCORING

(1) Cannot quiet self, makes no attempt, and intervention is always necessary
(2) A brief attempt to quiet self but with no success (less than 5 seconds)
(3) Several attempts to quiet self, but with no success (less than 5 seconds)
(4) One brief success in quieting self for period of 5 seconds or more
(5) Several brief successes in quieting self
(6) An attempt to quiet self which results in a sustained successful quieting with the infant returning to state 4 or below
(7) One sustained and several brief successes in quieting self
(8) At least 2 sustained successes in quieting self
(9) Consistently quiets self for sustained periods

Self-Quieting from Motoric Arousal (States 3, 5A, 6A)

SCORING

(1) Cannot quiet self, makes no attempt, and intervention is always necessary
(2) A brief attempt to quiet self but with no success (less than 5 seconds)
(3) Several attempts to quiet self, but with no success (less than 5 seconds)
(4) One brief success in quieting self for a period of 5 seconds or more
(5) Several brief successes in quieting self
(6) An attempt to quiet self which results in a sustained successful quieting
(7) One sustained and several brief successes in quieting self
(8) At least 2 sustained successes in quieting self
(9) Consistently quiets self for sustained periods

Consolability with Intervention

If the infant does not bring himself out of either states 6B or 5B or out of state 6A or 5A or state 3 in 15 seconds, the examiner then begins to intervene in a graded series of maneuvers. Some babies will quiet when they are being talked to softly; others need to be talked to very firmly and for prolonged periods, while still others need their arm and leg movements restrained, their hands and/or feet held, their bodies tucked, they need to be picked up, need to be rocked gently or even vigorously or they need to be allowed to suck in order to be consoled. Consoling is demonstrated when the infant quiets for at least 5 seconds. The BNBAS equivalent scales require a starting point of state 6B. If the infant does not achieve such a state, this scale is scored N. Only minor revisions of the original BNBAS scale are proposed here, as indicated.

BNBAS EQUIVALENT	SCORING (State 6B)
(1)	(1) Not consolable
(2)	(2) Picking up and rocking vigorously or picking up and sucking, in addition to dressing, holding, and rocking
(3)	(3) Picking up, dressing, swaddling, and sucking, or dressing, holding in arms and rocking softly
(4)	(4) Picking up and holding, dressing and swaddling or holding and rocking
(5)	(5) Picking up and holding
(6)	(6) Restraining of arms and legs or hand on belly and restraining one or both arms
(7)	(7) Restraining of arms or legs or placing hand on belly steadily, or against soles of feet
(8)	(8) Examiner's face and voice firmly or persistently
(9)	(9) Examiner's voice very softly or examiner's face alone

Consolability with Intervention from Motoric Arousal (States 3, 5A, 5B, or 6A)

SCORING. Same as above.

PEAK OF EXCITEMENT (BNBAS EQUIVALENT)

This is a measure of the overall amount of motor and crying activity observed by the examiner during the course of the whole examination and is scored similarly to the BNBAS. The examiner sees peaks of excitement and notes how the infant's behavior brings him back to a more responsive state. The kind of intense reactions which some infants demonstrate when they reach their peak of excitement makes them unavailable to the outside world and must be scored high. Others are hardly able to be jogged to respond at all, and their peak is very low. An optimal response would fall in the moderate, reachable range, in which the infant could be brought to respond to stimuli in spite of a high degree of upset or excitement, but then returns to a more moderate state.

Scoring

(1) Low level of arousal to all stimuli. Never above state 2, does not awaken fully
(2) Some arousal to stimulation—can be awakened to state 3
(3) Infant reaches either state 4 briefly, but is predominantly in state 3 or lower
(4) Infant reaches either state 5 or 6A, but is predominantly in state 4 or lower
(5) Infant reaches state 6B after stimulation once or twice, but predominantly is in state 5 or lower
(6) Infant reaches state 6B after stimulation but returns to lower states spontaneously
(7) Infant reaches state 6B in response to stimuli but with consoling is easily brought back to lower states
(8) Infant screams (state 6B) in response to stimulation, although some quieting can occur with consoling, with difficulty
(9) Infant achieves insulated crying state. Unable to be quieted or soothed

RAPIDITY OF BUILDUP

This is a measure of use of states from quiet to agitated state. It measures the timing and the number of stimuli which are used before the infant changes from his initially quiet state to a more agitated one. Since this implies that we start with an initially quiet infant, it measures the period of "control" which he can maintain in the face of increasingly demanding stimuli as well as the additive effect of these stimuli in changing his initially quiet state.

Since many infants do not achieve a full blown crying state, build-up to an upset state as evidenced by motor discharge or fussing should be scored separately, applying the same criteria to build-up from state 1 or 2 to state 5A, 5B, or 6A, or a motor arousal 3. N is scored on the rapidity of build-up to state 6 scale, which is equivalent to the BNBAS.

Rapidity of Buildup to State 6B

BNBAS Equivalent	Scoring
(1)	(1) No upset to state 6B at all
(2)	(2) Not until TNR, Moro, prone placement, and defensive reaction
(3)	(3) Not until TNR, Moro, prone placement, or defensive reactions
(4)	(4) Not until undressed
(5)	(5) Not until pulled to sit
(6)	(6) Not until low tactile maneuvers
(7)	(7) Not until uncovering him and placing him supine
(8)	(8) At first auditory and light stimuli
(9)	(9) Never was quiet enough to score this

Rapidity of Buildup to Motor Arousal

Scoring

(1) No motor arousal at all
(2) Not until TNR, Moro, prone placement, and defensive reactions
(3) Not until TNR, Moro, prone placement, or defensive reactions
(4) Not until undressed
(5) Not until pulled to sit
(6) Not until low tactile maneuvers
(7) Not until uncovering him and placing him supine
(8) At first auditory and light stimuli
(9) Never was quiet enough to score this; was motorically aroused at beginning of examination

IRRITABILITY

This measures the frequency and threshold with which the infant gets upset. Irritability in the sense of the BNBAS, i.e., audible fussing to specific stimuli listed, is scored on the BNBAS irritability scale. Some infants show irritability either motorically or with pained expression or also with fussing and to almost any stimulation presented, while others will only be irritable to the more massively aversive maneuvers. In the additional irritability scale, any degree of upset and discomfort is counted. It may be evidenced by grimace or cry faces or other signs of clear irritation often accompanied by motor upset or fussing or active crying.

Irritability (BNBAS)

Aversive Stimuli Considered

uncover	pinprick
undress	TNR
pull to sit	Moro
prone	defensive maneuvers

SCORING

(1) No irritable crying to any of the above stimuli
(2) Irritable crying to one of the stimuli
(3) Irritable crying to two of the stimuli
(4) Irritable crying to three of the stimuli
(5) Irritable crying to four of the stimuli
(6) Irritable crying to five of the stimuli
(7) Irritable crying to six of the stimuli
(8) Irritable crying to seven of the stimuli
(9) Irritable crying to all of the stimuli

Irritability (A)

SCORING

(1) The infant does not react irritably to any of the stimulation presented.
(2) The infant reacts irritably only to the high tactile and vestibular maneuvers, but not otherwise.
(3) The infant reacts irritably only to some of the medium tactile maneuvers, but not all of them, and not to the low tactile ones, or the distal ones in sleep or awake state.
(4) The infant reacts irritably to most or all of the medium tactile maneuvers, but not to the low tactile ones or the distal ones in sleep or awake states.
(5) The infant reacts irritably to some of the low tactile maneuvers, to all or most of the stronger tactile maneuvers, but not to distal stimuli in sleep or awake states.
(6) The infant reacts irritably to most or all of the low tactile maneuvers and to most all other tactile maneuvers, but not to distal stimulation in sleep or awake states.
(7) The infant reacts irritably to being uncovered and put into supine position and to any other tactile stimulation, but not to distal stimulation in sleep or in awake states.
(8) The infant reacts irritably to social stimulation or to object presentation and to any other tactile stimulation, except distal stimulation in sleep.
(9) The infant reacts irritably to distal stimulation in sleep and to any other stimulation.

ROBUSTNESS AND ENDURANCE

Many young infants have only limited energy resources available and need intermittent times out to refuel themselves. Exhaustion may be evidenced by increasing lethargy, passive unavailability, and low-keyedness, or at times by hyperarousal. Some infants "come into their own" in the course of the examination and focus and mobilize their energy resources by the interaction provided through the examination. They are robust and have much endurance. The ease with which the examiner can proceed in the course of the examination can be an index of this robustness.

SCORING

(1) The infant has no energy at all, or appears very fragile and the examination is inappropriate.
(2) The infant's energies are very limited, the infant is quite fragile and long "periods out" are necessary; the examination has to be shortened and the examiner has to be very gingerly or very carefully paced.
(3) The infant shows considerable exhaustion and fragility or hyperarousability, yet with prolonged periods out and slowed timing the examination can be completed.
(4) The infant repeatedly shows some exhaustion; he is moderately fragile or arousable but with times out, the examination can be completed fairly well.
(5) The infant repeatedly shows some exhaustion or is somewhat fragile or arousable, but with brief times out he can recover himself each time and finishes quite well.
(6) The infant starts out quite robustly, yet half way into the examination he needs some times out; he then can recover himself to some extent.
(7) The infant is fairly robust and energetic throughout the examination and needs only minimal time out because of diminishing energy resources, or the infant starts out somewhat fragile but becomes more energetic and robust as he goes along.
(8) The infant may have brief periods of mild exhaustion or of minimal fragility, but is generally quite energetic and robust throughout.
(9) The infant is robust and has good energy resources throughout the examination.

CONTROL OVER INPUT

The infant's ability to control environmental input varies with his organizational maturity. Often the inability to control stimulation has behavioral physiological manifestations such as hiccoughing, urinating, or bowel movement straining. Sometimes the control mechanisms for cutting down on stimulation are subtle, such as in sneezing and yawning. Some infants cannot control environmental input at all and deal with stimulation by gross autonomic responses such as gagging, apnea, tachypnea, or seizing. Soft and brief talking or brief presentation of soft inanimate objects are considered as mild stimuli, as are very gentle movements of the infant. More prolonged and complex social or inanimate object presentation as well as more firm tactile manipulation of the infant are considered as moderate stimulation. The more vigorous tactile maneuvers, such as passive movements and the items in packages IV and V, are considered stimulation unless they are administered in a very attentuated and abbreviated fashion. An overall assessment of the level of the infant's control should be made which summarizes the self-regulatory capacities throughout the examination.

SCORING

(1) The infant has no control over any input and always reacts with apnea or severe tachypnea, seizures, or severe tremulousness.
(2) The infant has minimal control over any mild stimulation; he shows eye-rolling, eye-floating, facial-twitching, mild apnea, considerable tremulousness, or spitting up quite easily, or color change or frantic diffuse activity or abrupt state changes, and needs to be interacted with very carefully.
(3) The infant has some control over mild stimulation, may show hiccoughing, color change, motor activity increase, rapid and repeated state change, considerable bowel movement straining and/or urination, or some respiratory unevenness.

(4) The infant has moderate control over mild stimulation; he may show some hiccoughing, some gaze aversion, some bowel movement straining, or considerable motor increase, or repeated state change.
(5) The infant has some control over moderate stimulation; he may show some respiratory unevenness, some hiccoughing and gaze aversion, mild bowel movement straining, and some motor increase or mild state change.
(6) The infant has moderate control over moderate stimulation; he may show slight respiratory unevenness, maybe some hiccoughing, or gaze aversion, only occasional bowel movement straining, and some motor increase or state change up.
(7) The infant has good control over moderate and possibly even some strong stimulation; he may show considerable sneezing and yawning, mild color change, mild motor arousal or state change, but in general he stays well together.
(8) The infant has considerable control even over strong stimulation; he may show some transient motor increase or state change but stays well together.
(9) The infant has good control over all stimulation presented; his adaptations are effective, and he is "on top" of the interaction and maneuvers presented.

NEED FOR FACILITATION AND USE OF STIMULATION

This is a measure of the infant's need for facilitation and ability to make use of the examiner's stimulation to enhance his own organization. Some infants are so fragile that any kind of stimulation is too taxing for them, and they need to be left alone. Other infants can deal with very mild stimulation for short periods, while others, given the appropriate, sensitive facilitation, can improve in organization considerably, even if not for long periods. Some infants are quite well organized on their own, and social interaction brings out the beginnings of real social competence.

Scoring

(1) The infant cannot tolerate any kind of handling or stimulation; he appears very fragile even when left alone.
(2) The infant can only poorly tolerate stimulation, and facilitation appropriately consists of maintaining him quietly at a low level.
(3) The infant will try to shut out most stimulation, yet at some cost to his regulation; his regulation improves with cessation of stimulation.
(4) The infant can shut out stimulation quite well; he actively tries to stay in low states and usually is successful; or the infant is very disorganized when aroused and it is barely possible to modulate him.
(5) The infant comes to arousal states and appears very disorganized when he is in the higher states, yet with careful facilitation he can usually be brought down to a more balanced level.
(6) The infant appears quite disorganized when he comes to higher states, yet with facilitation he can be brought to a quite well organized, balanced level, either for examination and/or social interaction, for considerable periods.
(7) The infant appears disorganized when in higher states, yet social stimulation and/or facilitation brings him into his own; he is then available at a behavioral level, either for examination and/or social interaction, for considerable periods.
(8) The infant is quite well organized most of the time; occasionally some facilitation enhances his organization and makes him more actively available to respond to social and inanimate object stimulation to perform in a balanced fashion during other manipulations.
(9) The infant is well organized; he actively seeks out stimulation to fuel himself and facilitation is not necessary.

Overall Summary Parameter

ATTRACTIVENESS

This is a measure of the infant's overall social attractiveness. The examiner rates how appealing the infant is to work and interact with. His physical appearance obviously is only a small part of this dimension. More critical is how well organized and integrated the infant stayed during the examination, but also how differentiated, subtle, and engaging he was in his behaviors, and how much positive feedback he gave to the examiner. The examiner needs to reflect on his own reactions to arrive at a rating.

How hard did he have to work to get this infant's best performance? How much did the infant come through for him? Or did the infant, in fact, bring out and enhance the examiner's interactive repertoire? This score, then, reflects a composite of robustness, differentiation, and social engaging.

SCORING

(1) The infant was very stressed and did not permit any handling or interaction.
(2) The infant had brief periods of potential availability and organization but on the whole was unavailable for relaxed interaction and was quite stressed much of the time.
(3) The infant showed some periods of organization and stability, although there was no or only very brief opportunity for relaxed social interaction with much work on the examiner's part.
(4) The infant showed repeated, although brief, periods of organization and stability, and there may have been some opportunity for social interaction with moderate amount of work on the examiner's part.
(5) The infant is quite well organized, even if not available for social interaction, or he can be brought to some degree of relaxed social interaction with the examiner's help.
(6) The infant has good self-organization and/or is appealing when in interaction, yet the examiner has to do much of the stage-setting and facilitating.
(7) The infant is usually well organized and can be quite engaging in interaction, at least for some periods.
(8) The infant is well organized, and is generally engaging in social interaction; he can maintain himself for quite some time.
(9) The infant is always well organized, stable, and modulated; he is always engaging in social interaction and can maintain himself for prolonged periods.

Interfering Variables

This is a measure of the amount of interference from the environment which detracts from an optimal administration on the examiner's part and an optimal performance on the infant's part. Different examiners may be disturbed to differing degrees by the same interferences. The examiner needs to be aware of his own reactivity and his energy expended in shutting out distractions and noxious events. He should use the ideal circumstance of a quiet, dimly lighted room where he can examine the infant without observers, as baseline for his performance. The degree of interfering variables is rated on the cover of each examination.

SCORING

(1) There was essentially no interference from the environment which would have detracted from the examiner's relaxed comfortable administration of the examination.
(2) There was some interference from the environment, such as noise from the street or hallway, some less than optimal lighting circumstance or some space constraints.
(3) There was moderate interference from the environment such as some noise nearby, difficult lighting conditions, other persons watching, and less than optimal space.

(4) There was considerable interference from the environment, such as considerable noise, interfering lighting, interruptions through extraneous events, other persons watching, and poor space.
(5) There was massive interference from the environment, such as very interfering noise, or light, several interruptions due to extraneous circumstances, highly anxious parents or interfering observers, and very cramped quarters, etc.

Acknowledgment

Our special thanks to F. H. Duffy, M.D., for his continuing and critical discussion from a neurologist's point of view.

3

A Biobehavioral Perspective on Crying in Early Infancy

BARRY M. LESTER AND PHILIP SANFORD ZESKIND

1. INTRODUCTION

The central role of crying in early infancy has been discussed by parents, pediatricians, and theorists. In Western cultures crying is the primary mode of communication through which the young infant's needs and wants are expressed. The affective messages transmitted by the cry tell the caregivers that the infant needs attention, and in most cases crying is terminated when the infant's needs are met. There are times, however, when the immediate cause of crying is not clear—so-called "unexplained fussiness"—when the infant cannot be easily soothed.

Difficulty in coping with crying in early infancy is probably the major complaint of parents to pediatricians and is one of the most frequent reasons for visits to the hospital emergency room in the first few weeks of life. Infants who cry often and are difficult to soothe may place extreme demands on their parents which may in turn raise doubts in the parents about their own competence, thus jeopardizing the early development of the infant–caregiver relationship. The often-cited role of crying in child abuse may be one indicator of such interactive failures in which infant and parent both contribute to the child's maltreatment. Infant crying that is excessive and particularly aversive-sounding may set the stage for the develop-

---

Barry M. Lester, Ph.D. • Child Development Unit, Children's Hospital Medical Center, Boston, and Harvard Medical School, 333 Longwood Avenue, Boston, Massachusetts 02115. **Philip Sanford Zeskind, Ph.D.** • Department of Psychology, Virginia Polytechnic Institute and State University, Blacksburg, Virginia 24061. The preparation of this chapter was partially supported by grant No. 3122 from the Grant Foundation, New York, and by grant HD 10889 from the National Institute of Child Health and Human Development.

ment of nonoptimal childrearing practices which, in an extremely stressed environment, could lead to abuse or neglect.

Crying is only part of the behavioral repertoire of the infant (although to some parents it may not seem that way); therefore, in asking what we can learn from the cry it is important that the cry be interpreted within the context of the individual infant and the particular infant–caregiver relationship. We view the infant as an organized complex of dynamic biobehavioral systems and subsystems (Lester, 1978, 1979) and the cry as the major expression of the young infant's affective system. From this psychobiological framework we assume that much if not all of the behavior of the infant is adaptive; the neonate emerges with a preprogrammed response repertoire designed to maximize the survival of the individual and the species.

The cry is the young infant's major mode of communication and is used to express different needs, states, and demands. The properties of the cry carry messages to the caregiver such as pain, hunger, and perhaps need for attention. The cry of the infant at risk may be unique in order to elicit special caregiving to facilitate the infant's recovery (Lester & Zeskind, 1979). As a sign of behavioral and perhaps central nervous system (CNS) compromise (Lester & Zeskind, 1978), certain acoustic features of the cry may alert parents to the kind of handling and care their infant needs. This leads to questions about the adaptive significance of the cry, the kinds of messages and meanings that may be encoded in variations in the cry, and how the cry is integrated into the overall behavioral organization of the infant. When viewed as part of the neonate's behavioral repertoire, the cry can be studied in the context of the development of infant–caregiver interaction. The neonate, by his or her range and form of adaption, shapes the dyadic interaction. At the same time, the infant is being shaped by the practices and expectations of the caregiving environment. The study of infant crying becomes one window into the development of this reciprocal feedback system and may facilitate our understanding of how infant and parent negotiate their relationship.

The importance of crying in the development of early infant–mother attachment was proposed by Bowlby (1969) from an ethological perspective as a proximity-promoting behavior, an "acoustical umbilical cord" (Sander & Julia, 1966) that keeps the infant close to the mother and thereby protected from danger. Because human infants, unlike some other species, are not locomotor at birth, the potential for exclusive dependency on vocal communication to signal distress and maintain proximity is heightened, although other adaptive mechanisms may evolve. For example, in some non-Western societies infants are carried and use kinesthetic cues to communicate their needs to the mother. Murray (1979) has argued that the cry does not fit the definition of the releaser concept from classical ethological theory as an adaptive, species-specific behavior that evolved as an emergency signal and should function as a sign stimulus to release reciprocal parental caregiving responses. She argues that a broader definition of the releaser concept in which cries may contain multiple messages and derive their meaning, in part, from contextual cues is more compatible with the research literature. As a "social releaser" the cry would include motivational and cognitive features that are phylogenetically adapted.

As an adaptive response, the cry may also facilitate the organization of the infant's biobehavioral systems. At a physiological level, the cry of the neonate facilitates the reorganization of the cardiorespiratory system, improves pulmonary capacity, and helps maintain homeostasis (Brazelton, 1962). Developmentally, this function of the cry could give way to or also serve to enhance the behavioral organization of the infant. The periods of unexplained fussiness that have been reported during the first few months (Bernal, 1972; Brazelton, 1962; Emde, Gaensbauer, & Harmon, 1976; Rebelsky & Black, 1972) may serve to promote time spent between infant and caregiver that provides opportunities for social interaction. The infant is biologically prepared to respond to these social interactions and elicit from them inputs that facilitate his behavioral organization. For the caretaker, these interactions promote early attachment as the caretaker becomes more attuned to the infant's rhythms and needs. The finding by Bell and Ainsworth (1972), although questioned by Gewirtz and Boyd (1977), that mothers who respond more quickly to infant crying early in life have infants who cry less in the second year may be due to the role of crying in the early negotiation of the development of infant–parent interaction.

The preceding discussion leads to two broad approaches to the study of infant crying that will form the basis for most of this chapter: the cry as an indicator of the developmental status of the organism and the cry as a social signal. Although this distinction is somewhat artificial, it does appear to be useful in organizing much of the literature and our own thinking. Before we discuss the literature, however, it may be helpful to start from a common understanding of what a cry is and how it is produced.

2. ANALYSIS OF INFANT CRY SOUNDS

The first attempt at ascribing meaning to the acoustic structure of cry patterns was in 1838 by William Gardiner, who used musical scores to illustrate the "puling cry of a spoiled child," as well as other human cries, bird calls, and animal vocalizations. Darwin (1855) used photographs and line drawings to describe crying and other emotional expressions. Following the invention of the Edison phonograph, Flateau and Gutzmann (1906) recorded the vocalizations of 30 infants on wax cylinders and used a phonetic alphabet and musical notation to describe features of the cry. Fairbanks (1942) introduced acoustic methods applied to the analysis of the pitch of infant hunger cries, and Lynip (1951) reported the first spectrograms of infant vocalizations.

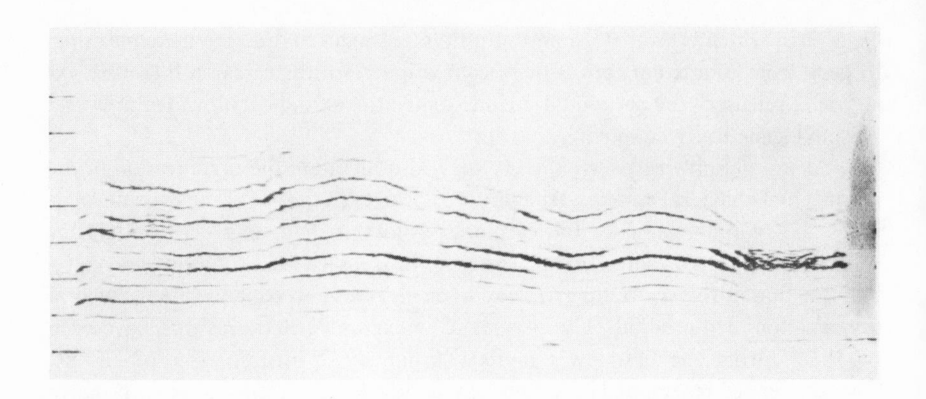

FIGURE 1. Narrow-band spectrogram of pain cry in a 3-day, term, healthy neonate. This is in seconds on the horizontal axis and frequency to 8000 Hz on the vertical axis.

The most important technological advance for the analysis of acoustic information was the invention of the sound spectrograph, which is still considered by many to be the most important instrument in the phonetics laboratory (Painter, 1979) and has certainly been the major tool for the analysis of infant cry sounds. The sound spectrograph was developed by Bell Laboratories, the trade name is Sonograph, and the visual record presented on paper is called a spectrogram, sonogram, or voice print. A spectrograph is a wave analyzer which produces a record of the distribution of energy in frequency and time (Figures 1 and 2). Frequency, which we perceive as musical pitch, is represented on the vertical axis of the paper and covers the speech range from 0–8000 Hz (cycles per second). Duration (time) is represented on the horizontal axis and covers 2.4 seconds. Intensity, which we perceive as loudness, is represented by the degree of darkness of the marks on the paper.

The spectrograph records a sound sample on pressure-sensitive paper mounted on a continuously rotating disc. With each revolution of the disc, a different portion of the sound spectrum is sampled as the signal is passed through a moving filter. A marker pen moves vertically along the revolving kymograph drum in synchrony with the moving filter and literally "burns" a mark on the paper whenever acoustic energy is present at a given frequency; the more energy present, the darker the markings.

Infant cry sounds contain both periodic and aperiodic (white noise) wave forms. The periodic wave forms are complex waves composed of sine waves as might be approximated by striking two or more different tuning forks at the same time. What the sound spectrograph essentially does is to analyze the complex waves into their sine wave components; that it, it sorts out the individual tuning forks in the sound we hear. Each cycle of vibration of the vocal fold produces a complex wave. A fundamental frequency of 400 Hz means 400 vibrations of the vocal folds per second or the production of 400 complex waves. The term *fundamental frequency* (written f_0) refers to the physical event of vocal fold vibration, whereas the term *pitch* refers to our perception of f_0.

The spectrograph offers two filter settings, narrow band (45 Hz) and wide band (300 Hz). Figures 1 and 2 show, respectively, a narrow-band and a wide-band spectrogram of the same cry from a 3-day-old, full-term healthy newborn infant. When a narrow-band filter is used and the fundamental tone is greater than the bandwidth of the filter, as in infant cry sounds, the resulting spectrogram displays the sine components of the frequency spectrum simultaneously according to their running intensities and thus as successive harmonics of the fundamental. When a wide band is used and the bandwidth of the filter is greater than the fundamental, the frequency spectrum appears as regions of energy concentration called formants instead of as a display of successive harmonics. While formants appear on both narrow-band and wide-band analysis, the resolution in narrow-band analysis is along the frequency axis, whereas in wide-band analysis resolution occurs along the time axis. Thus, the narrow-band filter is precise with respect to frequency and separates the harmonics on the print-out but imprecise with respect to time. The wide-band filter has the opposite features: it is precise with respect to time but imprecise with respect to frequency. The fundamental frequency can be measured using wide-band spectrogram by simply counting the number of vertical striations or vocal fold pulses. Since the harmonics are precise multiples of the fundamental frequency, with a narrow-band spectrogram the frequency of an unambigious higher harmonic can be located, such as the fourth or fifth harmonic, and the frequency of that harmonic is divided by four or five to calculate the fundamental frequency.

The sound spectrograph preserves both quantitative and qualitative information about the cry, but there is much information contained on a single spec-

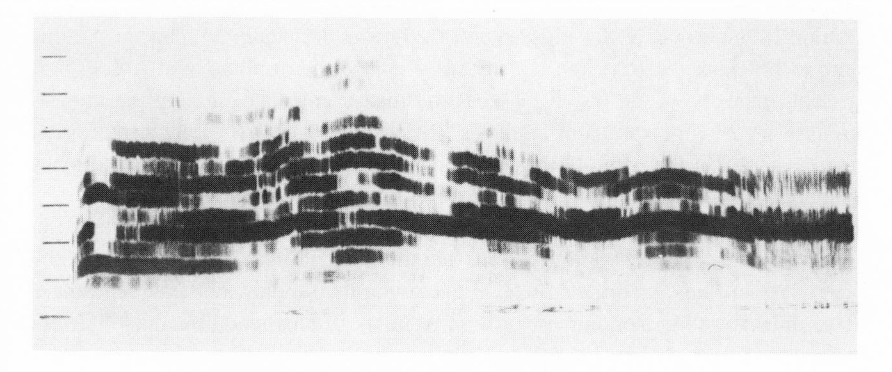

FIGURE 2. Wide-band spectrogram of same cry as in Figure 1.

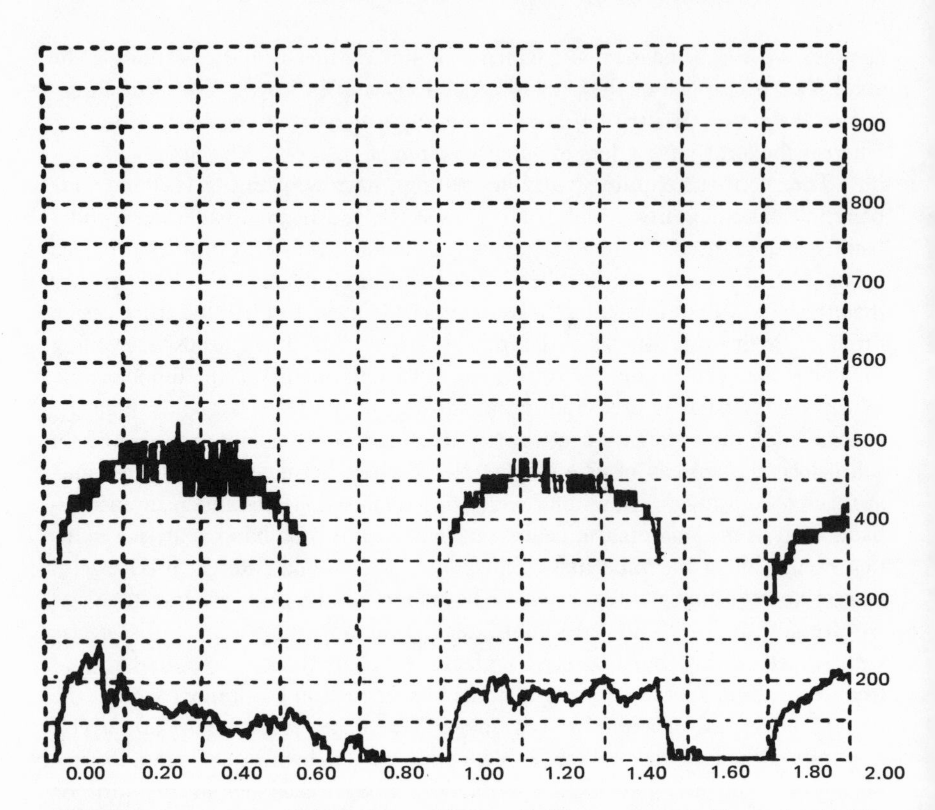

FIGURE 3. Computer plot of the fundamental frequency of the same cry as in Figures 1 and 2. The lower plot represents the amplitude of the signal. Time is shown in seconds on the horizontal axis and frequency in Hz on the vertical axis.

trogram, some of which is difficult to define and measure. Increasingly, computer extraction techniques are being used to process a large array of acoustic cry information. In our work, we have used spectrograms to decide on the relevant parameters to be extracted from the cry and to preserve the qualitative attributes and general "picture" of the cry. For the actual measurements of the cry parameters we used a real-time spectrum analyzer, which plots a Fourier transformation of a signal as an oscilloscopic-like display on a cathode ray tube with the frequency on the abscissa and the amplitude at each frequency on the ordinate. The spectrum analyzer provides a direct digital read-out of the duration of the signal and the amplitude and frequency at user-selected points within a specified frequency bandwidth but has the disadvantage of averaging the fundamental frequency over time; thus, the pitch contour or variability in the fundamental cannot be determined.

With high-speed computer technology, the Fourier transformation of the cry signal can be displayed as a running time series, as shown in Figure 3. This is a computer plot of the fundamental frequency of the same cry shown in Figures 1 and 2. Changes in the fundamental frequency are plotted every 25 msec, an interval which clearly shows the variability in the fundamental frequency. In Figure 4 a similar procedure (spectral analysis of the Fourier transformation) was used to show the full harmonic structure of the cry. This baby was full term and healthy but underweight for length by the ponderal index (Miller & Hassanein, 1973). This cry is an example of what Truby and Lind (1965) called *hyperphonation*, with a shift in pitch that fluctuates around 1800 Hz.

3. GLOSSARY OF CRY FEATURES

With this basic foundation we can now describe the parameters used in the acoustic analysis of infant cry sounds. We have attempted to organize the cry vocabulary into a set of working definitions which should provide a basis for com-

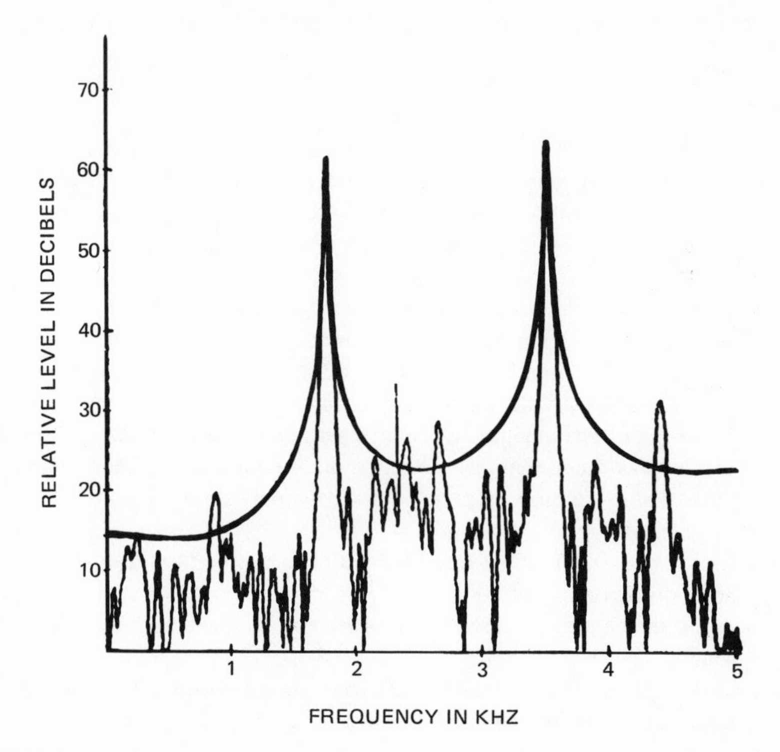

FIGURE 4. Computer plot of the harmonic frequencies up to 5000 Hz of the hyperphonated pain cry of a term, underweight-for-length neonate. This plot shows a single 25-millisecond window.

paring research findings. Rather than trying to define every possible parameter, we have chosen to restrict this glossary to features that have been used by investigators to discriminate among the cry sounds of normal infants or between normal and abnormal infants. In attempting to integrate and synthesize the work of many investigators, we will inevitably sacrifice a degree of specificity for the sake of generalization. Thus, this system should be viewed as an initial effort at establishing a common vocabulary. We have drawn most heavily on the work of the study groups in Helsinki (Sirvio & Michelsson, 1976; Wasz-Hockert, Lind, Vuorenkoski, Partanen, & Valanne, 1968) and Stockholm (Lind, 1965), for they have been the major contributors to the sound spectrographic analysis of infant cry sounds.

So that we can start from a common baseline, the referent for these terms is a narrow-band spectrogram of pain-induced cry that begins during the first expiratory phase of the respiration cycle following the stimulus. Figure 5 presents a schematic account of these acoustic features divided into three broad categories: durational features (designated by lower-case letters) that include both expiratory and inspiratory sounds, harmonic features (represented by numbers) which refer only to expiratory signals, and qualitative features (shown as upper-case letters).

3.1. Durational Features

Since the first cry following a pain stimulus seems to be unique, the durational features of this cry are distinguished from durational features of subsequent cry segments. These features pertain to the cry sounds within the first expiration–inspiration respiratory cycle following a pain stimulus.

- *Latency* (a–b, Figure 5)—time between the offset of the stimulus and the onset of the first expiratory cry phonation.
- *Length* (b–c)—time from the onset to the end of the first expiratory cry phonation. All signals that are judged to be part of the harmonic structure of the expiratory phonation are included, even though breaks in the harmonics may occur. Sounds such as grunts and moans which may appear at the end of the expiratory period but are not part of the main signal are not included.
- *First pause* (c–d)—time from the end of the expiratory phonation to the onset of the first inspiration.
- *Inspiration* (d–e)—time between onset and offset of the first inspiratory sound.
- *Second pause* (e–f)—measurement from the end of the inspiration to the beginning of the second cry expiration.

With subsequent cry cycles similar measures can be made. The length between concurrent cry expirations (first pause plus inspiration plus second

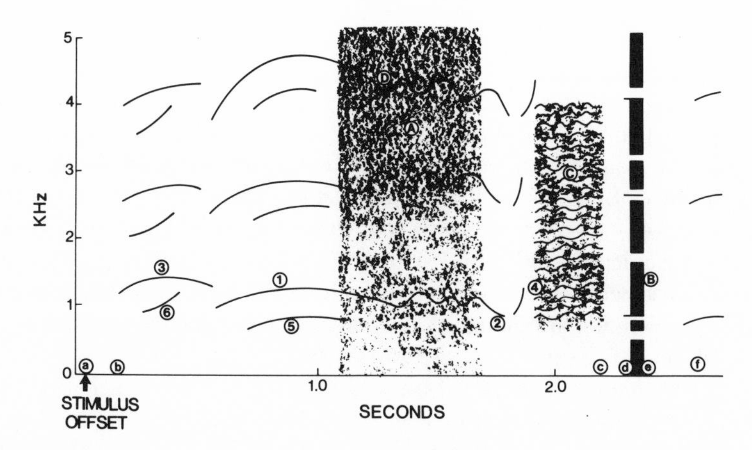

FIGURE 5. Schematic drawing of a narrow-band spectrogram of a pain-elicited cry showing three classes of cry features: durational, harmonic, and qualitative. Durational: latency (a–b), length (b–c), first pause (c–d), inspiration (d–e), second pause (e–f). Harmonic: maximum fundamental frequency (1), minimum fundamental frequency (2), shift (3), glide (4), double harmonic break (5), biphonation (6). Qualitative: distortion (A), voiceless (B), vocal fry (C), vibrato (D).

pause) can also be calculated and was found to be useful in one study (Lester, 1976) when measured from first to second cry expiration. Since the cry pattern does change over time, the cycle from which the durational measures are taken should be specified.

3.2. Harmonic Features

These features are based on the structure of the harmonics of the first expiratory cry phonation following the presentation of a pain stimulus. Although these features can also be measured from subsequent cry expirations, they may change as the cry pattern stabilizes; thus, the segment selected should be noted. In addition, a record should also be kept of when during the cry expiration a feature occurs.

- *Fundamental frequency* (f_0)—the frequency in Hz of the first harmonic. The f_0 can also be measured as the number of Hz between any two successive harmonics. The f_0 will change somewhat during the course of a given cry expiration, and this variability should be included. The *maximum fundamental* (2) is the lowest measurable point of the first harmonic. The *range of the fundamental* can also be computed. The *percentage change* and *percentage change per unit time* (i.e., divided by the length of the cry phonation) may also be important. These measures are not made from any of the splitting or breaks in the fundamental described below.

- *Shift* (3)—a clear break in the fundamental which is not a doubling or halving of the number of harmonics; for instance, from a fundamental frequency of 500 Hz to one of 700 Hz. The number, location, duration, frequency, and melody of the shift can be specified. This feature is more common in pain cries than in spontaneous cries (Stark & Nathanson, 1975; Truby & Lind, 1965; Wasz-Hockert *et al.*, 1968).
- *Glide* (4)—a very rapid continuous change in the f_0. The rate of change is on the order of 600 Hz in .1 second and is usually too rapid to be perceived by a listener. Glides tend to occur at the beginning or end of an expiratory phonation and are thought to be due to changes in vocal fold tension and subglottal pressure (Stark & Nathanson, 1975).
- *Double-harmonic break* (5)—a doubling in the number of harmonics in parallel with but shorter in length than the fundamental frequency and its harmonics. The cry sounds harsh but is without a lowering in pitch. This feature is frequently associated with pharyngeal constriction and turbulence caused by forcing the air stream through the pharynx.
- *Biphonation* (6)—also refers to the co-occurrence of two sets of harmonics but in this case there are two different sound sources. The second set of harmonics is not parallel with the fundamental frequency and its harmonics. The two fundamentals might move independently over time; they might go in opposite directions, or one might remain flat while the other rises or falls.

3.3. Qualitative Features

These features describe patterns in the harmonic structure and the presence of aperiodic noise that affect the quality of the cry sound.

- *Melody*—refers to the pattern of change of the fundamental and can be flat, rising, falling, rising-falling, or falling-rising.
- *Distortion* (A)—the presence of nonharmonic sound that masks the harmonic components. A harsh, rough, scratchy sound is perceived. Distortion is aperiodic sound produced at or above the level of the glottis. It can be due to muscle tension or mucous congestion and increases with the intensity of the cry.
- *Voice*—a voiced cry is produced by vibrations of the vocal cords and appears as a fundamental tone with its harmonics. A *voiceless* (B) phonation such as frication, the sound of air-rush in the vocal tract, appears as a blur of indistinguishable vibrations. A *halt-voiced* phonation includes both voiced and voiceless components. This judgment should be made separately for cry expirations and cry inspirations.
- *Vocal fry* (C)—or glottal roll, refers to aperiodic, low-frequency vibrations

of the vocal folds. Vocal fry usually appears at the end of a phonation and is seen as trembling, narrow harmonics of low intensity.

- *Vibrato* (D)—wave-like patterning of the fundamental frequency. Vibrato is related to considerable variation in pitch and is often heard as an irregular, warbly cry.

3.4. Additional Features

Some cry characteristics do not readily appear on a spectrogram but have been found to be useful in a number of studies.

- *Threshold*—the number of pain stimuli (e.g., number of snaps) required to elicit a cry of a minimum duration.
- *Intensity*—the amplitude of the fundamental frequency or the loudness (decibels) of the cry. On the spectrogram intensity is seen as the darkness of the energy. Amplitude can be measured from the tape-recorded cry but is also a function of the distance from the infant's mouth to the microphone and the recording level of the tape recorder. Intensity can be measured directly when the cry is elicited with a sound level recorder which records the sound pressure level for a given mouth–microphone distance.
- *Episode duration*—refers to the total amount of time the infant cried following a pain stimulus. In other words, this is a measure of the duration of the entire crying period. Activity can also indicate the number of cry sounds or phonations per unit time to indicate how much the infant cried.

4. MECHANICS OF CRY PRODUCTION

Crying is a complex motor response followed by sound. The cry sound is an acoustic phenomenon produced by excitation at the glottis causing the vocal cords in the larynx to oscillate. The sound is amplified by the surrounding resonating chambers—the oral, nasal, sinusal, laryngeal, and pharyngeal cavities. Unfortunately, little is known about the physiology of cry production and particularly about the role of the CNS in the production and modulation of the cry. Because of the appearance of stable cry patterns in virtually all healthy newborns, the cry is assumed to be at least partially controlled by autonomous central nervous system mechanisms (Lenneberg, 1967; Lieberman, 1967). Moreover, as the cry is a maximal type of response that originates in the CNS, it is reasonable to view the cry as indicative of the capacity of the nervous system to be activated and to inhibit or modulate that activation (Parmelee, 1962). Stark and Nathanson (1975) note that it is difficult to think of a structural lesion or mechanical defect that can account for the constellation of cry features found in neurologically impaired infants.

Crying is intimately related to respiration. The motor component of the cry or the cry act (Truby & Lind, 1965) originates in the CNS and requires the respiratory system to adapt to the generalized motor response. The respiratory pattern of crying dominates normal tidal respiration, converting it into a form of dyspnea in which the effort and duration of expiration are increased. Following the introduction of a pain stimulus, the newborn infant produces an expiratory sound followed by an inspiratory sound. These sounds correspond to the expiration–inspiration cycle of respiration. The inspiration phase of the cry is shorter than the inspiration of tidal respiration, while the expiratory cry phase is longer than normal respiratory expiration. According to Truby and Lind (1965), the infant response to a pain stimulus is complete occlusion of the respiratory passage at the larynx, followed by an expiratory cry sound. In children and adults, the expiratory cry is often preceded by a sharp inspiration with associated sound. Thus, an important consideration for the analysis of cry sounds is whether the sound is expiratory or inspiratory, which will be affected by the point during the respiration cycle when the stimulus is presented.

Much of the foundation for the work on infant cry sounds was provided by Truby and Lind (1965), who distinguished among three types of expiratory pain cries. The basic cry or phonation ranges in pitch from 200 to 600 Hz and is characterized by a smooth and symmetric spectral and intensity pattern. The infant does not seem to be in great discomfort or distress. The second cry type, turbulence or dysphonation, sounds "raucous" or "harsh," wherein overloading along the vocal tract causes noise or aperiodic sound that obscures all or part of the harmonics of the basic cry pattern. Hyperphonation, the third cry type, refers to a sudden shift in pitch, usually between 1000 and 2000 Hz, where it appears as if the fundamental frequency were suppressed and "replaced" by one of its overtones. Hyperphonated cries sound extremely high pitched and whistle-like and are thought to be due to muscular constriction of the vocal apparatus. The authors describe this kind of cry as a vocal response to intense distress.

Recently Golub (1979; Golub & Corwin, 1982) proposed a physioacoustic model of infant cry production based on the acoustic theory of sound production and the interaction between crying and respiration. According to acoustic theory (Lieberman, Harris, Wolff, & Russell, 1971), an infant's vocal tract functions acoustically as a one-quarter wavelength tube resonator of specified length which can be used to predict the formant frequencies of the cry. Golub suggests that neonates control tension in their muscles, especially the smaller muscles of the larynx, in a quantal or noncontinuous fashion and that differences in the acoustic properties of some cries can be explained by the relative tensions of two laryngeal muscles (vocalis and cricothyroid) and one respiratory muscle (abdominal). This model may provide a physiological explanation for the differences in pitch between the phonated and hyperphonated cry types described by Truby and Lind. As suggested by Golub, the sudden shift in pitch that characterizes hyperphonation may be due to shifting vocal registers by quantally controlling the appropriate laryngeal and respiratory muscles.

Using Golub's model, we may be able to explain how the cry is modulated by the nervous system, particularly as concerns parasympathetic (PNS) input from the autonomic nervous system (ANS). The sympathetic division (SNS) of the ANS is usually thought of as serving emergency functions. Stimulation of the SNS produces generalized physiological responses rather than discrete, localized effects. Pain and strong emotions, for example, produce massive SNS discharges that mobilize the body for action, the so-called "fight or flight" reaction. Most autonomically innervated organs receive fibers from both the SNS and PNS. When dual innervation occurs, the PNS has adaptive functions. It maintains homeostasis of the internal environment necessary for the organism to survive. Through inhibition the PNS assures homeostatic regulation or adaptation. For example, SNS fibers to the heart produce cardiac acceleration, whereas PNS fibers are cardiac decelerators. Graham and Clifton (1966), among others, argue that the cardiac deceleration often found to accompany the orienting response to a novel stimulus indicates PNS input.

The relevance of this distinction for infant cry production is that the larynx receives dual innervation from SNS fibers with PNS input to the cricothyroid muscle through the vagus nerve. If, as Golub (1979) suggests, tension on the cricothyroid muscle controls the frequency of vocal fold vibration, we may hypothesize that variability in the fundamental frequency of the cry may be due to input from the vagus nerve.

The vagus, cranial nerve X of the brain stem medulla, has six branches and contains parasympathetic and sensory fibers carrying feedback to and from the brain (especially with glossopharyngeal or nerve IX). One branch is the cardiac nerve responsible for the slowing of cardiac activity already mentioned. The pharyngeal branch is the chief motor nerve of the pharynx, the recurrent laryngeal branch supplies motor fibers to the intrinsic muscles of the larynx, and the superior laryngeal innervates the cricothyroid. Testerman (1970) showed that in cats stimulation of the cut vagus modified cricothyroid activity. The vagus can receive messages from the motor areas of the cerebral cortex through fibers in the corticobulbar tract which terminate in the brain stem on motor nuclei of the vagus. The primary motor area of the cerebral cortex, or Brodman's area 4, is a motor strip that controls the larynx and is near Broca's speech area, supporting the notion that early cry patterns may be related to the development of speech and language. Activity of the ANS may be governed by descending impulses from the brain stem and cerebral cortex or by local reflex stimuli. The primary integrative influence on the ANS is through the hypothalamus, which is often thought of as the "seat" of emotion.

The development of inhibition and the balance between inhibition and excitation is a basic element of central neurological development. The capacity of the nervous system to be activated and to inhibit or modulate that activation may be reflected in parameters of infant cry patterns. The generalized SNS responses following the introduction of a pain stimulus to the infant may include motor sympathetic input from the brain stem or cortex responsible for the initiation of the cry response. This sympathetic activation can be measured by the latency from stimulus to cry onset and by the threshold to cry production or amount of stimulation necessary to produce the cry. Modulation of the cry signal may be under parasympathetic control through vagal input to the larynx. PNS input may be expressed as variation in the fundamental frequency and spectral features of the cry. The dual SNS and PNS innervation of the larynx provides a balance of excitatory and inhibitory impulses to the vocal folds that produce the basic cry phonation with a smooth and symmetrical spectral pattern between 200–600 Hz. In this normal state, vagal inhibition controls the tension of the laryngeal musculature preventing contraction of the laryngeal (e.g., cricothyroid) muscles which maintain the production of a well-modulated cry signal. Decreased vagal tone or a dampening of vagal input may reduce the inhibitory effect of the vagus on the constriction of the laryngeal muscles. The contraction of these muscles can produce changes in the spectral features of the cry. The sudden shifts in pitch to 1000–2000 Hz that characterize hyperphonation, which according to Golub (1979) can be explained by shifting vocal registers by controlling the tension of the appropriate laryngeal muscles, may be due to decreased vagal tone, which allows the laryngeal muscles to contract. Parasympathetic or vagal hypoactivity may also explain the frequent shifts in pitch or high variability in the fundamental frequency of some cry phonations. Frequent and rapid changes in the fundamental frequency may be a reflection of the poor modulation of the cry signal caused by the autonomic imbalance that could arise from hypoactivity in the PNS. Low vagal input may disrupt the homeostatic adaptation of the cry signal to modulated phonation producing an imbalance or tension between SNS and PNS control systems. This autonomic imbalance may be expressed as fluctuations in the fundamental frequency of the cry. Clearly, one could also postulate parasympathetic hyperactivity as well as sympathetic hyperactivity and hypoactivity, as in Wenger's studies of autonomic balance in children and adults (cf. Wenger & Cullen, 1972). However, it seems more likely that PNS activity is primarily involved because the input is direct and localized, unlike SNS pathways, which are diffuse (mediated through the thoracic spinal nerves) and produce more massive and generalized physiological responses.

Consider the cry, as did Truby and Lind (1965) as a two-component phenomenon, act plus sound. The activation component, or initiation of the cry, is due to motor input from the SNS. The termination of the cry act, which we have not discussed, may well be due to afferent feedback from the vagus and possibly glossopharyngeal nerves to the medulla, although other mechanisms are undoubtedly involved. For example, the release of norepinephrine that accompanies SNS activation can augment arousal. The sound component of the cry, or the nature of the sound produced, is under PNS control following SNS activation. Vagal input to the laryngeal muscles determines how the vocal folds vibrate to produce the pitch of the cry. Decreased vagal tone may result in extreme variability of the fundamental frequency during phonation of shifting vocal registers altogether to produce hyperphonation.

As we shall see, changes in the pitch of the cry is the dominant feature associated with damage or insult to the nervous system in infants. To anticipate what follows, we shall propose that immaturity or insult to the nervous system may be seen in the autonomic regulation of the cry act and cry sound as differences in sympathetic (latency and threshold) and parasympathetic (fundamental frequency and spectral features) response systems. Infants who are compromised may differ in one response class and not the other (e.g., fundamental frequency and not latency) due to the different mechanisms involved depending, of course, on the nature of the insult. Lack of inhibition or decreased vagal tone is probably responsible for the increases in pitch that are almost ubiquitous in groups of infants at risk. These infants seem to alternate between phonation and hyperphonation, suggesting autonomic imbalance. This is most readily seen following a pain stimulus where the immediate response of the jeopardized infant is often hyperphonation as the muscles of the larynx constrict due to lack of parasympathetic input. The infant's cry then shifts registers into phonation as the nervous system recovers its homeostatic balance. But in many infants with CNS insult the cry shifts back and forth between phonation and hyperphonation virtually in step functions, probably indicating the struggle at the autonomic level caused by vagal hypotonia. Hyperphonation rarely occurs in normal healthy infants, as was seen in the modulated phonation of Figure 3. Figure 4 is an example of the hyperphonated cry in a term healthy infant underweight for length but recruited from the newborn nursery. The fundamental frequency of this 25 msec cry window is between 1000 and 2000 Hz. A different portion of this cry over a 2-second period is shown in Figure 6 and exhibits high variability in the fundamental frequency. Figure 7 shows the fundamental frequency contour of the cry of a premature infant born at 34 weeks gestation recorded at the baby's expected date of birth. The fundamental frequency remains within the range of phonation but is extremely variable, perhaps a less extreme manifestation of lack of parasympathetic inhibitory control. In premature infants this may be a sign of neurological immaturity in which higher-level systems of inhibition are less well developed than more basic sympathetic arousal systems.

The study of the relationship between crying and the cardiorespiratory system may provide additional support for the role of vagal control. Animal studies have shown that cooling or cutting of the vagal input to the heart increases cardiac variability (Chess, Tam, & Calaresu, 1975; Katona & Jih, 1975). Porges, Borher, Keren, Cheung, Franks, and Drasgow (1981) administered methylphenidate

FIGURE 6. Computer plot of the fundamental frequency of the cry of the same infant as in Figure 4.

(Ritalin) to hyperactive children hypothesizing that hyperactivity may be due to low vagal input (i.e., low inhibition) and found that the effects of the drug increased the coupling between heart rate and respiration. In a study discussed later (Lester, 1976), cardiac deceleration following a pure tone stimulus was related to the fundamental frequency of the cry. Malnourished infants who did not show the cardiac deceleratory orienting response cried at a higher fundamental frequency. Both increases in the fundamental frequency of the cry and decreases in the cardiac deceleration may be a result of vagal hypotonia.

As we learn more about the mechanisms involved in the production and control of the cry, we may gain a better understanding of the nervous system. We have concentrated on the autonomic nervous system because of its direct connection to the cry. Others (e.g., Porges *et al.*, 1981) have suggested that vagal input may also indicate CNS processes. Given what we know about the cry, it is likely that the cry may be a mural of higher CNS activity, including the regulatory functions of the medulla, ANS integration in the hypothalamus, or of the motor cortex. At

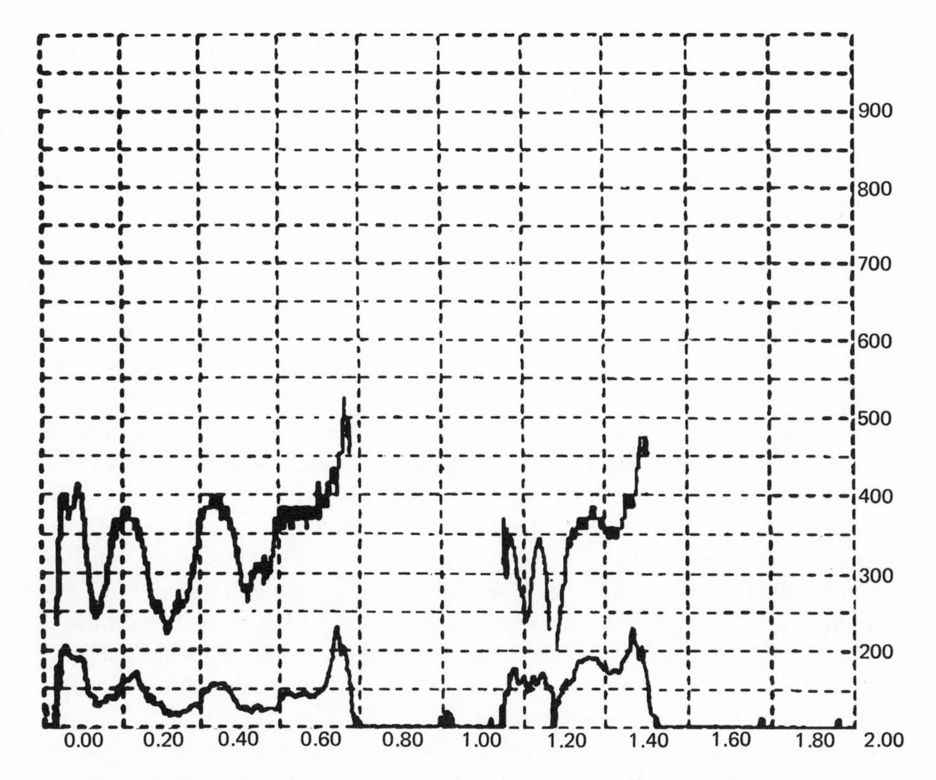

FIGURE 7. Computer plot of the fundamental frequency of the cry of a premature infant born at 34 weeks' gestation recorded during the administration of the Brazelton scale at term age.

a more general level, the cry may reflect the organization of the nervous system as seen, for example, in the development of inhibition and excitation. By studying the cry in relation to other biobehavioral and physiological response systems, the compensatory and recovery capacities that indicate the plasticity of development may be highlighted in the organization of the nervous system.

5. METHODOLOGICAL ISSUES

There are a number of methodological concerns that make it difficult to compare results among infant cry studies. A common strategy is to compare cry features between groups of infants labeled *normal* and *abnormal*. However, *abnormal* often includes infants suffering from a spectrum of recognizable syndromes, as well as infants presenting with symptoms suggestive of CNS impairment (Karelitz & Fisichelli, 1962). Ostwald, Phibbs, and Fox (1968) and Prechtl,

Theorell, Gramsbergen, and Lind (1969) used clinical judgment to divide their samples of 13 and 21, respectively, into normal, suspect, and abnormal groups. In addition to small sample sizes, the groups also represent a variety of different syndromes. Age effects are often confounded (Karelitz & Fisichelli, 1962; Michelsson, Sirvio, & Wasz-Hockert, 1977a; Ostwald *et al.*, 1968; Vuorenkoski, Lind, Partanen, Lejeune, LaFourcade, & Wasz-Hockert, 1966), and there are virtually no normative studies of age (or any other) changes or sex differences. Variables that affect the initial state of the infant, such as eliciting the cry before (Karelitz & Fisichelli, 1962; Prechtl *et al.*, 1969) or after (Lind, Wasz-Hockert, Vuorenkoski, Partanen, Theorell, & Valanne, 1966) feeding, or moving the infant onto a table and removing clothing (Karelitz & Fisichelli, 1962), are often not reported and may result in differential cry responses.

The comparability of results is also limited by differences in the operational definitions of cry features and by the instruments used to extract these features. For example, sound spectrographic and computer extraction techniques differ with respect to the kind of information they provide and the precision of measurement. Another problem is that investigators may include different numbers of cries per infant, thus confounding individual differences with group differences (Michelsson *et al.*, 1977a). Experimenter bias may also be a problem, since in most studies the condition of the infant is known when the cries are analyzed. Prospective studies are needed in which predefined cry features are used to predict developmental outcome performance.

The type of cry studies also varies. The cry can be elicited by the investigator or the eliciting stimulus may be unknown and presumed to be related to conditions such as hunger. The method of elicitation is important and techniques vary with respect to the intensity of the stimulus which can be graded from minimal to maximal (Lind *et al.*, 1966). Different procedures that have been used include a rubber band snap or finger flick to the sole of the foot, pinching the biceps, pulling the hair, vaccination or heel prick, rubbing the backbone or, for the birth cry, the traditional spanking. We can distinguish between stimuli of high intensity such as vaccination, which produces an immediate maximal response, and more graduated procedures such as flicking the sole of the foot which are less invasive and allow the baby to build up more slowly into a fully aroused state. Maximal stimuli provide better experimental control over the measurement of response latency, whereas with graduated procedures threshold levels can be studied.

Prechtl *et al.* (1969) found that in the same infants stochastic features of the cry were more stable when the cry was elicited with a pinch than when the cry was spontaneous. Individual differences in cry duration were also consistent so that infants who showed relatively long or short spontaneous cries showed similar durations for the pinch cry. We (Lester & Zeskind, 1979) found similar results comparing cries elicited with a rubber band snap with cries elicited by rubbing the baby's spine in preterm and full-term male and female newborns. Correlations among harmonic and durational features showed moderate intraindividual stability when the two methods of elicitation were compared. Zeskind (1980a) compared the cry elicited with a rubber band snap with a cry recorded eight hours later during the administration of the Brazelton scale in infants who varied in the weight for length ratio of the ponderal index. He also examined cry features of the first and last cry segments from the total cry episodes of the cry elicited by rubber band snap. Harmonic features of the Brazelton scale cry were highly correlated with the same features of the first and last segments of the rubber band snap cry, especially for the maximum fundamental frequency. The first and last rubber band snap cry segments were also highly correlated for the same features. Factor analysis showed that measures of the threshold to elicit the cry and total cry duration were generally unrelated to the harmonic features and may represent a separate dimension of neonatal cry behavior. The pattern of correlations from an earlier study (Lester & Zeskind, 1979) also indicates that durational and harmonic features are relatively independent.

Differences between initial and later cry segments and between different types of cries are important because most cry studies use the first expiratory cry phonation following a maximal pain stimulus. This cry segment seems to be sensitive to the status of the central nervous system (Lester & Zeskind, 1978) and may have some unique features. For example, Zeskind (1980b) found that the first segment of the pain cry more often exhibits the acoustic features which are related to the infant's risk status than either the last segment of the pain cry or the cry produced during the administration of the Brazelton scale.

In a study of birth cries, Blinick, Tavolga, and Antopol (1971) reported that the cries became lower in amplitude with more harmonic form as the vigor of the stimulation wore off. Wolff (1969) also alluded to a related phenomenon by stating that over time the pain cry takes on the rhythmic morphology of the basic cry.

Another indication of the status of the nervous system may be the capacity to recover from the initial response to a sudden painful stimulus by adopting a more rhythmic, harmonic, and presumably less costly (from an expense-energy model) cry pattern. We might expect that infants who do not shift to lower vocal registers by releasing tension in their laryngeal and respiratory muscles (Golub, 1979) and remain in hyperphonation (Truby & Lind, 1965) following the initial cry response may be at higher risk due to parasympathetic (vagal) hypotonia or lack of inhibition, which prevents the nervous system from maintaining homeostatic adaptation.

6. NORMATIVE STUDIES

Most of the research on infant cry has been focused on the clinical significance and meaning of abnormal cry sounds with less attention given to the ontogeny of crying in normal infants.

Earlier we mentioned the work of Truby and Lind (1965) in which the characteristics of phonated, dysphonated, and hyperphonated cries were described. Wasz-Hockert *et al.* (1968) identified four types of cries in infants ranging from birth to seven months: the birth, hunger, pain, and pleasure cries. The birth cry was characterized as about one second in duration, voiceless, tense, and with a flat or falling melody. The pain cry is longer (mean = 2.6 sec), tense, higher in pitch (mean = 500 Hz) with a flat melody form. Shifts in pitch, subharmonic break, and vocal fry occurred in one third to one half of the cries. Between the first and sixth month slight increases in length and pitch were noted. The hunger cry was shorter (mean = 1.3 sec) and primarily characterized by a rising-falling melody form. Pitch was variable although lower (mean = 470 Hz) than in the pain cry. Age changes were not marked. Pleasure cries were flat in form, nasal, and voiced with extreme variability in pitch.

Twenty-four cries, six from each category, were randomly selected and played to adults to compare with spectrographic identification of the cry types. The mean percentage of correct identification of the six cry samples for each cry type were birth, .48; pain, .63; hunger, .68; and pleasure, .85, although there was considerable variability in correct identifications within each cry type. The best identified cries from each class were spectrographically typical, whereas the worst identified cries showed atypical spectrographic features. Experience with infants also facilitated the identification of the cry type.

Spectrographic recordings were used by Wolff (1969) to describe the morphology of the basic or rhythmic cry as consisting of an initial cry proper (.6) followed by a brief rest period (.2 sec), a short inspiratory whistle (.1–.2 sec.), and another brief period before the next cry proper begins. Wolff argued that although this basic cry is heard when the infant is hungry, the term *hunger cry* is misleading if it implies a causal relation between hunger and the cry since this is the basic cry pattern to which the infant sooner or later reverts from other patterns and has no casual relation to hunger. A second type of cry was identified by Wolff as the mad or angry cry on the basis of parents' observations that their baby sounded angry. The morphology, or temporal segments, of the angry cry is the same as that of the basic cry. The angry cry is distinguished by turbulence or nonharmonic distortion created by forcing air through the vocal cords. The pain cry, Wolff's third type, has a different morphology from that of the basic cry. It is characterized by a sudden and loud onset, a long initial cry proper, followed by a long rest period, a brief inspiration and short rest before the next cry proper.

We may wish to consider the following classification schema. First, infant sounds can be broadly separated into distress and nondistress vocalizations (the pleasure cry is a nondistress vocalization). Distress vocalizations or crying can then be described in terms of the known or unknown eliciting stimulus. The basic or cry phonation has a known morphology and may be associated with hunger. Wolff reports a fundamental frequency of predominately 350–400 Hz with an initial cry phonation of .6 sec. For Wasz-Hockert *et al.* (1968), the mean fundamental frequency was higher (470 Hz) and the duration longer (1.3 sec). The rising-falling melody form suggested by Wasz-Hockert *et al.* can be seen in the spectrograms published by Wolff.

Some differences in the structure of the cry seem to be related to the intensity quality of the stimulus. A high-intensity stimulus such as a heel prick elicits the maximal response that is referred to as the *pain cry* (Wolff, 1969; Wasz-Hockert *et al.*, 1968). This cry has the distinct morphology described by Wolff and the harmonic features described by Wasz-Hockert *et al.* However, more graduated cry-eliciting procedures will probably produce cries that do not conform to the characteristics of the maximal pain cry response. Our speculation is that these changes in the cry lie on a continuum and that the less aversive the eliciting stimulus the more the cry approximates the morphology of the basic cry phonation. The maximal cry response to pain is characterized by longer and louder initial sound and higher fundamental frequency followed by a longer rest period before the first inspiration. This cry is also likely to be more tense and to show harmonic features such as shifts in pitch, subharmonic break, and vocal fry. Hyperphonation and dysphonation may occur following a maximal pain stimulus, depending on the condition of the infant.

From the baby's perspective, the initial state will probably influence the level of maximum response reached for a given amount of stimulation. Individual differences among normal neonates and perhaps areas of investigation for at-risk or damaged infants might include the variability in approximating the parameters of the maximal pain response for less than maximally intense stimuli and the infant's self-regulatory mechanisms for state control and modulation that enables the baby to shift from the maximal pain cry to the basic cry. Hypersensitive infants, for instance, may show a maximal pain cry response to less aversive stimuli than do normal infants and may have difficulty in the regulation of the cry to the basic pattern of phonation. The floppy, lethargic, hyposensitive infant may require more than the expected amount of stimulation to reach a maximal pain cry response or may not even show this response.

Within this system we can also study qualitative variants of the basic to maximal pain cry morphology such as the "angry cry" and the birth cry. The angry cry, as described by Wolff, has the morphology of the basic cry but is colored by turbulence of nonharmonic distortion (dysphonation). Nonharmonic distortion is found in varying degrees in many cries and is one of many attributes that affect the quality of the cry. However, it has no structural or antecedent causal properties. That this cry may have unique signal value or functional significance because parents perceive the cry as angry and react differently to it is a separate issue, and we may wish to investigate the effects of various cry attributes on infant–parent interaction. The birth cry, like the angry cry, shows considerable nonharmonic distortion but appears to be a variant of the basic cry depending on the intensity of the eliciting stimulus.

Developmental changes and individual differences in cry patterns have been investigated in a few studies. Lind *et al.* (1966) found that the length of the pause from the first to second cry segment decreased from birth to seven months and that the minimal length of the cry period increased with age. No age differences were found for the latency or length of the first cry signal. Ringel and Kluppel (1964) reported differences in the amplitude and fundamental frequency among but not within the pain cries of 10 newborns; that is, they found interindividual but not intraindividual differences. Both the discomfort cry and pain cry duration and interval decreased and became less variable between days 1 and 8 of life in a study of seven infants reported by Prechtl, Theorell, Gramsbergen, and Lind (1969). Comparisons of the two cry types over time revealed that the duration of the pain cry was more stable.

In an effort to study the genetic influence on vocal behavior, Ostwald, Freedman, and Kurtz (1962) recorded the spontaneous cries of 16 pairs of infant twins during the first month of life. Although many of the twin pairs showed similar cry patterns, these similarities were not related to the monozygosity of the pairs.

In a few studies the amount of daily crying at home during infancy was recorded. In one (Rebelsky & Black, 1972) 24-hour tape recordings of the vocalizations of seven infants were made every two weeks from 1–2 weeks of age to 3 months. They found considerable variation in the amount of crying both among and between infants. Overall, the amount of crying increased until 6 weeks and then decreased. By 12 weeks, the infants showed evidence of adaptation to cultural sleeping and eating patterns with peak crying times associated with feedings.

Bernal (1972) asked 20 primiparous mothers and 57 mothers of second babies to keep a diary of the amount of crying during the first 10 days of the babies' life. A peak in crying was found between 6 P.M. and midnight. Second babies cried less than first babies because mothers responded more quickly and more often by feeding to the cries of second babies. The diary method was also used by Brazelton (1962) by asking 80 mothers to keep daily records of their infants' crying for the first 12 weeks. Crying increased until the 6th week and then decreased with daily peaks in the evening. Brazelton emphasized patterns of individual differences related to environmental tension and stressed that a certain amount of crying is necessary for the infant's organization. A similar conclusion was reached by Emde, Gaensbauer, and Harmon (1976), who recorded episodes of unexplained fussiness, that is, crying that does not appear to be associated with hunger or pain that, in extreme cases, is labeled colic. They found that early fussiness occurred to a greater or lesser degree in all babies observed and was most common during the first 3 months, rarely occurring after 6 months. The apparent inborn readiness of such prolonged crying suggested an evolutionary explanation of crying to promote infant–caretaker interaction at times not taken up by feeding. In a follow-up study of normal newborns, Karelitz, Fisichelli, Costa, Karelitz, and Rosenfeld (1964) reported that the amount of newborn cry activity from an elicited pain cry was positively correlated with higher Cattell scale scores at 15–20 months and higher Stanford–Binet IQ scores at 3 years.

7. CRYING AND NEUROLOGICAL DIAGNOSIS

In pediatrics, crying is probably most associated with certain "classic" syndromes such as *cri du chat* (literally, "cry of the cat") and is part of the neurological examination of the neonate (cf. Prechtl, 1977). Clinical evaluations of the condition of an infant are often made based on the nature of the cry. In fact, most of the literature on infant crying is concerned with clinical populations. One of the goals of these studies has been to use cry features to aid in the differential diagnosis of specific syndromes. In most studies, the differentiation of so-called abnormal cry features is based on the retrospective analysis of the cry of an infant known to be abnormal or, in cases such as *cri du chat*, where the cry, although dramatic, is superfluous to making the diagnosis. Moreover, most cry features that have been described are not unique to a particular syndrome but are found in relations to a wide spectrum of CNS insult.

These studies are useful because they have enabled us to learn a great deal about variations in infant cry features and about the kind of CNS involvement that relates to differences in infant cries. As a result, these studies provide us with a window into the nervous system that may facilitate our understanding of normal and abnormal developmental processes, of how reorganization and recovery of the nervous system take place. At the diagnostic level, we may be able to identify infants who do not present with obvious clinical signs but are at risk for major catastrophe such as sudden infant death syndrome (SIDS), infants whose nervous system may be immature or stressed at birth and may be especially vulnerable to developmental insults, or infants who are at particular risk for social or interactional failure such as child abuse or failure to thrive. Our major interest in infant crying is in these latter kinds of situations, in using the cry as part of the field of behavioral pediatrics better to understand the infant and his developing interpersonal relationships. We shall first summarize the clinical studies (for a more detailed description of these studies, see Lester & Zeskind, 1979) and then our own work and the work of others on the behavioral pediatric side of crying.

Cry features have been described in relation to three chromosome aberrations. In a study of infants with *cri du chat*, Vuorenkoski *et al.* (1966) found that of the 13 cry features analyzed, a minimum pitch above 500 Hz differentiated 97% of the *cri du chat* from normal pain cries and 91% of *cri du chat* from normal hunger cries. Infants with Down's syndrome show shorter, less active, and less well-differentiated cry outbursts than do normal infants; more stimulation is needed to produce a one-minute cry response in infants with Down's syndrome and their latency from stimulus to cry onset is longer (Karelitz & Fisichelli, 1962;

Fisichelli & Karelitz, 1963). These features were also found in other infants suffering from brain damage. In a sound spectrographic study of infants with Down's syndrome, all eight features of the cry studied differentiated Down's syndrome babies from controls (Lind, Vuorenkoski, Rosberg, Partanen, & Wasz-Hockert, 1970). The cry in Down's syndrome infants was characterized as long, with a flat melody and lower fundamental frequency. The latency was longer and the Down's syndrome cries showed a higher frequency of biphonation, stuttering voice, and nasality. Especially intriguing was the finding that in five infants who were suspected of having Down's syndrome because of facial profile, muscular hyptonia, and no Moro reflex but who had a normal karotype, the cries were found to be normal. A case study of an infant with 13–15 trisomy (Ostwald, Peltzman, Greenberg, & Meyer, 1970) indicated a fundamental frequency at termination of the cry (100–250 Hz) and tonal instability or quaver in the cry. They noted the similarity between the cry in trisomy 13–15 and Down's syndrome. Both of these conditions are trisomy anomalies due to chromosomal reduplication in contrast to *cri du chat* which results from chromosomal deletion and has a wholly unique cry sound.

Wasz-Hockert, Koivisto, Vuorenkoski, Partanen, and Lind (1971) were interested in changes in cry features with signs of developing kernicterus and compared infants with Rh immunization, infants with ABO immunization, and infants with no blood group incompatibility. The cries of the total hyperbilirubinemia group had a higher maximum and minimum fundamental frequency, more shifts in the fundamental, a higher change in the fundamental, a shorter cry duration, and shorter cry latency. Splitting of the fundamental and biphonation occurred more frequently in jaundiced infant cries than in controls.

A cry score rating system developed from another sample by Vuorenkoski, Lind, Wasz-Hockert, and Partanen (1971) was applied to the bilirubin cry data and showed more abnormal cry ratings scores in the hyperbilirubinemia group. For the sample on which the cry score was developed, 95% of the normal infants and 4% of the abnormal infants were correctly classified.

The cries of babies with bacterial meningitis were found to have a higher minimum and maximum fundamental frequency; shorter duration with more voiceless periods; more rising, falling-rising, and flat melody forms; and a greater frequency of biphonation and gliding than the controls (Michelsson *et al.* 1977a). At a check-up between 4 and 18 months after the meningitis, 7 infants were normal and the other 7 had developed neurological sequelae. A comparison of the cries from the healthy infants with the cries of those with sequelae showed that the cries in the latter group were shorter, more voiced, continuous, and showed more variations in melody form. The fundamental frequency was not significantly different between these two groups. The possibility that changes in features of the cry may parallel the infant's recovery was suggested by spectrograms from two infants. For one infant, cries were recorded one and three days after the diagnosis and showed a change to a more typical pattern with recovery. The other infant developed hydrocephalus and showed no recovery in the cry pattern from the time meningitis was diagnosed.

A retrospective follow-up study of full-term asphyxiated newborns (Michelsson, Sirvio, & Wasz-Hockert, 1977b) showed that when compared with healthy controls at birth, the asphyxiated group had a shorter cry duration, higher fundamental frequency, more variability in melody type, and a more frequent occurrence of biphonation, vibrato, double harmonic break, and glottal roll. After follow-up at 2 and 8 years, the newborn cries of the damaged infants when compared with the newborn cries of normal infants were shorter in duration, had a higher fundamental frequency, and showed more unstable signals than those of the healthy infants who suffered asphyxia at birth. These findings were similar to those reported in an earlier study of premature and full-term asphyxiated newborns (Michelsson, 1971). Spectrograms were also presented to show changes in the cry with recovery from asphyxia in an infant later found to be healthy at follow-up and the persistence of the newborn cry pattern in an infant who later developed cerebral palsy.

From their model of cry production described earlier, Golub (1979) and Corwin and Golub (1980) developed computer extraction procedures to examine cries for eight acoustic features expected to represent abnormalities in the infant. Four groups of infants were studied: 55 normal term infants, 17 infants with multiple or severe abnormalities, 12 infants with hyperbilirubinemia, and a fourth group of 2 babies who later died of sudden infant death syndrome and an infant whose sibling had recurrent apnea. Of the normal infants, 45 had no cry abnormalities, with the remaining 10 infants each having one abnormality. All 17 abnormal infants had at least one abnormal cry feature with 14 having two or more. Eleven of the bilirubin infants showed the same abnormality of glottal instability; this feature occurred in only 8 of the 63 infants without jaundice. All three infants in the last group showed the same single abnormality, which was not found in any other infants, a constriction in the vocal tract which produced a higher fundamental frequency.

Colton and Steinschneider (1980) compared neonatal cries in 22 term siblings of SIDS victims with term and preterm control groups. The cry predictor that most clearly differentiated the siblings of SIDS infants from the other group was a higher first formant frequency. The author suggested that the cry may be related to SIDS and may reveal some of the underlying anatomical/physiological variants related to the cause of SIDS.

A retrospective case study of the cry of a 4-day-old full-term, full-birthweight, normal baby who at 6 months died suddenly and unexpectedly showed that when compared with the cries of four normal infants the cry of the SIDS baby had a higher fundamental frequency, more shifts of the fundamental, and more extremes in frequency than did those of the non-SIDS babies, indicating vocal tract constriction (Stark & Nathanson, 1975). No mechanical defects of the larynx were found on autopsy. A case study by Hopkins (1976) and data from the National Collaborative Perinatal Research Project (Anderson-Huntington & Rosenblith, 1976) also mentioned "abnormal" cries as well as other extremes in behavioral functioning in their retrospective reports of babies who died of SIDS. The Hopkins study used the Brazelton scale and found average performance on day 3 but extreme temperamental lability on day 30. An interview with the mother on day 30 revealed a baby who was unable to generate a consistent interpretable pattern of communication and had a cry that was weak and irritating. Interestingly, Naeye, Messmer, Sprecht, and Merritt (1976) found that SIDS victims' cries differed in pitch and that mothers had trouble differentiating between a victim's needs and moods by the cries based on a modification of the Carey questionnaire administered to the parents of 46 infants who died of SIDS. These studies suggest that unusual features of the cry may indicate CNS involvement and, as Stark and Nathanson (1975) suggested, infant cry analysis when combined with other measures might aid in the identification of infants who may be at risk for SIDS and in the understanding of some of the mechanisms underlying SIDS.

Durational patterns of the cry in normal, neurologically suspect, and abnormal infants during the first week of life were reported by Prechtl *et al.* (1969). They found greater variation in durational features, especially duration, in the suspect and abnormal infants as compared with the normal infants. A clinical rating of normal, suspect, or abnormal was also used by Ostwald *et al.* (1968), who found higher fundamental frequency cries in the suspect and abnormal groups. Michelsson (1971) reported cries of longer duration, a lower fundamental frequency but higher shifts in the fundamental, and more shifts in the glottal plosives in small-for-date babies. In premature, low-birthweight infants, the fundamental frequency and frequency of shift was higher, biphonation and gliding were more common, glottal plosives were less frequent. The birth cries of infants from narcotic-addicted mothers were reported by Blinick *et al.* (1971). Eleven percent of the cries of controls and 50% of the cries of infants of addicted mothers were categorized as abnormal. A high fundamental frequency was the abnormality most characteristic of the cries of infants of drug-addicted mothers.

These studies appear to suggest that certain conditions of the infant are associated with changes in features of the cry. There are a number of ways in which these findings may have clinical utility. Since listening for unusual-sounding cries is part of the pediatric examination of the newborn, the identification of certain cry features may help provide accurate descriptions of cry parameters and sharpen clinical sensitivity to the cry sound. For example, Partanen, Wasz-Hackert, Vuorenkoski, Theorell, Valanne, and Lind (1967) showed that physicians could correctly identify 15 out of 20 cries as abnormal or normal. The "hit" rate was better for abnormal than normal signals, although this level of accuracy was at the expense of labeling too many signals as abnormal. A spectrographic comparison of the best identified cries from the abnormal and normal infants suggested that the most striking difference between the cries was the higher fundamental frequency (1250 Hz vs. 450 Hz).

In a later study, clinicians were trained to discriminate above chance level normal cries from cries of infants with asphyxia, brain damage, hyperbilirubinemia, or Down's syndrome (Vuorenkoski, Wasz-Hockert, Lind, Koivisto, & Partanen, 1971). These studies support the practice of including the cry as part of the pediatric examination of the newborn and may suggest that the inclusion of some information on cry acoustics may be useful in clinical training.

8. CRY FEATURES IN THE INFANT AT RISK

Our own work on infant cry began somewhat fortuitously during a study of cardiac habituation to pure tones stimuli in malnourished Guatemalan 1-year-old infants. It was observed that the cry of these infants sounded high-pitched, arrhythmic, and strained. The cry was recorded following the habituation procedure and acoustic analysis performed with a spectrum analyzer. The malnourished infants showed lower cardiac orienting and dishabituation responses to the tones than did well-nourished infants (Lester, 1976). The cry of the malnourished infants had a longer duration, longer latency, higher fundamental frequency, lower amplitude of the fundamental, and fewer harmonics. These cry features were correlated with the heart rate data in the same infants, indicating that infants who showed lower cardiac orienting responses were accounting for the differences in cry features. Juntunen, Sirvio, and Michelsson (1978) reported similar alteration in the cries of children suffering from marasmus.

These and other studies of "suspect" babies (Ostwald, 1972; Prechtl *et al.*, 1969) led to the study of cry features in newborns who may be at risk due to prenatal and perinatal factors. Certain features of the pain cry seemed to be sensitive to CNS stress and could be expected to be related to the risk status of the neonate. If so, analysis of infant cry features may be useful in the identification and assessment of otherwise healthy infants who may be at risk. In addition, the effect of the cry, as part of the behavioral repertoire of the infant, on the development of infant–caregiver interaction may be especially important for the infant at risk when the infant's recovery may depend upon the sensitivity of the caregiving environment to the infant's needs. The crying of the high-risk infants may place greater demands on the ability of caretakers to interpret their infants' signals.

In one study (Lester & Zeskind, 1978) we studied the combination of measures of physical size and neonatal behavior in predicting acoustic features of the pain cry in 2-day-old neonates. Multiple regression analysis showed that poor performance on the Brazelton scale in conjunction with a low ponderal index (a

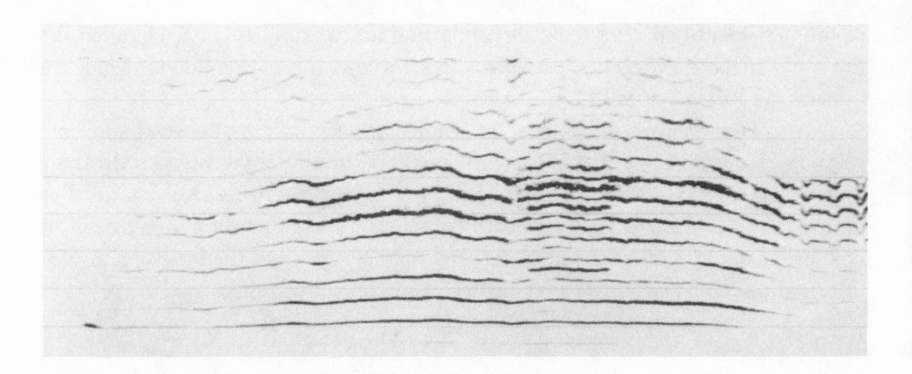

FIGURE 8. Spectrogram of pain cry in a term, average-weight-for-length neonate. From "A Synergistic Process Approach to the Study of Prenatal Malnutrition" by B. M. Lester, *International Journal of Behavioral Development,* 1979, *2,* 377–393. Copyright 1979 by the *International Journal of Behavioral Development.* Reprinted by permission.

weight-for-length ratio used to indicate fetal malnutrition) and short gestation was associated with infant cries of short duration, a high fundamental frequency, a high maximum frequency, and fewer harmonics.

We then wanted to take a closer look at the underweight-for-length infants because Als, Tronick, Adamson, and Brazelton (1976) had reported that infants below the 10th percentile of the ponderal index were more often scored clinically worrisome on the Brazelton scale. Twelve average weight-for-length and 12 infants below the 3rd percentile of the ponderal index from the previous study (Lester & Zeskind, 1978) were compared on the Brazelton scale and on the cry features (Lester, 1978, 1979a,b). Again, the infants were clinically healthy according to routine pediatric and neurological examinations. The underweight infants showed poorer performance on all four summary dimensions of the Brazelton scale: interactive processes, motor processes, organization of state, and physiological organization. On the cry features, the groups did not differ on latency to cry onset but did differ on the remaining measures, with the underweight infants showing a shorter first cry phonation, a longer expiratory period, a higher fundamental frequency, and fewer harmonics. Correlations between the Brazelton scale dimensions and cry features indicated that poor Brazelton scale scores were related to the differences in the cry features, most notably to the higher fundamental frequency. Figures 8 and 9 show spectrograms from a full-weight and an underweight-for-length infant from this study. The cry of the underweight infant shows more harmonic distortion, more variability or shift in the fundamental frequency resulting in a rising-falling melody form, in contrast to the generally flat melody form of the cry of the full-weight infant.

These findings were replicated and extended in two subsequent studies. In the first (Zeskind & Lester, 1981), the notion that intrauterine (and, presumably, CNS) stress affects the organization of neonatal crying was further elaborated by predicting that overweight-for-length infants would also show differences in cry features. Three groups of full-term normal, healthy infants were compared: infants below the 3rd percentile of the ponderal index, infants who were appropriate weight for length, and infants above the 90th percentile of weight for length.

Analysis of the pain-elicited cry indicated that whereas no differences were found between low and high PI infants on any of the cry features, both groups differed from the average PI group by requiring more stimuli to elicit the cry, a longer latency from stimulus to cry onset, and a higher fundamental frequency. Only the high PI infants cried, overall, for less time than the average PI infants. The cry features of the low and high PI groups also showed more variability than in the average PI group.

In the second study (Zeskind, 1980a) pain cry features, Brazelton scale performance, and the cry elicited during the administration of the Brazelton scale were investigated in normal, healthy groups of low, average, and high PI neonates. The highest dominant frequency of the initial pain cry, of the final pain cry segment, and of the Brazelton scale cry showed a higher mean and more variability for the underweight and overweight infants that the average weight infants. Underweight and overweight infants also required more stimuli to elicit the cry and cried for less time. Similarly, on the Brazelton scale the underweight and overweight infants did not differ from each other but differed from the average-weight babies on all four of the previously mentioned summary dimensions and on a fifth dimension of orienting to the environment. Cry features and Brazelton scale scores were also highly related, supporting previous findings.

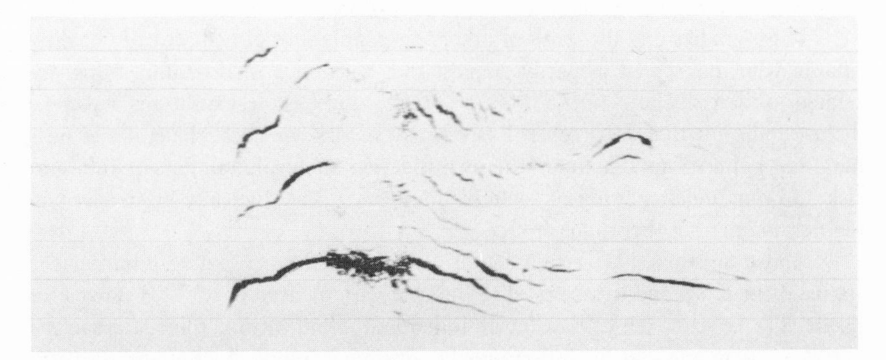

FIGURE 9. Spectrogram of pain cry in a term, underweight-for-length neonate. From "A Synergistic Process Approach to the Study of Prenatal Malnutrition" by B. M. Lester, *International Journal of Behavioral Development,* 1979, *2,* 377–393. Copyright 1979 by the *International Journal of Behavioral Development.* Reprinted by permission.

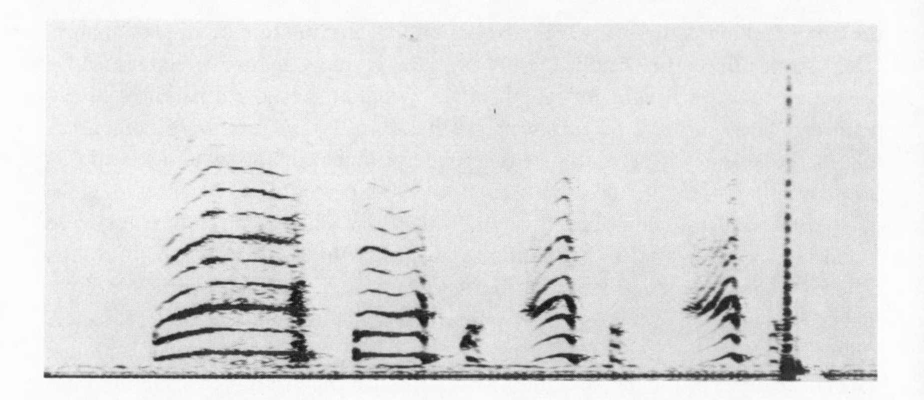

FIGURE 10. Spectrogram of pain cry in term neonate with fewer than three nonoptimal obstetric conditions. From "Acoustic Features and Auditory Perceptions of the Cries of Newborns with Prenatal and Perinatal Complications" by P. S. Zeskind and B. M. Lester, *Child Development*, 1978, *49*, 580–589. Copyright 1978 by *Child Development*. Reprinted by permission.

These alterations in cry features are not only associated with variations in physical growth. Infants at risk due to events associated with their obstetrical histories also differ in cry features (Zeskind & Lester, 1978). Using the maternal and parturitional list of nonoptimal obstetric conditions developed by Prechtl (1968), we compared cry features between full-term normal, healthy infants who were defined as low (0–2 nonoptimal conditions) and high (5–9 nonoptimal conditions) risk. The high-risk infants had a higher fundamental frequency, required more stimuli to elicit the cry, had a longer latency to cry onset and a shorter first cry expiration, and cried less overall than the low-risk group. Spectrograms from a low-risk and a high-risk infant in this study are shown in Figures 10 and 11.

These studies and the work of others on populations of pediatrically healthy infants who may be in jeopardy suggest that variations in neonatal crying are related to the risk status of the infant. The cry may have less utility as a specific disease indicator than as a general assessment of risk. As such, the study of neonatal cry patterns may facilitate the identification and evaluation of the infant at risk and our understanding of some of the mechanisms that may be involved in the development of the infant at risk.

It also appears that certain features of the cry or cry patterns mirror the status of the CNS and can be used to indicate various degrees of CNS trauma or insult. The cry may tell us that for an individual infant there is some stress to the nervous system and that the infant may be somewhat fragile. As a prognostic indicator, the cry may be useful as a way to help estimate which infants are at greater risk for CNS sequelae in situations where the risk status of the infant is ambiguous as, for example, in prematurity. Since the CNS is sufficiently plastic, such

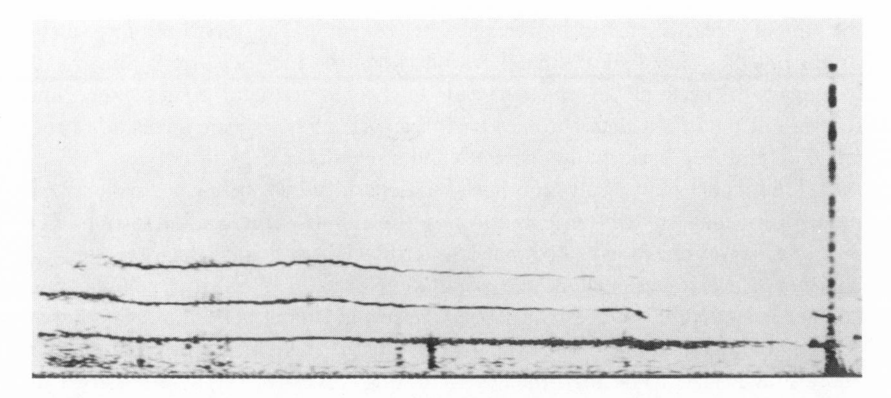

FIGURE 11. Spectrogram of pain cry in a term neonate with 5–9 nonoptimal obstetric conditions. From "Acoustic Features and Auditory Perceptions of the Cries of Newborns with Prenatal and Perinatal Complications" by P. S. Zeskind and B. M. Lester, *Child Development*, 1978, *49*, 580–589. Copyright 1978 by *Child Development*. Reprinted by permission.

that most premature infants do recover, we would expect repeated study of the cry to reflect the infant's recovery as seen in the organization of other behavioral and physiological systems.

9. EFFECTS OF CRYING ON CAREGIVERS

The cry not only reflects the biological integrity of the infant but may also play a role as part of the development of the infant–caregiver interactive system. That is, the cry may be one aspect of the infant's contribution to his or her own caretaking. This may be particularly important for the infant at risk when the sensitivity of the caretaking environment may be crucial for the infant's recovery. An illustration of this concept is shown in Figure 12 for the infant at nutritional risk and shows some of the major factors which may potentiate or magnify nutritional insult (from Lester, 1979b). Broadly speaking, the model includes the prenatal environment and the ecological-social milieu of the family. It shows that maternal obstetric and reproductive history interacts with variations in the intrauterine environment that may compromise the fetus and produce the infant at risk. These historical and intrauterine factors can act in concert to produce a stressed, underweight infant whose at-risk status may be indicated by disorganized behavior patterns. The poor eliciting behavior of these infants can exacerbate the effects of malnutrition. A poorly organized infant who has difficulty interacting with the environment and is a poor elicitor of maternal responses by being underdemanding of a caregiver who is already stressed and nutritionally depleted may not receive the kind of caregiving necessary for his or her recovery. The opportunity for catch-up during the immediate postnatal period is denied, and the stage is already set to reenact the cycle of poverty that leads to chronic postnatal malnutrition. The recovery of the infant then becomes part of a cycle of synergistic forces that capitalize on and feed back into the stresses and strengths of the culture.

This model may apply to other situations in which behaviors such as the infant's cry may contribute to failures in the infant–caregiver interaction. For example, surveys have indicated that low birthweight and small-for-gestation-age babies are overrepresented in the population of failure-to-thrive, abused, and adopted infants (Gil, 1970; Light, 1974). Parke and Collmer (1975) suggested that certain characteristics of the low-birthweight baby make him a prime target for abuse and neglect. The high-pitched, irritating, excessive cry of the abused or neglected infant has also been mentioned (Gil, 1970; Parke & Collmer, 1975; Ramey, Heiger, & Klisz, 1972). We may be witnessing a cycle of interactive failure in which an unattractive, poorly responsive, difficult-to-manage baby with a high-pitched, irritating cry violates the limits of an already stressed caregiver. This suggests that the way in which adults, especially parents, perceive and respond to infant cries may affect the early development of parent–infant interaction.

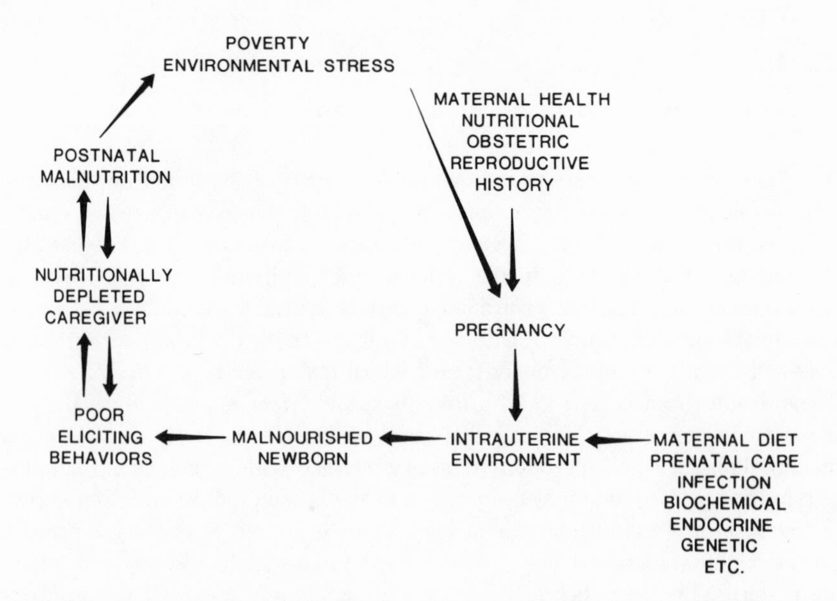

FIGURE 12. Synergistic model of the effects of prenatal malnutrition. From "A Synergistic Process Approach to the Study of Prenatal Malnutrition" by B. M. Lester, *International Journal of Behavioral Development*, 1979, *2*, 377–393. Copyright 1979 by the *International Journal of Behavioral Development*. Reprinted by permission.

Studies of adult responsiveness and ability to judge infant cry sounds reveal the functional effects and the social and signal value of the cry. A number of studies indicate that mothers can identify the cry of their own infant (Formby, 1967; Greenberg, Rosenberg, & Lind, 1973; Morsbach & Bunting, 1979; Valanne, Vuorenkoski, Partanen, Lind, & Wasz-Hockert, 1967) and wake when their own babies cry but not when other babies cry in a hospital (Formby, 1967) and that crying may raise the temperature of the breasts of lactating mothers 3–5 days postpartum (Vuorenkoski, Wasz-Hockert, Koivisto, & Lind, 1969). Wolff (1969) reported that mothers were more distressed when listening to a pain cry which they thought was from their infant than when the same cry was acoustically altered to sound more like a basic cry. Other studies indicate that even infants respond to the cries of other infants (Sagi & Hoffman, 1976; Simner, 1971).

The ability of adults to judge the reason why infants are crying is less apparent. Adults, independent of contextual cues, seem to have difficulty discriminating pain from hunger and other elicited cries (Muller, Hollien, & Murray, 1974; Murray, 1979; Sherman, 1927), even for mothers who can correctly identify their infant (Muller, Hollien, & Murray, 1974). Part of this difficulty may occur when the temporal patterning and/or fundamental frequency is similar among differently elicited cries. However, Wasz-Hockert *et al.* (1968) did find that adults could discriminate among some birth, hunger, pain, and pleasure cries. Others were more difficult to distinguish. They also found that women, regardless of experience with infants, were more accurate in their judgments than men. Using the same tapes, Berry (1975) found that children as young as 7 years of age could also distinguish the four cry types and that this ability improved from ages 7 to 13.

Our conceptions of the signal value of infant crying may also be influenced by cultural practices and expectations. In a study of foraging societies, Konner (1977) noted that the hunger cry of the infant was only attended to by the mother, whereas a pain cry elicited attention and immediate approach from other adults in the village. In a study of infant care practices in 186 societies throughout the world, Barry and Paxson (1971) found that the reward for crying was inversely correlated with the frequency of crying and that bodily contact was related to reward for crying. These findings could be interpreted to mean that crying is minimized by prompt attention to the infant's signals, such as body contact cues that are available when infants are carried, that enable the mother to begin soothing before crying begins. Thus, the meaning and signal value of the cry may be related to the adaptive significance of the cry for a culture. The role of maternal responsiveness to infant crying in Western societies is supported by the work of Bell and Ainsworth (1972), in which the frequency and duration of crying during the first year of life was negatively correlated with the promptness of the mother's response and positively related to the number of crying episodes ignored by the mother.

We became interested in whether adults could discriminate between the cries of low- and high-risk infants and, if so, what dimensions they were using to distinguish the cries. As part of a study described earlier (Zeskind & Lester, 1978), cries of low- and high-risk infants were randomly selected and rated on dimensions of cry attributes by adults with no professional child care experience or children of their own and by parents along dimensions of cry attributes. The adults were simply told that we wanted them to rate the cries, and no other information was provided about the study or the histories of the infants. The rating scales were generated from descriptions in the literature representing bipolar dimensions of urgent/nonurgent, pleasing/grating, sick/healthy, soothing/arousing, piercing/nonpiercing, comforting/discomforting, aversive/nonaversive, and distressing/nondistressing. Comparisons of the mean ratings for the low- and high-risk infants showed that adults, regardless of child care experience, rated the high-risk infant cries as more urgent, grating, sick, arousing, piercing, discomforting, aversive, and distressing. Also, parents rated all cries as less aversive than nonparents did, and females rated all cries as more arousing than males did. In order to determine whether low- and high-risk cries were perceived along separate dimensions, the cry ratings were factor-analyzed. This analysis showed that the ratings of the low-risk infant cries were represented by a single dimension composed of all of the rating scale scores, whereas high-risk infant cries were rated along two separate dimensions, one representing the unpleasant nature of the cry and the other indicating that these cries sounded "sick" and "urgent." In other words, the cry of the infant at risk carries additional information—a clear statement by the infant that special attention is required.

These same cries were used in another study (Zeskind, 1980b) to determine the kinds of caregiving behaviors that adults felt they would provide to these infants. Different groups of parents and nonparents, also naive to the purpose of the study, were asked to select the most appropriate caregiving response to each cry from the following list: feed, cuddle, pick up, clean, give pacifier, or wait and see. They were then asked to rank each caregiving response along two dimensions: how tender and caring the response is, and how immediately effective the response is at terminating the crying. Each cry trial was assigned the ranked score for each dimension that was given to the specific caregiving response to the cry. In effect, each adult created his or her own ordinal scale for his or her responses to low- and high-risk infant cries. In this study, parents but not nonparents rated responses to high-risk infant cries as more tender and caring and more immediately effective at terminating the cry. Also, more parents often gave the same responses to the high-risk infant cries than to the low-risk infant cries; that is, they were less ambivalent than nonparents about what they felt they should do when hearing the cry of the high-risk baby. The most common types of response that parents chose for the high-risk infants were contact–comfort responses, such as picking up and cuddling the baby.

The responses of adults to the high-risk infant cries in these two studies may be unique to infants with high-pitched cries, for Freudenberg, Driscoll, and Stern (1978) found that cries of infants with Down's syndrome, which are low-pitched and flat, were not found to be as unpleasant as normal infant cries, and normal infants were rated as in greater need of attention.

The effects of infant crying on adult physiological responses have been reported in a series of studies by Frodi in an attempt to determine some of the antecedents of child abuse. In one study (Frodi, Lamb, Leavitt, & Donovan, 1978) mothers and fathers watched a videotape of the same baby either smiling or crying, but the parents were told that the baby was either normal, difficult, or premature. The crying infant, especially when labeled premature, triggered the greatest autonomic arousal and was rated as more aversive. Another group of parents saw videotapes of either premature or term infants, but the cries were dubbed so that an independent assessment might be made of the cry effects from the effects of facial features (Frodi, Lamb, Leavitt, Donovan, Neff, & Sherry, 1978). Although the acoustic features of the cry were not analyzed, more autonomic arousal was elicited by crying in general, the cry of the premature infant elicited greater autonomic arousal and more negative emotions than the term infant cry, and the strongest autonomic and negative emotional ratings were to the combination of the premature face and premature infant cry. In a study of abusive mothers, a crying infant elicited greater autonomic arousal and more negative emotional ratings by the abusers than by controls (Frodi & Lamb, 1980). These studies may indicate how infant behaviors that are perceived as aversive by parents can elicit aggressive reactions and may contribute to failures in infant–parent interaction.

Infants perceived as temperamentally "difficult," some of whom are called "colicky," may also show cry patterns that differ from temperamentally "easier" infants. Lounsbury (1978) identified 4- to 6-month-old infants as difficult, average, and easy from maternal responses to the Infant Characteristics Questionnaire (Bates, Freeland, & Lounsbury, 1979) and recorded cries at home just prior to a feeding. A different group of primiparous mothers rated the infant cries on scales of caretaking responses, emotional reactions, and perceived cause of crying. Spectrographic analysis of the cries revealed that infants perceived as difficult paused longer between cries, paused longer overall during the total vocalization, and showed a higher fundamental frequency than infants perceived as easy or average by their mothers. Interestingly, the mean fundamental frequency for the difficult infants was approximately 530 Hz, which is between what most investigators report to be the fundamental frequency in normal infants (400–500 Hz) and what we found to be the average fundamental frequency in at risk infants (700–800 Hz) (e.g., Lester & Zeskind, 1978; Zeskind & Lester, 1978, 1981). The ratings of the cries showed that mothers were more angry and irritated at the cries of difficult infants and felt that the difficult infants were more spoiled. Difficult infants were also seen as crying for reasons such as fright, frustration, or wanting attention, whereas easy infants were perceived as crying from routine physical discomfort (hunger, wet diapers, etc.). The cry of the difficult infant seemed to carry a message to the mothers different from the cry of the easy infant.

The cry tapes from this study were used by Boukydis (1979) to determine parents' physiological responses and ratings of temperamentally easy, average, and difficult infants. Overall, the cries of the difficult infants elicited the greater changes in skin potential, followed by average and easy infant cries. Primiparous parents showed the highest levels of arousal to average cries, followed by difficult then easy cries. Both multiparous parents and nonparents showed the expected pattern of highest levels of arousal to difficult cries and lowest level to easy cries. Boukydis also administered the rating scales from the Zeskind and Lester (1978) study and found that difficult infant cries were rated as more grating, arousing, piercing, and aversive than the other two cry types. The means for the difficult infant cries fall between the means for the low and high complications groups in the previous study (Zeskind & Lester, 1978). Unlike the previous study, a factor analysis indicated that all three cry types were being perceived along similar dimensions (mostly grating, arousing, and discomforting). Difficult cries also received higher "irritation/anger" and "spoiled" ratings and were perceived to be less similar than average or easy infant cries to the cries of one's own infant.

The research reviewed in this section indicates that adults show differential responses to infant cries that vary in acoustic features and that are associated with other aspects of the infant, such as prematurity and temperament. While no study has yet looked at the relationship between cry features and parent–infant interaction, it seems likely that crying may represent one way in which the infant affects the caretaking environment. Thus, in a nonsupportive environment, the behavioral repertoire of the poorly organized infant with a cry that is perceived as grating and aversive may violate the limits of caregiver control behavior (Bell, 1971) and suppress the optimal caregiving patterns necessary to facilitate the recovery of the baby. These nonoptimal caregiving patterns could then exacerbate an already disorganized infant. In a supportive environment where the behavior of the baby can be integrated into the limits of caregiver control behavior, as in the vast majority of interactive dyads, the "sick" and "urgent" quality of the cry of the infant at risk may elicit caregiving behavior that facilitates the recovery and behavioral organization of the baby.

10. DISCUSSION

Parents, who always seem to be at least one step ahead of researchers in their understanding of children, have long been aware that there are vast individual differences in the behavior of infants, that infants differ in the clarity with which they signal their needs and wants and in the predictability of their eating, sleeping, and waking cycles; in short, how "readable" they are.

From a biological perspective, it makes sense that infants from birth must be able to solicit care from adults and that adults must be able to recognize and respond appropriately to these infant signals. As the most powerful infant signal, crying not only promotes physical survival in human and other newborn mammals but may also help establish a relationship that guarantees social interactions, thereby promoting the growth and development of the infant (Goldberg, 1982). For example, Korner and Grobstein (1966) found that of several strategies used, cuddling and holding a baby upright on the shoulder was the most effective in terminating crying. This strategy also elicits attention and visual scanning which in turn provides the opportunity for social interaction. The parent is doubly rewarded—not only does the crying stop but a mutually enjoyable interactive interchange can occur. In the process, the caretaker learns to know the rhythms and needs of the individual baby and the infant gains experience in interacting with and effecting change on both the animate and inanimate environment. Thus, as the major expression of the infant's affective system, it is not surprising that the cry signal contains encoded messages that represent the needs of the individual infant and that the functional significance of crying changes as part of the negotiation process of the developing infant–parent relationship.

We believe that infant vocalizations are adaptive, species-characteristic, communicative signals that convey information to the caretaking environment about the status and needs of the infant. While we have focused on crying, others have studied the communicative function of noncrying vocalizations (Morath, 1977). Converging evidence from the comparative literature indicates evolutionary continuity of vocal signals. Cry signals, in addition to relating to the organizational status of the infant, may be the result of selective pressures for acoustic signals tailored to interspecies and intraspecies communication. Low-frequency, harsh sounds are used by birds and mammals to signify hostility, whereas higher-frequency, more purely tonal sounds are used to convey fright, appeasement, or approach (Morton, 1977). For example, in altricial birds the young use high-pitched vocalizations to elicit parental care and to direct food toward the nest. Other work on vocal communication in animals suggests that variations in the amplitude, signal shape, and spectral composition of certain calls are related to the distance the calls cover and the ease with which the sender can be located in space. Distress calls, for example, are high-pitched, loud, and prolonged. Not surprisingly, as we move up the phylogenetic scale, the complexity of vocal signals increases, such that the higher monkeys and apes can vary several dimensions of the signal. This ability allows more flexible, less stereotypic responses and finer discriminations by the receiver. In the rhesus, lower frequency sounds are used to signify dominance and higher sounds submission.

In studies of neonatal kittens, Buchwald (1980) found that under conditions of stress, for instance, removal of the mother, the call emitted by kittens showed increases in the fundamental frequency, and she noted the similarity between the kitten "stress" call and the human neonatal "stress" cry (Wasz-Hockert *et al.*, 1968; Zeskind & Lester, 1978). In response to this cry of the cat *(cri du chat)*, the mother retrieves the kitten and carries it back to the nest. The highest levels of maternal retrieval responses occurred when the adult females were tested after, as compared with before, parturition and occurred more to a vocalizing than to a taped kitten call, or to a sine wave, square wave, or triangular wave, all at the fundamental frequency of the kitten call. A separate group of nonlactating, nonpregnant mothers primed to approximate hormonal changes prior to and immediately after birth showed almost no retrieval responses. This supports previous findings that parents responded differently from nonparents in their caretaking responses to high-risk infant cries (Zeskind, 1980b) and that autonomic responses to infant cries were related to parenting history and the temperament of the infant.

The notion that stress produces increases in pitch is also supported by adult studies in which stress was induced by exposing subjects to aversive stimulation, asking subjects to cope with a difficult task, or asking them to lie. In a review of these studies, Sherer (1979) concluded that increases in the fundamental frequency were the major vocal indicator of stress-induced, nonspecific arousal. Increases in the fundamental frequency were also found across studies of the activation dimension of discrete emotions such as fear and joy and for anxiety or tension states in psychopathological disturbances, particularly in cases of depression (Sherer, 1979).

Some of the changes in infant crying may be due to biobehavioral shifts in the organization of the nervous system. Emde, Gaensbauer, and Harmon (1976) postulate that there are two major biobehavioral shifts in development during the first postnatal year, one that occurs between two and three months, and a second between seven and nine months. They also noted that periods of unexplained fussiness waned at three months, increased slightly, and were negligible by six months, but that this curve was actually composed of three patterns: an early three-month pattern with a sharp decline, an intermediate pattern peaking at two to five months, and a prolonged pattern between one and four months with a slow decline to seven months.

These patterns could represent individual differences in the interplay between maturational and environmental factors in determining the course of these biobehavioral shifts. Periods of unexplained fussiness, by promoting attention and interaction from the caregiving environment, may be part of a neurophysiological feedback system which facilitates the two- to three-month shift at the CNS level; that is, the input from these interactions probably has an organizing influence on the nervous system. This would support the idea of Brazelton (1962) and others that periods of crying for no apparent reason are part of the normal development of infants and are, in fact, necessary. Brazelton also reported individual differences in the amount of crying at home during the first three months that were inversely related to the onset of other autonomous behavior such as sucking, and that mothers could distinguish this "paroxysmal" fussing from a hungry or uncomfortable cry.

It may be that the organization of crying in early infancy is a mirror that reflects the organization of the nervous system. Some infants whose excessive crying earns them the label of "colicky" or whose periods of unexplained fussiness are prolonged and do not decline until seven months may, for constitutional and interactional reasons, be less well organized at the CNS level; this affects the biobehavioral shift and causes them to be perceived as more temperamentally difficult by their parents. In the Lounsbury study (1978), the "difficult" dimension was derived from a factor analysis and found to be the first and most important factor and was defined by items relating to fussing, crying, changeability of mood, irritability, overall degree of difficulty, and intensity of protest. The cries of these infants were acoustically different (Lounsbury, 1978), and elicited heightened autonomic arousal and negative qualitative ratings by other adults (Boukydis, 1979).

Our own studies indicate that, at least in the newborn, prenatal and perinatal risk factors affect the infant's cry in ways that are related to the more global Brazelton assessment of the behavioral organization of the infant. This may explain some of the constitutional factors that contribute to later infant colic or to the temperamentally difficult infant by indicating early signs of CNS stress. Moreover, these cries seem to carry a unique message: Parental caretaking responses were more "tender and caring" and more immediately effective at "terminating the crying" for high-risk infants whose cry was rated as sounding more sick and urgent. The infant at risk signals the degree of jeopardy through the character of the vocal signal which sends the message to the caregiving environment that this infant requires special care. In addition to reflecting the biological integrity or level of organization of the infant, the cry may also contribute to the developmental process by influencing infant caretaking practices.

Stress appears to alter the character of the vocal signal and, most prominently, the fundamental frequency. However, the increase in the fundamental frequency caused by stress is probably due to mechanisms different from those influencing the high-frequency cries of brain-damaged infants. If, as Golub (1979) suggests, infants cry in one of two modes by shifting vocal registers, it may be that damaged infants cry in the high mode whereas the infant at risk alternates between the two modes due to poor modulation in the neurological control system that varies the tonicity of the vocal apparatus. As we discussed earlier, the mechanism for this might be parasympathetic hypotonia modulated by vagal innervation of the larynx. Our experience has been that in contrast to the cry of the damaged infant, in which the fundamental remains extremely high with little variability, infants at risk show many, and often dramatic, shifts in pitch, or hyperphonation (Truby & Lind, 1965), especially during the pain cry. It is as if the pain stimulus sent the stressed infant into hyperphonation, probably because the nervous system lacks the inhibition (due to low vagal input) which prevents the muscles in the larynx from contracting. Hyperphonation occurs from the heightened tension and tonicity in the larynx. What distinguishes the at-risk from the not at-risk infant is the frequency and duration of hyperphonation. Most low-risk infants rarely react to a pain stimulus with hyperphonation or quickly recover if they have that reaction. This is illustrated by the spectrograms in Figures 13 and 14 which show the first 10 seconds of the cry following a pain stimulus in a low-risk infant and a high-risk infant from the Zeskind and Lester (1978) study. In contrast to the phonated cry of the low-risk infant, the high-risk infant cry shows initial hyperphonation followed by other shifts in pitch, sometimes returning to hyperphonation.

Extreme variability in the fundamental frequency during phonation (Figure 6) may represent less severe autonomic imbalance as the nervous system struggles to maintain homeostatic adaptation. These different cry patterns may enable us to make better estimations of the integrity of the nervous system within at-risk populations. Damaged infants probably do not have this flexibility and are locked into the upper register of hyperphonation. While the high-risk and brain-damaged cries may appear similar on a spectrogram (Zeskind & Lester, 1978), the two cries do not sound the same, for more than the fundamental frequency changes. The cry in *cri du chat,* for instance, is hollow-sounding, flat, and monotonous; in fact, it sounds almost nonhuman and one wonders if adults might be driven to reject caretaking by these sounds as they were by the sounds of Down's syndrome babies (Freudenberg, Driscoll, & Stern, 1978). As an adaptive response, natural selection may not promote the caretaking of damaged members of the species, whereas infants in slight jeopardy are still considered within the range of the normal and can solicit the special attention they need. Moreover, the caretaking environment is prepared to listen. Humans are particularly responsive to higher-pitched sounds (Howard, 1973; Masterton & Diamond, 1973), with the maximum acoustic response of the ear above 800 Hz (Davis, 1959). Women, as the traditional caretakers of infants, have higher-pitched voices than males, and we all raise the pitch of our voices when we interact with babies. Infants as young as three months can reciprocate by imitating variations in vocal pitch (Kessen, Levine, & Wendrich, 1979).

The evolutionary continuity of neonatal cry signals suggests that crying is adaptive by eliciting parental caretaking behavior, that infant and environment are biologically prepared to respond to each other in ways that facilitate the development and organization of the infant. This biological synchrony (Zeskind, 1980b) may extend to the infant at risk when the acoustic properties of the cry of an infant with a stressed nervous system are uniquely suited to the sensory capacities of the parental caretaking environment, such that the distressed nature of the infant cry and need for special attention are communicated.

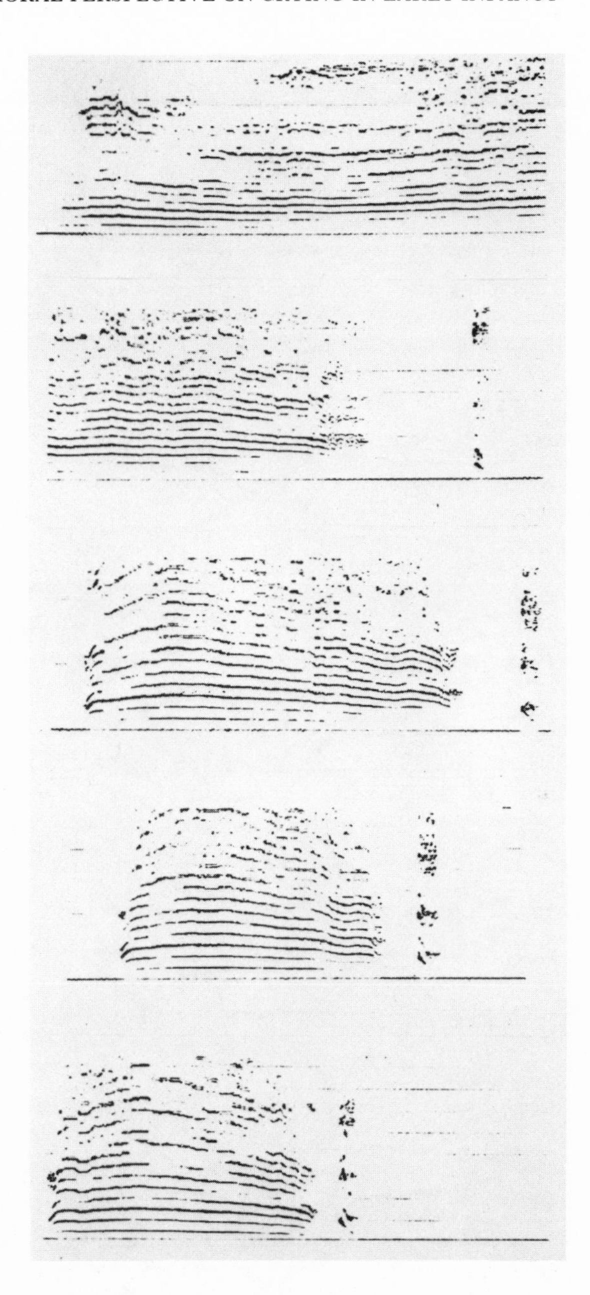

FIGURE 13. Spectrograms of approximately 10 seconds of pain cry in term neonate with fewer than 3 nonoptimal obstetric conditions.

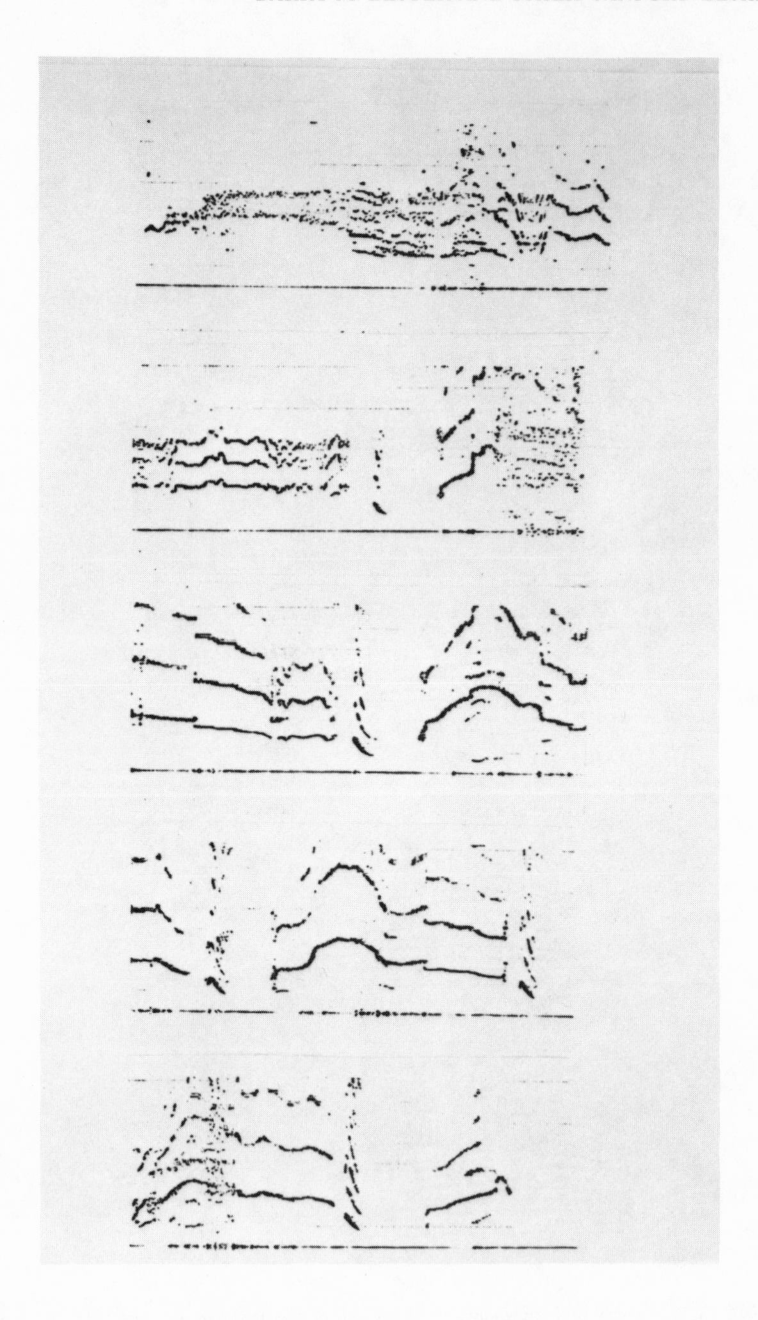

FIGURE 14. Spectrograms of approximately 10 seconds of pain cry in term neonate with more than 5 nonoptimal obstetric conditions.

11. CLINICAL IMPLICATIONS

Crying in infancy is a major problem in behavioral pediatrics. The literature reviewed in this chapter suggests that attention to the acoustic features of the cry may have clinical utility. Certain cry features are used to aid in the differential diagnosis of particular syndromes, and there is some evidence that clinicians can be trained in the discrimination of some of these cry patterns. The inclusion of acoustic cry information as part of the medical school curriculum seems warranted. This training could include the identification of infants who may be at risk but show no other clinical signs. What is striking about the infants we studied is that most were recruited from normal neonatal nurseries. They were clinically normal and healthy by routine physical and neurological examination, yet were behaviorally disorganized. Clearly, there are populations of neonates who because of some mild trauma, such as low weight at full term, may have specific needs to facilitate their optimal development. The cry may help in the identification of these infants who could then be treated to a full behavioral assessment, such as that provided by the Brazelton scale. This information could then be used to describe the strengths and weaknesses of the infant, which would help the pediatrician and parents together to identify the caretaking requirements for the optimal development of the infant. Early attention to infant cry patterns could help prevent later stress in the infant–parent relationship.

Such an assessment could also help describe the range of behavior in normal infants. Once we recognize the variation in behavior that occurs in the course of normal development, we can better advise parents by specifying the developmental issues the infant is negotiating—the strengths, vulnerabilities, coping strategies, compensatory mechanisms, likes and dislikes of the individual infant. The assessment of the infant's behavior tells us what the infant brings to the developing interaction with the parents. It provides a matrix for understanding the infant's side of the question and the problems that the parent is going to have to face. A high-pitched, irritating cry not only signals the sensitivity of the infant but is also an issue for the parents.

By understanding that much of the baby's behavior is due to his own organizational functions and not to their failure, parents can stop blaming themselves unnecessarily and feeling ambivalent about their baby. Excessive crying or aversive-sounding crying may be rendered easier to tolerate by legitimizing and supporting the observations and perceptions of parents. We do not want to minimize this behavior, which is real and difficult. At the same time, we can acknowledge that it is part of the process of normal development, and we can attempt to fit the crying behavior into the larger context of the total behavior of the infant and the infant–parent relationship. Feelings of failure about parenting and ambivalence toward the baby can lead to negative feedback which could drive the interactants further apart. This could exaggerate crying episodes and produce a "colicky" baby. On the other hand, the energy consumed by feeling inadequate and guilty could potentially be freed to facilitate the developing interaction and maximize the baby's development. For it is ultimately that interaction which will determine the infant's outcome and which we must support.

Acknowledgments

Thanks to C. F. Zachariah Boukydis for critical comments of the manuscript, to Kate Neff for her typing, suggestions, dedication, and cheerful disposition, and to Emily Burrows for her patience at the Xerox machine. We would also like to thank Howard Golub for use of the MIT acoustic laboratory computer facilities.

12. REFERENCES

Als, H., Tronick, E., Adamson, L., & Brazelton, T. B. The behavior of the full-term but underweight newborn infant. *Developmental Medicine and Child Neurology*, 1976, *18*, 590–602.

Anderson-Huntington, R. B., & Rosenblith, J. F. Central nervous system damage as a possible component of unexpected deaths in infancy. *Developmental Medicine and Child Neurology*, 1976, *18*, 480–492.

Barry, H., & Paxson, L. M. Infancy and early childhood: Cross-cultural studies. *Ethnology*, 1971, *10*, 467–508.

Bates, J. E., Freeland, C. A., & Lounsbury, M. L. Measurment of infant difficulties. *Child Development*, 1979, *50*, 794–803.

Bell, R. Q. Stimulus control of parent or caretaker behavior by offspring. *Developmental Psychology*, 1971, *4*, 63–72.

Bell, S. M., & Ainsworth, M. D. S. Infant crying and maternal responsiveness. *Child Development*, 1972, *43*, 1171–1190.

Bernal, J. Crying during the first ten days of life and maternal responses. *Developmental Medicine and Child Neurology*, 1972, *14*, 362–372.

Berry, K. K. Developmental study of recognition of antecedents of infant vocalization. *Perceptual and Motor Skills*, 1975, *41*, 400–402.

Blinick, G., Tavolga, W. N., & Antopol, W. Variations in the birth cries of newborn infants from narcotic addicted and normal mothers. *American Journal of Obstetrics and Gynecology*, 1971, *110*, 948–958.

Boukydis, C. F. Z. *Adult response to infant cries.* Unpublished doctoral dissertation, Pennsylvania State University, 1979.

Bowlby, J. *Attachment and loss.* Vol. 1: *Attachment.* New York: Basic Books, 1969.

Brazelton, T. B. Crying in infancy. *Pediatrics*, 1962, *29*, 579–588.

Buchwald, J. Development of acoustic communication functions in an experimental model. In S. L. Friedman & M. Sigman (Eds.), *Preterm birth and psychological development.* New York: Academic Press, 1980.

Chess, G. F., Tam, M. K., & Calaresu, F. R. Influence of cardiac neural input on rhythmic variation of heart period in the cat. *American Journal of Physiology*, 1975, *228*, 775–780.

Colton, R. H., & Steinschneider, A. Acoustic characteristics of first week infant cries: Some relationships to the sudden infant death syndrome. In T. Murry & J. Murry (Eds.), *Infant communication: Cry and early speech.* Houston: College Hill Press, 1980.

Darwin, C. *The expression of emotion in man and animals.* New York: Philosophical Library, 1855.

Davis, H. Excitation of auditory receptors. In J. Field (Ed.), *Handbook of Physiology: Section I. Neurophysiology* (Vol. 1). Baltimore: Williams & Wilkins, 1959.

Emde, R. N., Gaensbauer, T. J., & Harmon, R. J. Emotional expression in infancy: A biobehavioral study. *Psychological Issues*, 1976, *X*(1), Monograph 37.

Fairbanks, G. An acoustical study of the pitch of infant hunger wails. *Child Development*, 1942, *13*, 227–232.

Fisichelli, V., & Karelitz, S. The cry latencies of normal infants and those with brain damage. *The Journal of Pediatrics*, 1963, *62*, 724–734.

Flateau, T. S., & Gutzmann, H. Die Stimme des Saüglings. *Archiv für Laryngologie und Rhinologie*, 1906, *18*, 139–151.

Formby, D. Maternal recognition of infant's cry. *Developmental Medicine and Child Neurology*, 1967, *9*, 293–298.

Freudenberg, R. P., Driscoll, J. W., & Stern, G. S. Reactions of adult humans to cries of normal and abnormal infants. *Infant Behavior and Development*, 1978, *1*, 224–227.

Frodi, A. M., & Lamb, M. E. Child abusers' responses to infant smiles and cries. *Child Development*, 1980, *51*, 238–241.

Frodi, A. M., Lamb, M. E., Leavitt, L. A., & Donovan, W. L. Fathers' and mothers' responses to infant cries and smiles. *Infant Behavior and Development*, 1978, *1*, 187–198.

Frodi, A. M., Lamb, M. E., Leavitt, L. A., Donovan, W. L., Neff, C., & Sherry, D. Fathers' and mothers' responses to the faces and cries of normal and premature infants. *Developmental Psychology*, 1978, *14*, 490–498.

Gardiner, W. *The music of nature.* Boston: Wilkins & Carter, 1838.

Gewirtz, J. L., & Boyd, E. Does maternal responding imply reduced infant crying? A critique of the 1972 Bell and Ainsworth report. *Child Development*, 1977, *48*, 1200–1207.

Gil, D. G. *Violence against children: Physical child abuse in the United States.* Cambridge, Mass.: Harvard, 1970.

Goldberg, S. Some biological aspects of early parent–infant interaction. In S. G. Moore & C. R. Cooper (Eds.), *The young child: Reviews of research.* Washington, D. C.: National Association for the Education of Young Children, 1982.

Golub, H. L. A physioacoustic model of the infant cry and its use for medical diagnosis and prognosis. In J. J. Wolf & D. H. Klatt (Eds.), *Speech communication papers presented at the 97th meeting of the Acoustical Society of America.* 1979.

Golub, H. L., & Corwin, M. J. Infant cry: A clue to diagnosis. *Pediatrics*, 1982, *69*, 197–201.

Graham, F., & Clifton, R. Heart rate change as a component of the orienting reflex. *Psychological Bulletin*, 1966, *65*, 305–320.

Greenberg, M., Rosenberg, I., & Lind, J. First mothers rooming in with their newborns: Its impact upon the mother. *American Journal of Orthopsychiatry*, 1973, *43*, 783–788.

Hopkins, J. B. *A comparative study of sensorimotor development during the first six months of life.* Unpublished doctoral dissertation, University of Leeds, 1976.

Howard, I. P. The spatial senses. In E. C. Carerette & M. P. Friedman (Eds.), *Handbook of perception: Vol 3. Biology of perceptual systems.* New York: Academic Press, 1973.

Juntunen, K., Servio, P., & Michelsson, K. Cry analysis in infants with severe malnutrition. *European Journal of Pediatrics*, 1978, *128*, 241–246.

Karelitz, S., & Fisichelli, V. R. The cry thresholds of normal infants and those with brain damage. *Journal of Pediatrics*, 1962, *61*, 679–685.

Karelitz, S., Fisichelli, V. R., Costa, J., Karelitz, R., & Rosenfeld, L. Relation of crying activity in early infancy to speech and intellectual development at age three years. *Child Development*, 1964, *35*, 769–777.

Katona, P. G., & Jih, F. Respiratory sinus arrhythmia: Non-invasive measure of parasympathetic cardiac control. *Journal of Applied Physiology*, 1975, *39*, 5.

Kessen, W., Levine, J., & Wendrich, K. A. The imitation of pitch in infants. *Infant Behavior and Development*, 1979, *2*, 93–99.

Konner, M. Infancy among the Kalahari Desert San. In H. Leiderman, S. Tulkin, & A. Rosenfeld (Eds.), *Culture and infancy*. New York: Academic Press, 1977.

Korner, A., & Grobstein, R. Visual alertness as related to soothing in neonates: Implications for maternal stimulation and early deprivation. *Child Development*, 1966, *37*, 867–876.

Lenneberg, E. H. *Biological foundations of language*. New York: Wiley, 1967.

Lester, B. M. Spectrum analysis of the cry sounds of well-nourished and malnourished infants. *Child Development*, 1976, *47*, 237–241.

Lester, B. M. The organization of crying in the neonate. *Journal of Pediatric Psychology*, 1978, *3*, 122–130.

Lester, B. M. Behavioral assessment of the neonate. In E. S. Sell (Ed.), *Follow-up of the high-risk newborn: A practical approach*. Springfield, Ill.: Charles C Thomas, 1979. (a)

Lester, B. M. A synergistic process approach to the study of prenatal malnutrition. *International Journal of Behavioral Development*, 1979, *2*, 377–393. (b)

Lester, B. M., & Zeskind, P. S. Brazelton scale and physical size correlates of neonatal cry features. *Infant Behavior and Development*, 1978, *1*, 393–402.

Lester, B. M., & Zeskind, P. S. The organization and assessment of crying in the infant at risk. In T. M. Field, A. M. Sostek, S. Goldberg, & H. H. Shuman (Eds.), *Infants born at risk*. New York: Spectrum, 1979.

Lieberman, P. *Intonation, perception and language*. Cambridge, Mass.: MIT Press, 1967.

Lieberman, P., Harris, K. S., Wolff, P., & Russell, L. H. Newborn infant cry and non-human primate vocalization. *Journal of Speech and Hearing Research*, 1971, *14*, 718–727.

Light, R. S. Abused and neglected children in America: A study of alternative policies. *Harvard Educational Review*, 1974, *9*, 198–240.

Lind, J. (Ed.). Newborn infant cry. *Acta Pediatrica Scandinavia*, Supplement 163, 1965.

Lind, J., Wasz-Hockert, O., Vuorenkoski, V., Partanen, T. J., Theorell, K., & Valanne, E. Vocal response to painful stimuli in newborn and young infants. *Annales Paediatriae Fenniae*, 1966, *12*, 55–63.

Lind, J., Vuorenkoski, V., Rosberg, G., Partanen, T. J., & Wasz-Hockert, O. Spectrographic analysis of vocal response to pain stimuli in infants with Down's syndrome. *Developmental Medicine & Child Neurology*, 1970, *12*, 478–486.

Lounsbury, M. L. *Acoustic properties of and maternal reactions to infant cries as a function of infant temperament*. Unpublished doctoral dissertation, Indiana University, Bloomington, Indiana, 1978.

Lynip, A. The use of magnetic devices in the collection and analysis of the preverbal utterances of an infant. *Genetic Psychology Monographs*, 1951, *44*, 221–262.

Masterton, B., & Diamond, I. T. Hearing: Central neural mechanisms. In E. C. Carterette & M. P. Friedman (Eds.), *Handbook of perception. Vol. 3. Biology of perceptual systems*. New York: Academic Press, 1973.

Michelsson, K. Cry analysis of symptomless low birth weight neonates and of asphyxiated newborn infants. *Acta Paediatrica Scandinavica,* Supplement 216, 1971.

Michelsson, K., Sirvio, P., & Wasz-Hockert, O. Sound spectrographic cry analysis of infants with bacterial meningitis. *Developmental Medicine and Child Neurology*, 1977, *19*, 309–315. (a)

Michelsson, K., Sirvio, P., & Wasz-Hockert, O. Pain cry in full-term asphyxiated newborn infants correlated with late findings. *Acta Paediatrica Scandinavica*, 1977, *66*, 611–616. (b)

Miller, H., & Hassanein, K. Fetal malnutrition in white newborn infants: Maternal factors. *Pediatrics*, 1973, *52*, 504–512.

Morath, M. Differences in the non-crying vocalizations of infants in the first four months of life. *Neuropaediatrie*, 1977, *8*, 543–545.

Morsbach, G., & Bunting, C. Maternal recognition of their neonates' cries. *Developmental Medicine and Child Neurology*, 1979, *21*, 178–185.

Morton, E. S. On the occurrence and significance of motivation-structural rules in some bird and mammal sounds. *American Naturalist*, 1977, *111*, 855–869.

Muller, E., Hollien, H., & Murray, T. Perceptual responses to infant crying: Identification of cry types. *Journal of Child Language*, 1974, *1*, 89–95.

Murray, A. D. Infant crying as an elicitor of parental behavior: An examination of two models. *Psychological Bulletin*, 1979, *86*, 191–215.

Naeye, R. L., Messmer, J., Sprecht, R., & Merrit, T. A. Sudden infant death syndrome temperament before death. *Journal of Pediatrics*, 1976, *88*, 511–515.

Ostwald, P. The sounds of infancy. *Developmental Medicine and Child Neurology*, 1972, *14*, 350–361.

Ostwald, P., Freedman, D. G., & Kurtz, S. H. Vocalization of infant twins. *Folia Phoniatrica*, 1962, *14*, 37–50.

Ostwald, P., Phibbs, R., & Fox, S. Diagnostic use of infant cry. *Biology of the Neonate*, 1968, *13*, 68–82.

Ostwald, P., Peltzman, P., Greenberg, M., & Meyer, J. Cries of a trisomy 13–15 infant. *Developmental Medicine and Child Neurology*, 1970, *12*, 472–477.

Painter, C. *An introduction to instrumental phonetics*. Baltimore: University Park Press, 1979.

Parke, R., & Collmer, C. W. Child abuse: An interdisciplinary analysis. In E. M. Hetherington (Ed.), *Review of child development* (Vol. 5). Chicago: University of Chicago Press, 1975, 509–590.

Parmelee, A. H. Infant crying and neurologic diagnosis. *Journal of Pediatrics*, 1962, *13*, 68–82.

Partanen, T. J., Wasz-Hockert, O., Vuorenkoski, V., Theorell, K., Valanne, E. H., & Lind, J. Auditory identification of pain cry signals of young infants in pathological conditions and its sound spectrographic basis. *Annales Paediatriae Fenniae*, 1967, *13*, 56–63.

Porges, S., Borher, R., Keren, G., Cheung, M., Franks, G., & Drasgow, F. The influence of methylphenidate on spontaneous autonomic activity and behavior in children diagnosed as hyperactive. *Psychophysiology*, 1981, *18*, 42–28.

Prechtl, H. F. R. Neurological findings in newborn infants after pre- and para-natal complications. In J. Jonix, H. Visser, & J. Troelstra (Eds.), *Aspects of prematurity and dysmaturity: A nutricia symposium*. Leiden: Stenfert Kroese, 1968, pp. 303–321.

Prechtl, H. F. R. The neurological examination of the full-term newborn infant. *Clinics in Developmental Medicine*, No. 63. Spastics International Medical Publications. Philadelphia: Lippincott, 1977.

Prechtl, H. F. R., Theorell, K., Gramsbergen, A., & Lind, J. A statistical analysis of cry patterns in normal and abnormal newborn infants. *Developmental Medicine and Child Neurology*, 1969, *11*, 142–152.

Ramey, C. T., Heiger, L., & Klisz, D. Synchronous reinforcement of vocal responses in failure to thrive infants. *Child Development*, 1972, *43*, 1449–1455.

Rebelsky, R., & Black, R. Crying in infancy. *The Journal of Genetic Psychology*, 1972, *121*, 49–57.

Ringel, R. L., & Kluppel, D. D. Neonatal crying: A normative study. *Folia Phoniatrica*, 1964, *16*, 1–9.

Sagi, A., & Hoffman, M. L. Empathic distress in the newborn. *Developmental Psychology*, 1976, *12*, 175–176.

Sander, L., & Julia, H. Continuous interactional monitoring in the neonate. *Psychosomatic Medicine*, 1966, *28*, 822–835.

Scherer, K. R. Nonlinguistic vocal indicators of emotion and psychopathology. In C. E. Izard (Ed.), *Emotions in personality and psychopathology*. New York: Plenum Press, 1979, pp. 495–529.

Sherman, M. The differentiation of emotional responses in infants. *Comparative Psychology*, 1927, *7*, 335–351.

Simner, M. L. Newborn's responses to the cry of another infant. *Developmental Psychology*, 1971, *5*, 136–150.

Sirvio, P., & Michelsson, K. Sound spectrographic cry analysis of normal and abnormal newborn infants. *Folia Phoniatrica*, 1976, *28*, 161–173.

Stark, R. E., & Nathanson, S. Unusual features of cry in an infant dying suddenly and unexpectedly. In J. F. Bosma & J. Showacre (Eds.), *Development of upper respiratory anatomy and function*. Washington, D.C.: U. S. Department of Health, Education and Welfare, Publication No. 75-941, 1975.

Testerman, R. L. Modulation of laryngeal activity by pulmonary changes during vocalization in cats. *Experimental Neurology*, *29*, 281–297.

Truby, H. M., & Lind, J. Cry sounds of the newborn infant. In J. Lind (Ed.), Newborn infant cry. *Acta Paediatrica Scandinavica*, Supplement 163, 1965.

Vuorenkoski, V., Lind, J., Partanen, T. J., Lejeune, J., LaFourcade, J., & Wasz-Hockert, O. Spectrographic analysis of cries from children with maladie du cri du chat. *Annales Paediatriae Fenniae*, 1966, *12*, 174–180.

Vuorenkoski, V., Wasz-Hockert, O., Koivisto, E., & Lind, J. The effect of cry stimulus on the temperature of the lactating breast of primiparas: A thermographic study. *Experientia*, 1969, *25*, 1286–1287.

Vuorenkoski, V., Lind, J., Wasz-Hockert, O., & Partanen, T. J. Cry score: A method for evaluating the degree of abnormality in the pain cry response of the newborn and young infant. *Speech Transmission Laboratory Progress Report*, 1971, *1*, 68–75.

Vuorenkoski, V., Wasz-Hockert, O., Lind, J., Koivisto, M., & Partanen, T. J. *Training the auditory perception of some specific types of the abnormal pain cry in newborn and young infants*. Progress report, Speech Transmission Laboratory, Royal Institute of Technology, Stockholm, Sweden, 1971.

Wasz-Hockert, O., Lind, J., Vuorenkoski, V., Partanen, T., & Valanne, E. The infant cry. *Clinics in Developmental Medicine*. Spastics International Medical Publications. London: Lavenham Press, 1968.

Wasz-Hockert, O., Koivisto, M., Vuorenkoski, V., Partanen, T. J., & Lind, J. Spectrographic analysis of pain cry in hyperbilirubinemia. *Biologia Neonatorum*, 1971, *17*, 260–271.

Wenger, M., & Cullen, T. Studies of autonomic balance in children and adults. In N. Greenfield & R. Sternbach (Eds.), *Handbook of psychophysiology*, New York: Holt, Rinehart & Winston, 1972, 535–570.

Wolff, P. The natural history of crying and other vocalizations in infancy. In B. M. Foss (Ed.), *Determinants of infant behavior (IV)*. London: Methuen, 1969.

Zeskind, P. S. *Differences in the cries and behaviors of infants varying in fetal growth*. Unpublished doctoral dissertation, University of North Carolina, Chapel Hill, 1980. (a)

Zeskind, P. S. Adult responses to cries of low and high risk infants. *Infant Behavior and Development*, 1980, *3*, 167–177. (b)

Zeskind, P. S., & Lester, B. M. Acoustic features and auditory perceptions of the cries of newborns with prenatal and perinatal complications. *Child Development*, 1978, *49*, 580–589.

Zeskind, P. S., & Lester, B. M. Analysis of cry features in newborns with differential fetal growth. *Child Development*, 1981, *52*, 207–212.

4

The Role of Physical Appearance in Infant and Child Development

KATHERINE A. HILDEBRANDT

1. INTRODUCTION

Physical appearance cues convey much information about another person during a social encounter. Not only are such characteristics as gender, age, attractiveness, and body type immediately obvious, but knowledge of these characteristics may elicit a host of assumptions about the person's personality characteristics and expected behavior patterns. Recent social psychological research (Adams, 1977; Berscheid & Walster, 1974) attests to the salience and power of physical appearance in social interactions among adults. Our society's preoccupation with physical appearance, as reflected by the cosmetic, diet, and fashion industries, also points to the significance of this variable in everyday life.

The landmark review of the physical attractiveness literature by Berscheid and Walster in 1974 established physical attractiveness as a legitimate and promising focus of scientific research on interpersonal relations. Although their paper emphasized adulthood, a few studies were described which indicated that the physical attractiveness of a child could influence interactions with peers and adults. More recently, Adams (1977) reviewed the physical attractiveness literature from a life-span perspective, providing both an updating of Berscheid and Walster's review and a theoretical model of the process by which physical attractiveness can influence development. Adams's model will be employed here to structure consideration of the role of physical appearance (especially facial attractiveness and body type) in infancy and childhood. The model will then be extended

Katherine A. Hildebrandt, Ph.D. • Department of Psychology, State University of New York at Buffalo, 4230 Ridge Lea Road, Buffalo, New York 14226. Preparation of this manuscript was supported in part by a grant from the University Awards Committee, State University of New York.

to support the suggestion that some physically atypical children encounter difficulties in social interaction due at least in part to their physical appearance, rather than as a simple result of their functional disabilities. Suggestions for the alleviation of the adverse effects of undesirable or atypical physical appearance will also be considered.

The information provided in this chapter should be particularly relevant to parents, teachers, day-care workers, pediatricians, nurses, and other adults who have frequent contact with infants and children. Since the effects of physical appearance can be subtle, developing an awareness of the potential biasing effect of children's appearance on adults' expectations and behavior is an important first step in assuring fair treatment and optimal development of all children.

Adams's major premise is that outer appearance and inner behavioral processes become interdependently related through four reciprocally interacting mechanisms. First, variations in appearance are assumed to elicit different expectations from others. Second, as a result of these expectations, people behave differently toward individuals who vary in appearance. Third, as a result of being perceived and treated differently, individuals develop differing self-concepts and personality attributes. Finally, these internal characteristics are manifested in different overt behavior patterns.

The evidence for each of these processes during infancy and childhood will be reviewed in turn. However, it is first necessary to consider the discriminability of appearance variations in infants and children.

2. DIFFERENTIATION OF CHILDREN VARYING IN APPEARANCE

In order for a particular variation in physical appearance to influence development, it must be discriminable. Thus, the first topic to be addressed is the differentiation of children on the basis of their facial and bodily appearance. The primary issue is whether or not these differentiations can be made reliably. Although many physical features are objectively measurable, individual judgments of the attractiveness or likableness of a child with particular features may be entirely subjective. If this is the case, we can expect few consistent differences in perceptions or treatment of children with specifiable physical characteristics. On the other hand, if these judgments can be made reliably, we can justifiably proceed to the questions of whether or not different-appearing children are perceived or treated differently.

2.1. Judgments of Physical Attractiveness

The stimuli used to elicit judgments of physical attractiveness are typically facial photographs. Only a few studies have used full-body photographs or have obtained live judgments. The judgments have taken the form of ratings, rank orderings, and paired comparison choices. The range of attractiveness in the stimuli, the method of obtaining judgments, and the method of calculating reliability all can influence the level of agreement obtained from judges (e.g., Sugarman, 1979). Nonetheless, studies which report adults' judgments of children's attractiveness (either as a focus of the study or as part of the procedure for selecting stimuli or assigning children to groups) have found moderate to high reliability among judges (Dion, 1973, $r = .81$; Langlois & Downs, 1979, $W = .46$; LaVoie & Andrews, 1976, $r = .72$; Marwit, Marwit, & Walker, 1978, $r = .96$; Salvia, Sheare, & Algozzine, 1975, 36–69% agreement; Styczynski & Langlois, 1977, $W = .42$; Trnavsky & Bakeman, 1976, 84% agreement). In addition, Lerner and Lerner (1977) found significant agreement between live ratings and ratings made from photographs.

Children's judgments of the attractiveness of other children also have been found to agree well with judgments by adults (Cross & Cross, 1971; Langlois & Stephen, 1977). Even preschool children agreed with adults when asked to choose which of a pair of age-mates they considered to be cuter, prettier, or more handsome (Dion, 1973; Styczynski & Langlois, 1977). Each pair contained a child of high attractiveness and one of low attractiveness as judged by a group of adults.

When a more complex ranking procedure was used by Cavior and Lombardi (1973), kindergarten children neither agreed among themselves nor with older children. This failure could have been due to the children's inability to understand the rank ordering test, or to the fact that they were asked to judge substantially older children (Dion, 1973). Inter-judge reliabilities were significant for slightly older children, and agreement with rankings made by fifth- and eleventh-grade children improved with age from kindergarten to third grade (Cavior & Lombardi, 1973). Dion (1973) suggested that children first learn to discriminate extreme levels of attractiveness of children their own age before developing a broader concept of physical attractiveness standards.

A few very recent studies have tested adults' abilities to judge reliably the attractiveness of infants. Contrary to the expectations of those (typically nonparents) who believe all infants look alike, significant reliabilities have been obtained. For example, Corter, Trehub, Boukydis, Ford, Celhoffer, and Minde (1978) reported that nurses reliably agreed on the attractiveness of premature infants, whether or not they regularly worked with prematures. Hildebrandt and Fitzgerald conducted a series of studies in which subjects were asked to judge infant photographs on a "cuteness" dimension. College students showed significant agreement in rank-ordering photographs of 4- and 8-month-old infants (Hildebrandt & Fitzgerald, 1978), and the ratings given a group of 60 photographs of infants ranging from 3 to 13 months were highly correlated across different groups of students (Hildebrandt & Fitzgerald, 1979a). Live ratings of 3-month-old infants also reflected substantial agreement among college student judges (Hildebrandt,

1980). Mothers of toddlers agreed with college women on rankings and ratings of unfamiliar infants (Hildebrandt & Fitzgerald, 1981).

2.2. Influences on Judgments of Attractiveness

Judgments of attractiveness potentially can be influenced by characteristics of the judges and characteristics of the individuals being judged. The effects of age, sex, familiarity, and physical characteristics will be considered.

2.2.1. Age

Although the reliability of judgments has been shown to vary with age of judge, the actual magnitude or pattern of judgments does not appear to be influenced by judge's age (e.g., Cross & Cross, 1971). In addition, adults' ratings do not vary with the age of child being rated, at least between the ages of 3 and 9 years (LaVoie & Andrews, 1976). Their ratings of infants, however, are influenced by age. Hildebrandt and Fitzgerald (1979a) reported that within the range of 3 to 13 months, 11-month-old infants received the highest cuteness ratings. Ratings increased from 3 months to a peak at 11 months, followed by a slight drop-off at 13 months. Age interacted with sex, however, since female infants reached the "peak" of rated cuteness earlier (at 9 months) than males (at 11 months). This was hypothesized to reflect the relative developmental maturity of female infants compared with same-age males.

2.2.2. Sex

The sex of the judge frequently has been found to have little or no effect on attractiveness judgments (Cavior & Dokecki, 1973; Cavior & Lombardi, 1973; Dion, 1973; Styczinski & Langlois, 1977). The only report of a simple and consistent effect of judges' sex was a study by Cross and Cross (1971), where female judges of various ages gave higher ratings to both 7- and 17-month-old faces than male judges did. Another type of sex difference appeared in the Hildebrandt and Fitzgerald (1979a) study. Although there was no difference in the average cuteness ratings obtained from male and female judges, the female judges used the ends of the rating scale more, which may have reflected better discrimination among the infant stimuli.

Attractiveness judgments also can be influenced by the sex of the person being judged. Some investigators simply avoid this issue by having male and female stimuli ranked or rated separately (e.g., Cavior & Dokecki, 1973). Moreover, the label for the physical attractiveness dimension has differed in some cases for judgments of boys and girls, making comparability of the ratings difficult. For example, in Dion's (1973) study, children chose which girl in a pair was "prettier" and which boy in a pair was "cuter." Styczynski and Langlois (1977) labeled girls as "prettier or cuter" and boys as "cuter or more handsome." Neither of these studies report any effect of the sex of the stimulus child on attractiveness ratings. The most obvious explanation for these findings is that there is no difference in the perceived attractiveness of male and female children and that the exact term used to describe the attractiveness dimension is irrelevant. This interpretation is supported by studies which report no sex of stimulus effect when the same label is used for judgments of boys and girls (LaVoie & Andrews, 1976; Lerner & Lerner, 1977), and by the finding that, for rating infants at least, it makes no difference whether the rating scale is labeled according to "cuteness" or "physical attractiveness" (Hildebrandt & Fitzgerald, unpublished data).

Only Cross and Cross (1971) have reported a significant sex difference in ratings, with girls generally receiving higher ratings than boys (the effect was complicated by interactions with the sex and race of the judges and the age and race of the stimuli). However, the possibility that there was a similar sex difference in relative *perceptions* of the attractiveness of girls and boys in the studies reported above that was obscured in the ratings by the operation of an attractiveness sex stereotype is suggested by Hildebrandt and Fitzgerald's (1979a) study of adults' ratings of infant cuteness. Because adults were not very accurate at guessing the sex of the stimulus infants (Hildebrandt & Fitzgerald, 1977), it was possible to look at differences in ratings when the infant's sex was known and when it was not known. When the sex was not known, female infants tended to receive slightly higher ratings than males. In addition, the cuter infants were more likely to be guessed to be female than were the less cute infants. When the sex was known (by telling the subjects either the correct or incorrect sex), labeled males received significantly higher ratings than labeled females. This finding was interpreted as evidence for a sex stereotype in cuteness ratings: females are expected to be cuter than males, so when an "average" infant is labeled male, it is perceived as slightly cuter than the average male, and when it is labeled female, it is perceived as slightly less cute than the average female. Since the sexes of the stimulus children in the studies reported above were probably known to the judges, because of hair or clothing cues, even if they perceived the girls to be more attractive than the boys, the operation of this attractiveness sex stereotype could have lead to the lowering of their actual ratings of the girls and raising of their actual ratings of the boys.

Further support for sex stereotyping in judgments of physical appearance can be found in a study of parents' responses to their newborn infants (Rubin, Provenzano, & Luria, 1974). The primary source of information these parents had about their infants was their sex, although most of them had also had some brief contact with the baby. On an adjective scale, daughters were rated as significantly softer, finer featured, and littler than sons. When asked to describe their infants, the terms "beautiful," "pretty," and "cute" were more frequently used by parents of daughters than by parents of sons. This finding seems to contradict that of Hildebrandt and Fitzgerald (1979a); however, Rubin *et al.* suggested that the parents were describing their infants in terms of their expectations for an infant of a particular sex rather than the actual features of their infants. There were no sex differences in the infants' birth weight, birth length, or Apgar score.

2.2.3. *Familiarity*

This last study raises the issue of the influence of familiarity on attractiveness judgments. Although most studies of physical attractiveness involve first impressions (thus, unfamiliar stimuli), a few have investigated judgments of familiar individuals. For example, Styczynski and Langlois (1977) found that preschool children's choices of attractive agemates were not affected by whether they knew the children in the stimulus pairs. The results of a study of fifth- and eleventh-graders by Cavior and Dokecki (1973), however, suggest that one reason no difference was found was that judgments were being made of children at the extremes of the attractiveness dimension. Although the rankings by knowers and nonknowers in their study were generally significantly correlated, there was less agreement on the rankings of individuals of average attractiveness than on individuals of high or low attractiveness. Moreover, there was slightly better agreement among nonknowers than among knowers, and knowers gave higher overall ratings than nonknowers.

Other studies also suggest that familiarity raises attractiveness ratings. Corter *et al.* (1978) found that nurses who worked with premature infants gave higher ratings to unfamiliar premature infants than did nurses who worked with full-term newborns, and a particular premature infant was ranked higher within a standard set by nurses who had actually worked with him or her than by other nurses in the same premature ward. Two additional studies (Hildebrandt, 1980; Hildebrandt & Fitzgerald, 1981) found that mothers judged their own infants to be cuter than either college students or other mothers judged them to be. There was an element of objectivity in the mothers' judgments, though; in both studies the mothers' judgments were moderately but significantly correlated with those of the college students. Thus, if college students gave a particular infant an above-average cuteness rating, the infant's mother was likely to rate him or her as very cute, whereas an infant rated below average by college students would likely be rated average or only slightly above by his or her mother.

2.2.4. *Physical Features*

Since there is at least some agreement among judges about who is attractive and who is not, we can proceed to ask what characteristics make one infant or child more attractive than another. This question has not been systematically studied for judgments of children, although Kleck, Richardson, and Ronald (1974) asked 9- to 14-year old boys why they chose particular photographs of unfamiliar male age-mates as more attractive. The following features (and the percent of the sample who named each) were mentioned: configuration of the face (73%), hair (53%), nonfacial cues (53%), affect cues (53%), eyes (47%), teeth (47%), mouth (40%), lips (26%), ears (20%), and action tendencies (20%). However, no attempt was made to determine whether any of the objectively measurable features were actually related to attractiveness judgments.

Tangential evidence that hair style influences attractiveness comes from a study by Dion (1974) in which children's attractiveness was manipulated. One of the ways in which the girl in the study was made unattractive was to pull her hair back from her face. Other manipulations included the use of makeup. Whether or not the hair style change was a primary factor is unknown, but a manipulation check attested to the effectiveness of the group of changes in creating an unattractive appearance.

Race, cleanliness, and clothing are also potential influences on attractiveness judgments. Although Cross and Cross (1971) reported that white children received higher attractiveness ratings than did black children from children and adults of both races, Epstein, Krufat, and Obudho (1976) found that facial cleanliness had a stronger influence than race on the preferences of second-, third-, and fourth-grade students for facial photographs. Attractiveness ratings of fourth-grade children, but not kindergarten children, were found by Langlois and Stephen (1977) to be higher for stimulus children of the same race as the judges. However, judgments of liking were more strongly influenced by attractiveness than race. No systematic information is available on the influence of children's clothing on judgments of their attractiveness, although the role of clothing in impression formation has been discussed (Haley & Hendrickson, 1974).

An analysis similar to that of Kleck *et al.* (1974) was conducted to determine what facial features college students considered important in their judgments of infant cuteness (Hildebrandt, 1976). The features mentioned most frequently (and the number of students out of 43 who named each) were facial expression (32), eyes (32), hair (22), fatness (19), facial proportions (9), and ears (9). A further study was conducted to determine if cuteness ratings could be predicted from objectively measurable facial features (Hildebrandt & Fitzgerald, 1979b). A number of facial features were measured from slides and then entered into a multiple regression equation predicting cuteness. It was found that about half the variance in cuteness ratings could be accounted for by the measured facial features, and cuter infants were more likely to have short and narrow features, large eyes and pupils, and a large forehead. A third study in this series (Power, Hildebrandt, & Fitzgerald, 1982) indicated that cuteness ratings also are influenced by facial expression, with smiling infants receiving higher ratings than crying infants.

2.3. Judgments of Body Type

Efforts to classify individuals on the basis of body type appear to date back at least to Hippocrates (McCandless, 1967). In more recent times, the work of Sheldon (1940, 1942) represents one of the first attempts to measure and describe body configurations objectively. Sheldon primarily studied adolescent and adult men, and his complicated method of "somatotyping" rarely has been used in studies of children (see Walker, 1962, 1963, for exceptions). The three dimensions on which he rated individuals were endomorphy (fatness), mesomorphy (muscularity), and ectomorphy (slenderness). A particular body received a score on each dimension. Recent studies (e.g., Lerner & Gellert, 1969) have simplified the system by categorizing children as chubby, average, or thin, although the terms *endomorphic, mesomorphic,* and *ectomorphic* are still frequently used to refer to these general categories.

Adults appear to be able to discriminate reliably among children on the basis of body type. Lerner and Gellert (1969) reported that the body type classifications of a group of kindergartners made by experimenters and teachers were significantly correlated. Even using a rating system similar to that of Sheldon, high reliability among adults rating preschool children has been obtained (Walker, 1962).

Children from as young as 5 years of age also show that they discriminate among the three basic body types by differentially choosing body type stimuli in response to questions about preferences and attributions of behavioral and personality characteristics (Brenner & Hinsdale, 1978; Lerner & Korn, 1972; Lerner & Schroeder, 1971; Staffieri, 1967, 1972; Young & Avdzej, 1979). An aversion toward the chubby body type is especially evident. Children overchoose a chubby figure when asked which of three silhouette figures they would not want to look like (Lerner & Gellert, 1969; Lerner & Korn, 1972; Lerner & Schroeder, 1971). The salience of chubbiness for even young children is further reflected in the finding by Lerner and Gellert (1969) that kindergartners could correctly name a peer who was chubby, but only girls were accurate above chance at identifying average and thin peers.

When children are asked which body type they would prefer to look like, they show an overall preference for mesomorphy (Brenner & Hinsdale, 1978; Lerner & Gellert, 1969; Lerner & Korn, 1972; Lerner & Schroeder, 1971; Staffieri, 1967, 1972). However, the ectomorphic body type is often chosen by a substantial minority of the children, while the endomorph is almost never chosen. This occasional tendency to prefer ectomorphy over mesomorphy is more evident in girls than in boys (Staffieri, 1967, 1972). Ectomorphy may be desirable to girls because of exposure to thin fashion models, and mesomorphy may be desirable to boys because of the importance of muscularity to successful performance in sports. In any event, these studies indicate that children differentiate chubby and nonchubby body types by at least kindergarten age, and they differentiate between average and thin body types by the middle of elementary school.

2.4. Summary

Adults have been shown to be capable of discriminating reliably among children on the dimensions of facial attractiveness and body type. Evidence suggests that familiarity raises adults' ratings of children's attractiveness, and that adults' ratings are influenced by such physical characteristics of children as facial features, race, cleanliness, clothing, and hair style. Older infants receive higher ratings than younger infants, but ratings of children do not vary with age. Ratings of male and female children typically do not differ, although it has been suggested that girls are perceived as more attractive than boys but are rated equally due to the operation of a physical attractiveness sex stereotype which implies a higher standard of attractiveness for girls than for boys.

Children can discriminate variations in attractiveness and body type among their peers, although preschool and kindergarten children only show the ability when the discrimination task is very simple and the appearance variations are obvious. Chubbiness can be identified at an earlier age than thinness. The reliability of children's judgments of attractiveness and body type improves with age. Sex differences rarely are reported.

3. STEREOTYPING OF DIFFERENT-APPEARING CHILDREN

Adams (1977) proposed that the most central assumption underlying the development of a relationship between outer appearance and psychological characteristics is that different appearances elicit different expectations from others. If expectations are systematically related to particular appearance variations, the existence of a stereotype is implied. If the stereotype functions in some way to produce behavior congruent with expectations, the operation of a self-fulfilling prophecy is evident. In this section we will consider the evidence supporting the existence of stereotypes based on physical appearance.

3.1. Adults' Perceptions of Children Varying in Appearance

Much of the research on self-fulfilling prophecies has been conducted within an educational framework, especially since the appearance of Rosenthal and Jacobson's (1968) book describing the effects of experimenter-manipulated teacher expectancies on pupils' academic performance. Although failures to replicate Rosenthal and Jacobson's findings have been reported (cited by Kehle, Bramble, & Mason, 1974), further research on the elicitation of differential teacher expectancies by various student characteristics, including physical attractiveness, has proliferated. Although differential expectancies have been found in many of these studies, whether they have consequences for the teachers' or their pupils' behavior is still open to question.

The typical paradigm for these studies involves asking a large number of teachers to evaluate the personal characteristics and expected future achievement of a child on the basis of information presented in a school folder containing a photograph (which apparently is a standard practice in many school systems). A number of student characteristics, physical attractiveness, gender, home background, IQ, race, school performance, and previous teachers' comments on the child's conduct or abilities, have been manipulated by varying the contents of the folder.

An unfortunate problem with some of these studies is the use of too few photographs. Several studies (Kehle *et al.,* 1974; Marwit *et al.,* 1978; Ross & Salvia, 1975) employed only one photograph per experimental condition, which severely restricted assessment of the highest order interaction and sometimes meant that only two photographs were used to represent each level of attractiveness. It is difficult in such cases to determine if the obtained effects are due to attractiveness variations or to idiosyncratic facial features. The other studies of this type reported here used at least three photographs per condition.

Clifford and Walster (1973) reported a strong effect of attractiveness on teacher expectations. An identical report card was paired with a photograph of either an attractive or an unattractive fifth-grade student. The 404 fifth-grade teachers in the study rated the attractive student as having a higher educational potential, a higher IQ, parents more interested in academic achievement, a greater likelihood of more future education, and better peer relations. A replication of the study with first-grade teachers (Clifford, 1975) revealed similar attractiveness effects on rated IQ, expected parents' attitudes toward achievement, and expected future education.

Several more studies have replicated the basic finding that teachers have higher expectations for and more favorable current perceptions of attractive children than they do for unattractive children, although in some cases attractiveness interacted with other student characteristics (Adams & Cohen, 1976a,b; Rich, 1975; Ross & Salvia, 1975). Parents also were found to perceive unfamiliar attractive children more favorably than unfamiliar unattractive children (Adams & LaVoie, 1975). Female college students evaluated an unattractive child who had committed a severe transgression more negatively than an attractive child who had committed the same transgression (Dion, 1972).

In general, these studies indicate that attractive children are perceived more favorably by adults than are unattractive children. This conclusion is consistent with the physical attractiveness stereotype reported in the adult literature. For example, Dion, Berscheid, and Walster (1972) found that attractive individuals depicted in photographs were rated more positively than unattractive individuals on a wide variety of dimensions. They suggest that these trait attributions are made following the general rule that "what is beautiful is good."

Before concluding that the physical attractiveness stereotype is universally applied when adults evaluate children, it is important to consider studies in which no attractiveness effect was found, or in which attractiveness affected perceptions in a seemingly contradictory fashion. In the first case, no attractiveness effect was found when female college students rated children who committed mild transgressions (Dion, 1972) or children who had not transgressed (Dion, 1974), when student teachers rated the severity of a child's transgression (Marwit *et al.,* 1978), when teachers rated a child's personality after having read the child's essay (Kehle *et al.,* 1974), and when teachers rated a child after being given information about the child's general conduct (LaVoie & Adams, 1974). In addition, few of the studies which report attractiveness effects found significant effects for all the dimensions assessed.

Results which seem to contradict the operation of a physical attractiveness stereotype include the findings that teachers rated the work habits of highly attractive children as lower than those of moderate and low attractive children (Adams & LaVoie, 1974), that college men rated unattractive children as more pleasant than attractive children (Dion, 1974), and that teachers rated as more severe the transgressions of attractive than unattractive children (Marwit *et al.,* 1978). Although it is not unreasonable to find three contradictory effects among all the studies reviewed here by chance alone, it is also possible that high attractiveness may be associated with some negative perceptions. One problem in comparing studies such as these is that it is not possible to ascertain whether the same range of attractiveness was used in all studies.

The large number of characteristics which were not affected by attractiveness also requires further examination. Adults may be influenced by children's attractiveness, in a manner consistent with the physical attractiveness stereotype, only when more concrete information about the child's personality, achievement, or conduct is lacking. Thus, the more ambiguous the information about the child the more likely it will be that an adult will be influenced by physical attractiveness. This hypothesized process may account for some, but not all, of the failures to find attractiveness effects. Further research of a more systematic nature will be needed to determine more accurately the effective limits of the physical attractiveness stereotype.

At least in some circumstances, attractive children appear to be perceived more positively by adults than are unattractive children, although this is by no means a universal effect. Before proceeding to the issue of whether adults treat children differently on the basis of attractiveness, we can ask if children hold a similar physical attractiveness stereotype.

3.2. Children's Perceptions of Unfamiliar Peers Varying in Attractiveness

The best evidence for a physical attractiveness stereotype in children comes from those studies in which children evaluate unfamiliar persons varying in attractiveness. Perceptions of familiar persons may reflect the stereotype, but are likely to be strongly influenced by the actual behavior of the stimulus persons. On the assumption that perceptions of familiar children are based primarily on exposure to those children, studies on this topic will be considered in the section on differences among children varying in appearance.

Studies of preschool children's evaluations of unfamiliar peers (Dion, 1973; Styczynski & Langlois, 1977; Trnavsky & Bakeman, 1976) have found that more attractive children are expected to behave more prosocially and less antisocially than unattractive children. In Dion's study, for example, attractive children were more frequently chosen when subjects were asked to pick a child who is "very friendly to other children," "doesn't like fighting or shouting," "doesn't hit, even if someone else hits first," and were less likely to be chosen as a child who "hits without a good reason," "scares you," and "might hurt you." Older children (kindergarten and fourth grade) also rated more attractive children as smarter, friendlier, nicer, and less mean than less attractive children in a study by Langlois and Stephan (1977). Attractiveness proved to have a stronger and more consistent influence than race, which was also manipulated in this study. Although main effects of attractiveness were found in all of these studies, interactions between attractiveness and other factors also occurred. However, findings contrary to those expected on the basis of the physical attractiveness stereotype were rare.

Attractive children are not only expected to behave prosocially, but they may be more likely than unattractive children to be chosen as potential friends. Although this effect was clearly evident in Dion's (1973) study and in Langlois and Stephen's (1977) study, it was more difficult to discern in the study by Styczynski and Langlois (1977). Here, the effect appeared only with one of the derived measures of popularity and only reached significance for boys.

Only one study was found which considered individual differences in the strength of the physical attractiveness stereotype. Trnavsky and Bakeman (1976) hypothesized and found that more attractive children held the attractiveness stereotype more strongly than less attractive children. Unfortunately, there have been no reported attempts to replicate this finding with older children.

3.3. Body Type Stereotypes

Stereotypes related to body type primarily have been studied by presenting subjects with photographs or drawings representing each of the three standard body types: endomorphic, mesomorphic, and ectomorphic. Then they are asked to assign each of a list of characteristics to one of the body types or, alternatively, to rate each body type stimulus on each characteristic. The findings from these studies have been consistent: positive traits are assigned to the mesomorphic stimulus and negative traits are assigned to the endomorphic and ectomorphic stimuli. This general pattern has been found with children and adolescents (Brenner & Hinsdale, 1978; Lerner, 1969; Lerner & Korn, 1972; Staffieri, 1967, 1972), mentally retarded adolescents (Staffieri, 1968), Japanese adults and adolescents (Iwawaki

& Lerner, 1974, 1976; Lerner & Iwawaki, 1975), and Mexican children (Lerner & Pool, 1972). In several of these studies the endomorphic stereotype was more markedly negative than the ectomorphic stereotype (Brenner & Hinsdale, 1978; Lerner & Korn, 1972; Staffieri, 1972).

A very recent study (Young & Avdzej, 1979) employed a somewhat more sophisticated assessment technique. Children viewed two videotapes of either obese or normal-weight boys interacting either obediently or disobediently with an adult. They then assigned adjectives to and chose a playmate from the two boys they had seen. Although an obese boy received a more negative evaluation and fewer friendship choices than a normal-weight boy did when their behavior was similar, an obedient obese boy was evaluated more favorably than a disobedient normal-weight boy. This finding implies that an individual's behavior can exert a stronger influence than a physical appearance stereotype. The consistency in the literature on body type stereotypes noted above may be due to the use of a procedure which provides no additional information about the stimulus person other than body type. If more information had been provided, which would have lessened the ambiguity of the situation, the stereotype might have proved less strong. This process appears to have lessened the impact of attractiveness on evaluations of children, as described in a previous section.

No age differences were obtained in Lerner's (1969) study of 10- to 20-year-old males, Young and Avdzej's (1979) study of third-, fourth-, and fifth-grade children, or in Lerner and Pool's (1972) study of fourth-, fifth-, and sixth-grade rural Mexican children, and no analyses for age effects are reported in Staffieri's (1972) study of 7- to 11-year-old girls. In Staffieri's (1967) study of 6- to 10-year-old boys (which also reports pilot data for 4- and 5-year-olds), the stereotypes were established by 6 years but were less evident at 4 and 5 years. Lerner and Korn's (1972) study of 6-, 15-, and 20-year-old males (which is discussed further by Lerner, 1972) revealed that, although the positive mesomorphic stereotype and negative ectomorphic and endomorphic stereotypes were present at all ages, the strength of the stereotypes was related to age. Between 6 and 15 years, the mesomorphic stereotype became increasingly positive and the endomorphic and ectomorphic stereotypes became more negative. This occurred because the adolescents assigned more good attributes to the mesomorph (and therefore fewer to the other body types) than did the younger children (Lerner, 1972). In addition, Lerner and Korn (1972) reported that the older subjects assigned a greater number of negative attributes to the endomorph than the younger subjects did. A replication of Lerner and Korn's (1972) study with females (Brenner & Hinsdale, 1978) confirmed this general finding. The youngest subjects had a more favorable view of the ectomorphic and endomorphic figures than did the older subjects.

The only evidence for a sex difference in body-build stereotypes comes from a comparison of Staffieri's studies (1967, 1972). School-age boys and girls differed in their perception of characteristics associated with ectomorphic body type. Girls attributed weakness and quietness to the ectomorph but otherwise did not associate ectomorphy with negative traits (Staffieri, 1972). Boys also attributed weakness and quietness to the ectomorph but additionally attributed other undesirable traits such as tendencies to be lonely, sneaky, afraid, and sad (Staffieri, 1967). Recall that there was also a substantial group of the girls who expressed a preference for ectomorphy, whereas only a few of the boys preferred ectomorphy over mesomorphy. A problem arises in interpreting the results of these two studies. The boys were shown male body-type stimuli and the girls were shown female body-type stimuli. Therefore it is unclear whether the difference in the stereotype is due to the sex difference between the stimuli or the sex difference between the raters. Unfortunately, insufficient research reports are available to clarify this issue. Although Lerner and Pool (1972) studied both male and female children's responses to the same stimuli, only adult male stimuli were used and the sample was obtained in a rural Mexican town. No sex differences were reported.

Observed cultural differences in body-build stereotypes have thus far been minor and related more to the strength of particular stereotypes than to their overall content. For example, Japanese adolescents rated ectomorphy and endomorphy similarly negatively (Lerner & Iwawaki, 1975), whereas American adolescents attributed a greater number of negative traits to the endomorphic stimulus than to the ectomorphic stimulus (Lerner & Korn, 1972). The data from Lerner and Pool's (1972) study of Mexican children were compared with 15-year-olds' data in the Lerner and Korn (1972) study. Although the direction of the stereotypes was the same in both groups, the difference in valence between mesomorphy and endomorphy was smaller for the Mexican children than for the American adolescents. More studies will be required to determine the universality of body-build stereotypes. The studies reported here suggest that it may be reasonable to expect some variation in the strength of the stereotype across cultures, although the general pattern should remain the same.

The final variable which has been assessed with respect to its relationship to body-build stereotypes is that of the body build of the individual holding the stereotype. One might expect that individuals whose body types are typically associated with a negative stereotype would hold a less negative stereotype for that particular body type. This does not seem to be the case. Lerner and Korn (1972), Brenner and Hinsdale (1978), and Staffieri (1967) categorized their subjects by body type and reported that there were no differences in the stereotypes held by children varying on this dimension.

3.4. Summary

From the information available, it appears that people do hold stereotypes related to the physical appearance of children. Both adults and children hold positive expectations and perceptions of attractive children as compared to unattractive children. This physical attractiveness stereotype appears to be particularly influential when little other concrete information about a stimulus child is available. Children perceive mesomorphic persons positively and endomorphic and ectomorphic persons negatively. This difference in stereotypes related to body types increases with age. Girls seem to perceive ectomorphy as less negative than boys perceive it.

4. DIFFERENTIAL TREATMENT OF CHILDREN VARYING IN APPEARANCE

Given that there are stereotyped expectancies associated with variations in children's physical appearance, the next important consideration is whether these stereotypes influence interactions with children. Evidence for this process will be reviewed in this section.

4.1. Teachers' Behavior toward Children Varying in Attractiveness

Very little research attention has been directed to the issue of teachers' treatment of children who vary in physical appearance. This is surprising in light of the number of studies comparing teachers' perceptions and expectations of attractive and unattractive children, but not so surprising when one considers the methodological difficulties inherent in directly observing teacher–student interactions.

Only one study (Adams & Cohen, 1974) has attempted directly to relate student attractiveness to teacher–student interactional behavior, and it is unfortunately limited to observations of only three teachers, each at a different grade level (kindergarten, fourth, and seventh grades). Interactions between male students and teachers were recorded on the third and fourth day of a new school year. Although a significant interaction between facial attractiveness and age suggested that teachers were more influenced by the attractiveness of older children, this interpretation is flawed by the confounding of student age with individual differences among teachers. (It should be noted that the teachers were matched on teaching style and some personality variables in an attempt to control individual difference variables.) All in all, this study provides little information about the effects of appearance on teachers' interactions with students.

Another observational study of teachers in the classroom reported a relationship between students' appearance and teacher behavior (Rist, 1970). In this case it was noted that certain children were consistently placed in low ability groups during kindergarten, first, and second grade. One of the dimensions on which these children differed from the children placed in the higher ability groups was physical appearance. The low ability children wore older and dirtier clothing and were more likely to have a very dark skin color and matted or unprocessed hair (all the children in the classes were black). Unfortunately, the children in the high and low ability groups also differed on at least three other dimensions: interactional behavior, use of language, and social status. It is therefore impossible to assess any independent influence of physical appearance on teacher behavior. This study illustrates a persistent problem in assessing appearance effects in a natural setting using simple correlational methodology. Certain aspects of appearance, if not all, are likely to have naturally occurring covariates which may in fact be causally related to any response behaviors observed. Attractiveness of clothing, hair style, and cleanliness are likely to be associated with social class, for example, and the possibility must be considered that even more physiologically based aspects of appearance (such as facial attractiveness and body type) have influential correlates.

The final study on this topic (Barocas & Black, 1974) reported that the more attractive children in four third-grade classrooms were more likely to have been referred by their teachers for supplemental psychological and educational services. The authors interpreted this finding as an indication that more attention is paid to attractive children and therefore their need of help is more likely to be noticed and responded to. The obvious alternative explanation, that the attractive children had more problems, was presumed to be unlikely. Barocas and Black observed that most of the referrals were of a "helpful" nature, rather than being in response to disciplinary problems, and suggested that unattractive students may be subject to more "controlling" referrals.

4.2. Other Adults' Behavior toward Children Varying in Attractiveness

None of the available studies of adults' responses to children varying in attractiveness has used direct observations of adult–child interactions. Rather, they have used laboratory simulations in which the children are represented in photographs or videotapes and only a limited number of behaviors (or reports of expected behaviors) are obtained from the adult subjects.

Dion (1972, 1974) conducted a series of studies of this type, directed at the issue of whether children varying in attractiveness are evaluated and punished differently for the same transgression. In the first study (Dion, 1972), female college students rated attractive and unattractive children who had committed either a mild or severe transgression, as described in a written paragraph. Although the results of the study were in line with the physical attractiveness stereotype (as described in the previous section), the one measure that tapped the potential behavior of the subjects toward a child who had transgressed did not show an attractiveness effect. The intensity of punishment advocated for children in the different attractiveness conditions did not vary. Dion proposed that a more detailed assessment of the type of punishment recommended for the different children (and not just the intensity) would reveal an attractiveness effect in line with the attributional inferences and trait ratings obtained.

Dion's second study (1974) more directly assessed the question of differential adult behavior toward children varying in attractiveness. In separate experiments with female and male college students, subjects first viewed a videotaped interaction between a child and an experimenter and then penalized the child for errors on a purported picture-matching task. The child viewed was either male or female and either attractive or unattractive. Women showed what Dion labeled a "cross-sex leniency effect mediated by attractiveness." The attractive male child was penalized less than either the unattractive male child or the attractive female child. The results for the men were different, with no sex or attractiveness effects on their use of penalties.

Dion's interpretation of these findings rested on the assumption that men are more task-oriented and women more interpersonally oriented in the type of situation she used in her study. The women would thus be expected to be more influenced than the men by such social cues as sex and attractiveness.

Another set of studies assessed adults' behavioral responses to photographs of infants varying in cuteness. In the first study (Hildebrandt & Fitzgerald, 1978), male and female college students viewed photographs of infants singly and in pairs, while measures were taken of their looking time, smiling (zygomaticus muscle activity), and skin conductance responses. Smiling was not related to cuteness, but was shown to occur in response to infants as opposed to other types of social and nonsocial photographs and to be more likely in women than in men, at least during the early presentations. Looking time differentiated the photographs during both single and paired presentations, with no sex difference. Infants rated cuter were looked at longer. These findings were interpreted as reflecting two processes in adults' responses to infant physical appearance: an initial positive response to "babyishness" and an individual cognitive preference response to "cuteness."

A subsequent study (Hildebrandt & Fitzgerald, 1981) was conducted with mothers. They also smiled at the infants unselectively and looked longer at infants they perceived as cuter. An additional feature of this study was that a photograph of each mother's own infant was included. The mothers looked longer at and showed a larger skin conductance response to their own infant than to the unfamiliar infants, and smiled more at their own infant than other mothers smiled at their own infants if their infants were especially cute. These reactions could have been mediated by perceived cuteness, since the majority of mothers judged their own infant to be cuter than the other infants they viewed, or by familiarity. Whatever the reason, most mothers responded positively to their own infants. The observation that both college students and mothers looked longer at unfamiliar infants perceived as cuter implies that infant attractiveness may elicit increased attention from caregivers, particularly in group care settings, where adults must decide how to distribute their attention among several or many equally familiar infants.

4.3. Children's Behavior toward Peers Varying in Attractiveness

There is no direct observational evidence available concerning children's differential treatment of attractive and unattractive peers. It has frequently been found that more attractive children have more friends and receive more social acceptance (Cavior & Dokecki, 1973; Cavior, Miller, & Cohen, 1975; Dion & Berscheid, 1974; Kleck *et al.,* 1974; Lerner & Lerner, 1977; Salvia *et al.,* 1975). The only exceptions to this general finding have occurred with young preschool children (Dion & Berscheid, 1974; Styczynski & Langlois, 1977). It can only be assumed that these popular, attractive children are treated favorably by their peers. These findings may also reflect different expectations for behavior from attractive and unattractive peers, as well as actual behavior differences between them.

An experimental study by Dion (1977) suggests that preschool children prefer to look at and are more rewarded by the view of an attractive peer. Children whose lever presses illuminated a photograph of an attractive child pressed a lever more than those shown an unattractive photograph. This finding was interpreted as support for the idea that physical attractiveness influences the behavior of others because it is intrinsically rewarding to the viewer. The study also supports the hypothesis that more attractive individuals receive more attention from others. Discovering whether or not the findings of these studies can be generalized to natural interpersonal interactions will require further investigations.

4.4. Behavior toward Children Varying in Body Type

Lerner and his colleagues conducted a series of studies on children's use of personal space with others who vary in body type. They used a felt board simulation technique which involved asking children to place a felt marker or figure representing themselves as close as they considered appropriate or comfortable to a figure representing a same- or opposite-sexed child of either ectomorphic, mesomorphic, or endomorphic body build. Both Japanese and American children, ranging in age from kindergarten through sixth grade, were tested, and in all cases the major finding was that children placed themselves farther from endomorphs than from either mesomorphic or ectomorphic individuals (Iwawaki, Lerner, & Chihara, 1977; Lerner, 1973; Lerner, Karabenick, & Meisels, 1975; Lerner, Venning, & Knapp, 1975; Lerner, Iwawaki, & Chihara, 1976). The only possible exception to this generalization occurred when kindergarten children were tested; in one case the effect occurred (Lerner, Karabenick, & Meisels, 1975), whereas in two others it did not (Lerner, 1973; Lerner *et al.,* 1976). In any event, the avoidance of physical closeness with endomorphs shows clearly by first grade and is consistent across both sexes and at least two cultures.

Although several of these studies uncovered interactions among the factors, in no case was the finding that more space was used with the endomorph contradicted. There does appear to be an increase in the strength of the aversion toward endomorphy with age, in that the difference in the use of space between the endomorph and the other body shapes increased with age (Lerner, Karabenick, & Meisels, 1975). Less space was used with same-sex than opposite-sex figures, although this effect was attenuated when the figures were endomorphic. This latter finding suggests that body type may be a more potent signal in interpersonal interaction than gender.

The lack of a significant difference in the use of personal space with mesomorphic and ectomorphic figures is somewhat surprising in light of the previously reported differences in the stereotypes for these body types. The differences in use of space in the Lerner, Karabenick, and Meisels (1975) study with kindergarten through third-grade children were in a direction consistent with the stereotypes and nearly significant. The authors suggested that there may be a time lag between the acquisition of some stereotypes and evidence of their influence on behavior. However, even older children did not use different amounts of space with mesomorphs and ectomorphs (Iwawaki *et al.,* 1977; Lerner, Venning, & Knapp, 1975). These findings raise the issue of the validity of generalizing from attitudinal measures to behavioral measures. Although children hold a predominantly negative stereotype of ectomorphic people, this may have little or no influence on their behavior toward them. Alternatively, the negative stereotype of the ectomorphic person may be reflected in interpersonal behaviors other than the use of personal space. However, the possibility that the stimulation technique used by Lerner does not assess how children behave in actual interpersonal interactions must also be considered. Although Lerner, Karabenick, and Meisels (1975) cite evidence that personal space use as assessed by the felt board technique is predictive of personal space use in real life, they did not validate their findings with observations of children's interactions with endomorphic peers.

4.5. Summary

There have been no good observational studies designed to assess whether children varying in appearance are treated differently by peers or adults. However, anecdotal evidence and experimental studies which simulate social interchanges provide some evidence for differential treatment. For example, photographs of more attractive children elicit more looking from children and adults, implying that attractive children elicit more attention in natural interactions. In addition, children place themselves farther from endomorphic stimuli than from mesomorphic and ectomorphic stimuli in a felt-board simulation, implying that they avoid contact with chubby peers. Obviously, more direct observational evidence is needed in this area.

5. DIFFERENCES AMONG CHILDREN VARYING IN APPEARANCE

5.1. Self-Perception

Do children who vary in appearance perceive themselves differently? It might be expected that they would at least show differences in their perceptions of their own appearance, as a result of objective evidence or exposure to peers' judgments of their appearance, or both. Only Cavior and Dokecki (1971) reported children's judgments of their own attractiveness. The correlations between fifth- and eleventh-grade children's rankings of themselves within their own age and sex group and the ranking given them by other children were positive, but only 3 of 16 were significant, possibly due to small sample sizes. This finding is consistent with the adult literature in which correlations between ratings by self and others are rarely high and only occasionally significant (Berscheid & Walster, 1974). One of the reasons why self-ratings were so poorly related to external ratings for fifth-grade girls, at least in Cavior's study (1970, cited by Berscheid & Walster, 1974), was that three fourths of them ranked themselves as the least attractive girl in their class. There are numerous reasons why these girls might have considered themselves unattractive: modesty, age-related changes in concern about appearance, and unfamiliarity with their appearance due to recent puberty-related changes, to cite a few. Overall, the children in Cavior and Dokecki's (1971) study based their judgments of their own attractiveness more on somatic variables (such as coloration and height) than on the perceptions of their peers. It would be interesting to determine the age at which children first develop an attractiveness self-concept, whether it is based more on objective physical characteristics or the perceptions of significant others, and how it changes as children's appearances change.

Several studies have investigated children's ability to identify their own body type. The evidence indicates that kindergarten children are not highly accurate (Lerner & Gellert, 1969; Lerner & Schroeder, 1971) but that accuracy does improve with age (Gellert, Girgus, & Cohen, 1971; Lerner & Korn, 1972; Staffieri, 1967). There is some tendency for girls to be more accurate than boys (Gellert *et al.,* 1971; Lerner & Gellert, 1969) and for nonchubby children to be more accurate than chubby children (Gellert *et al.,* 1971; Lerner & Gellert, 1969; Lerner & Korn, 1972). Although the majority of average body-type boys correctly identified themselves at all ages (5, 15, and 20 years) in the Lerner and Korn (1972) study, the majority of chubby subjects were accurate only at the two older ages. Only 26.7% of the chubby 5-year-olds were correct, whereas 63.3% of the nonchubby 5-year-olds were. These results suggest that chubbier children have a distorted image of their own body. Another possibility, with the younger children at least, is that they misunderstood the question and thought they were being asked which body type they preferred, rather than which they looked like. This is unlikely to have occurred with the older subjects though, and even at this age only 60% of the chubby subjects were accurate while 81.7% of the average subjects were.

More general measures of psychological self-concepts have been found to be related to appearance, although typically only for children at the extremes of the appearance dimensions. Salvia *et al.* (1975) compared the self-concept scores of the upper and lower 10% of a group of third- to fifth-grade children who had been rated on the facial attractiveness dimension. The unattractive children had lower self-concepts than the attractive children. Obese children in this same age range also have been reported to have lower self-concepts than nonobese children (Sallade, 1973).

LaVoie and Andrews (1976) sought relationships between physical characteristics and sex role self-concepts in 3- to 9-year-old children. No consistent relationships were found, although endomorphs scored lower on two of the six measures of sex role self-concept: gender recognition and sex role preference. The results of the study by Salvia *et al.* (1975) suggest that the inclusion of a large number of children of average appearance may have obscured differences between children with more extreme appearance variations, especially on the attractiveness dimension.

Cortes and Gatti (1965) studied late adolescents' self-descriptions and found a specific pattern of attributes associated with each of the three major body types. Subjects in Lerner and Korn's (1972) study and Brenner and Hinsdale's (1978) study also were asked to assign characteristics to themselves. Both chubby and average subjects described themselves as having more of the attributes they had assigned to the mesomorphic and ectomorphic stimuli than those attributed to the endomorph. Unfortunately, the authors do not report any analyses of the content of the attributes assigned by chubby and average subjects to themselves. Their reported results may simply be due to the reluctance of subjects to assign socially undesirable characteristics to themselves in contrast to the ease of assigning them to a stereotypical chubby person. On the basis of Cortes and Gatti's findings, one might expect the chubby and average subjects to identify different socially desirable adjectives as descriptive of themselves.

5.2. Overt Behavior

Do children who vary in appearance behave differently? Two recent studies have directly observed children and then attempted to find an association between attractiveness and behavior. Trnavsky and Bakeman (1976) observed preschool children during a free play period. The only behavior which differentiated the attractive and unattractive children was time spent in solitary play, with attractive children engaging in this activity less than unattractive children. The authors interpreted this finding as showing that the attractive children lacked independence and were unable to entertain themselves when alone. However, it is unclear from their report what the attractive children were doing when they were not playing alone. There were no significant differences for nonnegative or negative time spent with peers (analyzed separately), time spent with teachers, time spent in large groups of children, or time spent with other attractive children. Although "unoccupied" was one of the categories used in coding, it cannot be derived from the report whether the attractive children were spending more time unoccupied or in social interaction (but distributed across the different types of interaction so that none of them was statistically significant).

Langlois and Downs (1979) observed play interactions between pairs of preschool children. The children were either both attractive, both unattractive, or one was attractive and one was not. More affiliative behaviors were seen when the children were from the same attractiveness level, possibly indicating awareness of their own attractiveness. Unattractive children were more aggressive and active than attractive children, in line with behavioral expectancies.

The social behavior and personality characteristics of attractive and unattractive children also have been assessed indirectly through the perceptions of peers. As mentioned previously, attractive children tend to be more popular than unattractive children, which may indicate that attractive children behave in a more friendly manner. It may also, of course, reflect perceivers' differential expectations of children varying in appearance.

Two studies reported that children perceived their attractive, familiar peers positively. Dion and Berscheid (1974) found that unattractive male preschoolers were considered more aggressive than attractive male preschoolers, unattractive female preschoolers were perceived as more fearful than attractive female preschoolers, and unattractive preschoolers were more likely to be chosen as "someone who scares you." Lerner and Lerner (1977) found that attractive fourth- and sixth-graders were more likely than their unattractive peers to receive nominations for desirable traits and less likely to receive nominations for undesirable traits from familiar peers.

Findings which partially contradict these were reported by Styczynski and Langlois (1977). They found that attractive preschoolers were perceived by peers as exhibiting more antisocial behavior as well as more prosocial behavior. These authors suggested that children may pay more attention to attractive peers, thereby being more likely to notice both their positive and negative behaviors. In addition, attractive children may be more self-confident and assertive, thus leading to their behaving both more prosocially and antisocially.

Several attempts have been made to find a relationship between physical attractiveness and cognitive achievement. These attempts were instigated by the previously reviewed studies of teachers' stereotyping of attractive students. Only one study has been found which reports a significant relationship between physical attractiveness and grades. Lerner and Lerner (1977) found low but significant positive correlations between children's facial attractiveness and their grade point average in both the current year and two years before. In addition, students' attractiveness was positively related to the teachers' perceptions of their academic ability and academic adjustment. Clifford (1975) and Maruyama and Miller (cited by Adams, 1977) both failed to find a relationship between attractiveness and grades. However, the latter study did find a positive relationship between attractiveness and performance scores on the Wechsler Intelligence Scale for Children (WISC). This relationship did not hold for WISC verbal IQ scores.

It is possible that more attractive children are more cognitively competent, either for genetic reasons or through the operation of a third variable such as social class. However, the argument that teacher expectancies (based on stereotypes) operate to influence, rather than determine, cognitive performance is not adequately tested by the simple correlational approach. It would seem more reasonable to predict a child's grades from some valid measure of cognitive competence (IQ or achievement test scores) and then to assess whether attractive children's grades are higher than expected and unattractive children's grades are lower than expected.

Reports of several attempts to find relationships between physique and behavior in childhood are available (Davidson, McInnes, & Parnel, 1957; Kagan, 1966; Walker, 1962, 1963). Walker's studies were the most successful in showing that behavior patterns of preschoolers (as rated by parents and teachers) were related to physique measurements made using a modification of Sheldon's (1940, 1942) technique. For example, in the first study, 73% of the 292 predictions made on the basis of Sheldon's work with adults were confirmed in direction, and 21% were statistically significant. Only 3% were significantly disconfirmed. Unfortunately, parents' ratings of behavior (Walker, 1963) were not highly correlated with teachers' ratings (Walker, 1962), so it is unclear exactly what child characteristics were reflected in the ratings. Judging from the near absence of subsequent studies on this topic, it appears that either the approach proved to be not very fruitful or not to have elicited much interest. In light of the recent work by Lerner and his colleagues on body-build stereotypes, a revival of interest in the relationship between body build and both self-concept and overt behavior seems imminent.

Mention should be made of studies which have found a relationship between physical appearance and delinquent behavior, even though they are concerned with adolescents and adults rather than children. Glueck and Glueck (1956) first attempted to find determinants (actually correlates) of delinquency for different body types. A major finding was that the mesomorphic body type was overrepresented in the delinquent sample. This finding has recently been replicated by Shasby and Kingsley (1978). Behavior-disordered males were found to have a higher degree of mesomorphy and lower amount of total body fat than a randomly selected sample of boys from the same vicinity.

Cavior and Howard (1973) looked at the relationship between facial attractiveness and different types of delinquency. College students' ratings of facial photographs discriminated among five different categories of delinquency. In a second study, black delinquents were judged as having a darker skin color and as being less attractive than a comparison group of black high school students.

5.3. Summary

Attractive and nonobese children have been found to have better self-concepts than unattractive or obese children. The observations that children are not very accurate at perceiving their own attractiveness and that obese children are not very accurate at identifying their own body type imply that these differing self-concepts derive from differential treatment by others.

A few studies report that children differing in appearance show different patterns of behavior, although this topic has not been studied in detail, and findings of no differences in behavior are also common. Attractive preschool children have been observed to spend less time in solitary play and to be less aggressive than unattractive preschoolers. Attractive children show superior cognitive performance in some studies but not others. Different patterns of behavior and personality characteristics associated with body type variations have been reported.

6. DEVELOPMENT OF PHYSICALLY ATYPICAL CHILDREN

The studies of facial attractiveness and body type testify to the significance of physical appearance during early development. Children varying in appearance are, at least at times, perceived and treated differently by significant others. They may also develop different personality characteristics and behavioral styles. These processes are not thought to be unidirectionally related, but rather to reflect circular functions in ontogeny (Lerner, 1976). For example, an attractive child may be treated favorably because of cultural stereotypes, but the favorable treatment may also produce desirable behavior in the child which in and of itself elicits further favorable treatment.

One of the implications of the studies of facial attractiveness and body type is that the effects of appearance on development are likely to be more pronounced for children whose appearance is at one of the extremes rather than in the middle of an appearance dimension, as is the case with highly attractive or obese children. Children whose appearance deviated dramatically from the average might be expected to be subject to especially potent differences in treatment and effects on self-concept development. A number of congenital and medical disorders have visible manifestations or correlates. We will consider the effects on child development of some of these visible abnormalities. It must be acknowledged, of course, that most of these disorders produce functional limitations and other cognitive or behavior effects in addition to the effects on physical appearance. As with children within the normal range of appearance, no claims are being made that appearance is the sole or even the primary determinant of developmental outcome. The contention is rather that physical appearance can mediate social relations and personality development and therefore must be considered in any attempt to understand individual development.

6.1. Sources of Atypical Appearance

A newborn infant may fall outside the range of normal appearance due to prematurity or the presence of a congenital abnormality. The most obvious distortions of appearance result from limb deformities or deficiencies and facial deformities such as cleft lip. The various physical anomalies associated with such disorders as Down's syndrome may or may not be noticed by the untrained eye at birth. Subsequent to the newborn period, atypical appearance can result from accidents, injuries, or medical care resulting in amputations, or from the manifestations of a congenital disorder.

6.2. Differentiation of Atypical-Appearing Children

When attractiveness is the criterion measure, handicapped children are usually rated lower than nonhandicapped children (e.g., Down's syndrome children: Siperstein & Gottlieb, 1977). On a preference measure, again, handicapped children are rated lower than nonhandicapped children by other children (Richardson, Goodman, Hastorf, & Dornbusch, 1961) and by adults (Goodman, Richardson, Dornbusch, & Hastorf, 1963; Richardson, 1970). Siperstein and Gottlieb (1977) even suggest that there may be a separate attractiveness dimension for children with physical handicaps. Richardson and his colleagues have investigated this dimension by asking various groups of children and adults to rank-order a standard set of line drawings of children varying in type of handicap.

The first study in this series (Richardson *et al.*, 1961) reported a high degree of uniformity in 10- to 11-year-old children's rankings of the drawings. In general, there was no effect of rater's sex, disability, race, home setting (urban or rural), SES, or place of testing (school or camp) on their rank orderings, and there was significant agreement within each group of children. The most prevalent rank ordering was: (1) no handicap; (2) crutches and brace on left leg; (3) sitting in wheelchair with blanket over both legs; (4) left hand missing; (5) facial disfigurement on left side of mouth; and (6) obese.

Subsequent studies have investigated variations in this ordering by specific groups or individuals, and the contribution of race to changes in the rank ordering. One fairly consistent finding has been that girls rank children with cosmetic handicaps (such as obesity and facial disfigurement) lower, whereas boys rank children with functional disabilities lower (Richardson, 1970, 1971b; Richardson *et al.*, 1961; Richardson & Royce, 1968). Girls also seem to be slightly more responsive to racial differences in the pictures, leading Richardson and Royce to suggest that visual appearance cues are more important for girls than for boys. Changing the race of a picture in this study generally had only small effects on its ranking. However, the study was conducted at a racially integrated summer camp where racial prejudice was strongly discouraged. The procedure was repeated in a Southern city with black girls in two largely segregated schools (Richardson & Emerson, 1970). In this case the effect of race on rankings was more prevalent, but since the effects were not consistent across the two schools, they are difficult to interpret. The authors concluded that there was some evidence for a white preference. As in other studies, these girls chose the obese picture as the least liked, a choice unaffected by the obese child's race.

The age of the rater was investigated in another study (Richardson, 1970). Rankings were significantly concordant from first grade on; kindergartners may have found the ranking task too difficult. Agreement dropped temporarily at sixth grade, indicating a possible transition point in preferences toward handicaps. With increasing age, children were more likely to prefer children with functional handicaps and less likely to prefer children with cosmetic handicaps, although this trend was modified by the sex difference noted earlier. Richardson suggests that younger children prefer drawings which are closer to expected physical appearance and older children prefer children whose disfigurements are located farther from their face.

The process by which children acquire preferences for handicaps also has been explored. Simple exposure to handicapped peers does not seem to change rank orderings (Richardson, 1971a) but may change evaluations of individual peers. More popular boys (who are assumed to have greater exposure to peer values) are more likely than unpopular boys to give normative rank orderings of a standard set of drawings (Richardson & Friedman, 1973). Overall, it appears from this set of studies that children and adults show fairly definite preferences for children with some handicaps over children with other handicaps. Whether these preferences are predictive of attitudes and behavior toward physically atypical children will be explored below.

6.3. Stereotyping of Atypical-Appearing Children

The stereotypes held about children with atypical appearances may be similar to the stereotypes assigned to the unattractive. Physical abnormality of any type may cause perceivers simply to categorize a person as a member of the class of unattractive individuals. The ensuing stereotyped expectancies then follow the typical "attractiveness is good" and "unattractiveness is bad" classification system. On the other hand, specific appearance deviations may elicit specific assumptions about etiology and expectations for behavior (Longacre, 1973). Several studies have investigated the operation of stereotypes in perceptions of physically atypical children.

Both Aloia (1975) and English and Palla (1971) asked adults to rate photographs of normal and atypical children. A severely retarded child elicited more negative responses on a semantic differential than a mildly retarded child (English & Palla, 1971), and children with various types of physical deformities or stigmata (at least one of the five had Down's syndrome) were perceived as mentally subnormal to a greater extent than normal-appearing children (Aloia, 1975). In the latter study, judgments were not influenced by labeling the children "mentally retarded" or "normal." The author suggested that the children represented the positive and negative ends of the attractiveness continuum, and thus judgments would be expected to be in line with the physical attractiveness stereotype.

A study by Podol and Salvia (1976) uncovered a stereotyping effect specific to a particular disorder. Advanced speech pathology students were asked to rate the adequacy of the somewhat nasal speech of a girl following surgical correction of a prepalatal cleft. Half of the students saw a picture of the girl showing postoperative disfigurement and half saw the same picture, retouched to cover the disfigurement. When the disfigurement was visible, the girl's speech was judged to be more nasal and the students were more likely to suggest that she receive speech therapy. Judgments of the adequacy of the girl's speech on other dimensions were not affected by the visual manipulation within the nasal speech condition. The authors proposed the operation of a "cleft palate stereotype" which leads to the expectation of nasal speech.

A final study of this type was conducted with children by Siperstein and Gottlieb (1977). They were concerned with the effects of the physical appearance and competence of a child on other children's perceptions of and acceptance of that child. Subjects listened to a tape recording of either competent or incompetent spelling performance by the target child and viewed a picture of that child (either normal-appearing or Down's syndrome). After listening to the incompetent spelling performance, all the children responded that they were better spellers than the target child, regardless of the child's appearance. After the competent performance, however, 22% of the children perceived themselves as better spellers than the normal-appearing stimulus child, and 55% of them perceived themselves as better spellers than the Down's syndrome child. On an adjective checklist, there were main effects of both competence and appearance. Again, the less attractive individual (the authors reported that pictures of a Down's syndrome child were always rated as least attractive in a pilot study) was perceived and expected to behave in a less desirable manner.

6.4. Reactions to Atypical-Appearing Children

Evidence exists that visibly handicapped children are treated differently from nonhandicapped children. It is difficult, though, to determine to what extent these behaviors are elicited by the physical appearance of the child and to what extent by the child's real or expected functional limitations or behavior. There are several ways in which a physical handicap can affect social relations (Richardson, 1969). The effect can be direct, in that the handicap may put limits on certain functions, such as the use of nonverbal communicative cues. In addition, the handicap can directly influence the responses of others by violating their expectations of normal appearance. Finally, a physical abnormality can indirectly influence social relations because of the tendency to associate physical appearance cues with certain personal characteristics.

6.4.1. The Role of Visibility of the Handicap

One source of evidence that physical appearance is an important elicitor of reactions to an atypical child is the comparison of reactions to children with visible and nonvisible handicaps. For example, Tisza and Gumpertz (1962) reported that parents were more traumatized initially by the birth of a child with a cleft lip, which was immediately visible, than of one with a cleft palate only. They may even prefer to wait until the cleft lip is repaired to take the infant home. The infant with a cleft palate, on the other hand, looks like a normal infant. The cleft palate takes on more significance later, when feeding and speech difficulties occur.

Another study of parents' reactions to the birth of a child with a congenital abnormality (Johns, 1971) also revealed that parents seemed more concerned with visible abnormalities than with those which could be hidden easily. In addition, head and neck abnormalities, and especially facially disfiguring lesions, were regarded as particularly embarrassing and significant in terms of the child's future development.

Teachers' perceptions of the academic abilities of cleft lip and palate children in their classes also appear to be affected by the visibility of the child's deformity (Richman, 1978a). The children were divided into two groups, those with relatively normal facial appearance and those with noticeable disfigurement. The teachers accurately assessed the ability of the normal-appearing children. Within the visibly disfigured group, however, the teachers underestimated the ability of the high-ability children and overestimated the ability of the low-ability children. This finding was thought to be due to the operation of both a stereotypic expectancy of a certain ability level for disfigured children and a sympathetic response to the low-ability disfigured children.

Children also notice and respond to the visibility of physical handicaps. Richardson (1971b) found in one study that drawings of children with amputations were liked more often if they included a prosthesis. Even with a prosthesis, however, amputee children were perceived more negatively than nonamputee children. Centers and Centers (1963) found that amputee children were more likely to be named by classmates as the saddest, the least liked, the least nice-looking, and the least fun child in class. Another study by Richardson and his colleagues (Richardson, Ronald, & Kleck, 1974) suggested that such attitudes are expressed behaviorally but also that exposure to a visibly handicapped peer can increase liking of that child. The boys in an integrated summer camp were categorized as nonhandicapped, visibly handicapped, or nonvisibly handicapped, and their sociometric status was assessed both within their bunk group and outside of it. The visibly handicapped boys had lower status than the other two groups outside of their bunk groups but were rated no differently from the nonhandicapped boys within their bunk group. The nonvisibly handicapped boys received lower ratings within than outside their bunk groups. Observations at meals showed that boys tended to sit near those they had named as their best friends. Thus, it seems that sociometric choices were reflected in proxemic behavior. The implication from this study is that visible handicaps serve as a cue for attitudes and friendship choices to a greater extent when the handicapped child is not well known. One might expect proxemic behavior toward the visibly handicapped to change following intensive interaction with them. The study by Centers and Centers (1963) suggests that even after continued exposure, many visibly handicapped children may still be perceived negatively. This may, of course, have as much to do with the handicapped children's behavior as with the effect of their visible appearance.

6.4.2. Physical Appearance and Atypical Adult–Child Relationships

Physical appearance is commonly mentioned as a potential contributor to certain atypical adult–child relationships. For example, premature infants are more likely to be abused by their parents than are full-term infants (Klein & Stern, 1971). There are a number of dimensions on which these infants differ, including their physical appearance. It has been suggested (Frodi, Lamb, Leavitt, Donovan, Neff, & Sherry, 1978) that premature infants fail to elicit the same positive caregiving response as full-term infants because they have fewer of the "babyish" characteristics thought by ethologists (Hess, 1970) to be critical in this regard. Thus, the lack of a babyish appearance, the difference between the expected appearance of the baby and the premature's actual appearance (Boukydis, 1977), or simply the unattractiveness of the premature infant may combine with the premature's behavioral tendencies to push parents into abuse.

The parents' perceptions of and acceptance of their child's appearance may be more critical than its actual appearance. Attempts at predicting child abuse (e.g., Gray, Cutler, Dean, & Kempe, 1977) have at times included observations of parents' comments on the infant's physical appearance immediately or soon after birth. Although no one parental response during the neonatal period seems to predict child abuse, a negative attitude toward the child at this point on a cluster of dimensions, including physical attractiveness, might predispose them to abuse.

Certainly most parents of premature infants do not neglect or abuse their infants. Even in nonabusive families, however, the atypical, unattractive appearance of the premature infant may contribute to minor adjustment difficulties on the part of the parents (Blake, Stewart, & Turcan, 1975). For example, the parents may be discouraged from paying attention to the infant because of its appearance (Bidder, Crowe, & Gray, 1974). In addition, the aversiveness of the cry of a premature infant has been found to be compounded by the combination with the facial configuration of a premature (Frodi *et al.*, 1978). Undesirable behaviors emitted by a normal-appearing infant might be more easily tolerated than the same behaviors emitted by an atypical-appearing infant.

Physical appearance may also act to improve relations between adults and atypical children in some cases. It has been noted (DesLauriers & Carlson, 1969) that autistic children tend to be quite attractive. It may be their attractiveness which accounts for the disproportionate amount of attention paid to the syndrome as compared with other, similarly devastating disorders. In addition, children with other functional disorders with attractiveness in the normal to high range may be more likely to be kept at home or adopted earlier than less attractive, similarly disabled children. This hypothesis is supported by the difficulty in finding adoptive parents for a child with a facial deformity, despite the fact that the deformity itself is not physically disabling (Longacre, 1973).

6.5. Differences among Atypical-Appearing Children

Differences in social behavior and personality between atypical-appearing and normal-appearing children abound, but again it is difficult to know to what extent they can be attributed to appearance variations. It has been suggested that physically atypical children are frequently socially rejected because of their appearance and that this rejection can produce deviant social and personality development (Goffman, 1963; Longacre, 1973; Richardson, 1969). In many cases, however, such deviations are nonexistent or slight. For example, Lambert, Hamilton, and Pellicore (1972) found that most of the amputees they studied, who had been fitted with prostheses before the age of 6, showed a normal social adjustment at the age of 21. Other reports and case studies (Aitken, 1972; Woods, 1975) indicate that the incorporation of the prosthesis into the child's body image is an important aspect of healthy psychological development for the amputee child. It also has been reported that concern with appearance increases as children get older and that children may wish to start using a prosthesis during adolescence for the purpose of improving their appearance, even though they previously preferred to function without one. Unfortunately, it may be difficult for them to learn to use the prosthesis after functioning without it (Richardson, 1969), and it may be difficult for them to incorporate it into their body image. Early fitting of prostheses is frequently recommended for both psychological and medical reasons.

Richman (1976) stated that most studies of cleft palate children report no significant emotional maladjustment. Several studies of infants and children with cleft lip or palate (Richman, 1976, 1978b; Starr, Chinsky, Canter, & Meier, 1977) have found that these children are more passive and inhibited than normal children, but not to the extent of creating a serious problem. Richman (1976) suggested that cleft palate children may be avoiding situations in which they might receive negative responses from others. Their passivity in childhood seems to be limited to the school setting; their parents describe their behavior as being very similar to that of normal children on the passivity dimension (Richman, 1978b).

Longacre (1973) contends that facial disfigurement has little effect on a child's social or academic adjustment if surgical correction takes place before the age of four, but that problems may arise as the child encounters adverse reactions from peers and adults outside the home. Some common responses of disfigured children which may be detrimental to their emotional and mental health are preoccupation with the disfigurement and the acquisition of defense mechanisms. Some other characteristics of disfigured children and adults observed by Longacre (1973) include underproductivity, lack of self-acceptance, and feelings of dependence, isolation, hostility, and depression. Both Longacre (1973) and Kalick (1978) report that, for most individuals, social relations improve dramatically following surgical correction of the deformity.

6.6. Summary

Children's appearance may be outside the normal range due to congenital defects, accidents, or illnesses. Evidence suggests that atypical-appearing children are perceived as less attractive than normal-appearing children and that negative stereotypes are held about them. Unfortunately, it is difficult to differentiate between the effects of atypical appearance and the effects of atypical behavior or functional disabilities which are associated with a child's handicap. However, it was proposed that the visible manifestation of a defect could influence others' reactions independently.

In general, it can be concluded that atypical appearance may adversely affect a child's social relations and self-concept, although in most cases the effects will not be severe. Improving the child's appearance, through the fitting of prostheses or surgical correction, will help minimize these undesirable effects.

7. DISCUSSION

The studies reviewed above support the conclusion that the physical appearance of a child can influence early psychological development. It is not an all-powerful variable, determining outcome single-handedly, but rather a subtle modifier of the child's own and others' perceptions and behavior. Despite the fact that the momentary influence of appearance may be minor or dependent on an array of other factors, its influence occurs repeatedly and persistently, leading to observable outcomes. In cases where a child's appearance is particularly distinctive, the effects may be more immediately obvious.

Children are differentiated on the basis of physical appearance by adults and peers. Even very young children can identify obese and unattractive peers. Although agreement among judges of children's appearance is typically statistically significant, it must be kept in mind that agreement is frequently only moderate, especially for children in the average range on the attractiveness dimension. There is a strong subjective factor in many of these judgments, as well as influences of age, sex, and familiarity.

Perceptions and expectations of children who vary in appearance are often in line with the physical attractiveness stereotype. Attractive children and children of average body build are perceived more favorably and expected to engage in more desirable behavior than unattractive, obese, or very thin children. It is reasonable to propose that average body build is perceived as more attractive than obesity or thinness, leading to the generalization of the physical attractiveness stereotype to body type.

Presumably because of stereotyping, children varying in appearance are sometimes treated differently by others. Direct observational evidence is sparse, but simulation studies suggest that attractive and nonobese children are treated more favorably by both peers and adults.

As a result of either differential treatment or awareness of cultural stereotypes, children varying in appearance may come to behave and think of themselves differently. The evidence presented here implies that these effects are most likely to occur for children who are especially attractive or especially unattractive.

7.1. Directions for Further Research

More research is needed to determine (1) what makes a child attractive (to everyone or to particular individuals, such as teachers, who frequently interact with children), (2) under what conditions appearance is most likely to be influential, (3) to what extent a child's actual behavior strengthens, modifies, or negates the operation of physical appearance as a determinant of the reactions of others, (4) how and when children develop a physical appearance self-concept, and (5) to what extent physical appearance and physical appearance self-concepts are continuous during the childhood years.

Although there is some evidence that physical features, cleanliness, and hair style influence attractiveness judgments, most studies in this vein have used standardized facial photographs which eliminate some of the cues potentially important to attractiveness assessments in real-life settings. For example, children with braces on their teeth or children wearing eyeglasses are usually not included in these studies. These children might be more affected than others by the effects of their appearance on others and on their self-concept. There also is a certain discontinuity in the research in that facial attractiveness and body type are considered separately. Both cues are simultaneously present in actual social interactions. It would seem important to determine whether these cues are related or independent. Finally, the effects of clothing on attractiveness judgments must be considered more systematically.

In general, more observational studies need to be conducted in real-life settings to test the conclusions from well-controlled laboratory studies. It seems likely that the effects of appearance will be stronger in some settings (such as school, where comparisons among children are probable), than in others (such as the home, where only one highly familiar child of a particular age is usually present). In addition, the possibility must be seriously considered that some of the findings from laboratory studies will not be replicated in real-life settings because of the powerful influence of behavioral differences among children. Studies should be designed, however, to detect subtle influences of appearance, rather than hoping to uncover dramatic, unitary effects.

More information is needed on how children come to perceive their own appearance. Distortions in their perceptions may be especially important to an understanding of how appearance comes to affect psychological and behavioral characteristics. In fact, self-perceptions of physical appearance could prove to be more critical to psychological development than actual physical appearance.

Finally, a critical consideration is whether an attractive child at one age remains attractive at subsequent ages. Certain aspects of appearance necessarily change with age, and it may be that children in general are perceived as more attractive at some ages than others. Perhaps more important to an understanding of individual development is the relative position of a child on the attractiveness dimension at progressive ages. Adams (1977) reported a pilot study in which children's facial attractiveness was found to remain fairly stable from the first to the sixth grade. He also suggests, however, that psychological reorganization may take place at times of appearance changes. Systematic information on when (or if) discontinuities in appearance occur is needed. In addition, the effects on social relations and personality of appearance changes attributable to such interventions as vision corrections, weight loss, or correction of a facial deformity should be considered.

7.2. Implications for Child Care

From a humanistic point of view, many people would prefer that physical appearance not influence a child's development. It seems undemocratic for certain children, through accidents of nature, to be benefited or disadvantaged by their appearance (Berscheid & Walster, 1974). Despite overt contentions that "beauty is only skin deep," however, external appearance and internal psychological characteristics may become concordant through the processes described here. The final issue to be considered is whether the adverse effects of undesirable appearance can be minimized.

Again, it must be recognized that appearance is likely to be more influential in some settings than others. Initial encounters are probably more susceptible to the influence of physical appearance than subsequent encounters, and appearance will probably be more critical in group than individual settings. Thus, appearance is most likely to matter when a child's family moves, when a child first joins a group of other children (such as starting to attend preschool, school, day-care centers, or religious schools), or when a child is involved in short-term interactions with someone who has contact with many children (such as doctors and dentists). Many parents and children recognize the importance of appearance at these times, and most children arrive for the first day of school, for example, dressed in nice new clothes, with clean and neatly combed hair. Some parents may be less aware of the effects of physical appearance, and their children might benefit if the parents were informed of the potential influence of appearance and the ways in which they can make their children appear more attractive. In addition, medical interventions which improve a child's appearance can sometimes be timed to precede a child's first group social experience, and interventions which will worsen a child's appearance can sometimes be delayed until a child's social status and relations with significant others have been established.

Adults who interact with children in groups should be made aware of the possibility that they and the children in the group can be influenced by the physical appearance of a particular child. This effect can be especially evident when the group contains physically or mentally handicapped children. By monitoring their own and other children's reactions to children varying in attractiveness, they may be able to minimize adverse effects.

7.3. Conclusion

Physical appearance is an immediately salient variable in social encounters. It has the potential to modify perceptions of and behavior toward children, as well as children's own behavior and self-perceptions. Awareness of its subtle but pervasive power can aid both in an understanding of the process of individual development and in optimizing the development of all children, regardless of their overt appearance.

Acknowledgments

Thanks are due to Roger Burton, Roy Ford, Carol Oliva, and Ed Whitson for a critical reading of an earlier version of this chapter.

8. REFERENCES

Adams, G. R. Physical attractiveness research: Toward a developmental psychology of beauty. *Human Development,* 1977, *20,* 217–239.

Adams, G. R., & Cohen, A. S. Children's physical and interpersonal characteristics that affect student–teacher interactions. *Journal of Experimental Education,* 1974, *43,* 1–5.

Adams, G. R., & Cohen, A. S. An examination of cumulative folder information used by teachers in making differential judgments of children's abilities. *Alberta Journal of Educational Research,* 1976, *22,* 216–225. (a)

Adams, G. R., & Cohen, A. S. Characteristics of children and teacher expectancy: An extension to the child's social and family life. *Journal of Educational Research,* 1976, *70,* 87–90. (b)

Adams, G. R., & LaVoie, J. C. The effect of student's sex, conduct, and facial attractiveness on teacher expectancy. *Education,* 1974, *95,* 76–83.

Adams, G. R., & LaVoie, J. Parental expectations of educational and personal-social performance and childrearing patterns as a function of attractiveness, sex and conduct of the child. *Child Study Journal,* 1975, *5,* 125–142.

Aitken, G. T. (Ed.). *The child with an acquired amputation.* Washington, D.C.: National Academy of Sciences, 1972.

Aloia, G. Effects of physical stigmata and labels on judgments of subnormality by preservice teachers. *Mental Retardation,* 1975, *13,* 17–21.

Barocas, R., & Black, H. Referral rates and physical attractiveness in third grade children. *Perceptual and Motor Skills,* 1974, *39,* 731–734.

Berscheid, E., & Walster, E. Physical attractiveness. In L. Gerkowitz (Ed.), *Advances in Experimental Social Psychology* (Vol. 7). New York: Academic Press, 1974.

Bidder, R., Crowe, E., & Gray, O. Mothers' attitudes to preterm infants. *Archives of Disease in Childhood,* 1974, *49,* 766–770.

Blake, A., Stewart, A., & Turcan, D. Parents of babies of very low birthweight: Long-term followup. In *Parent–Infant Interaction,* Ciba Foundation Symposium 33. Amsterdam: Elsevier, 1975.

Boukydis, Z. C. F. *Infant attractiveness and the infant–caretaker relationship.* Presented at the International Conference on Love and Attraction, University College of Swansea, Wales, U. K., September 1977.

Brenner, D., & Hinsdale, G. Body build stereotypes and self-identification in three age groups of females. *Adolescence,* 1978, 551–562.

Cavior, N., & Dokecki, P. R. Physical attractiveness and self-concept: A test of Mead's hypothesis. *Proceedings of the Annual Convention of the American Psychological Association,* 1971, *6,* 319–320.

Cavior, N., & Dokecki, P. R. Physical attractiveness, perceived attitude similarity, and academic achievement as contributors to interpersonal attraction among adolescents. *Developmental Psychology,* 1973, *9,* 44–54.

Cavior, N., & Howard, L. R. Facial attractiveness and juvenile delinquency among black and white offenders. *Journal of Abnormal Child Psychology,* 1973, *1,* 202–213.

Cavior, N., & Lombardi, D. Developmental aspects of judgment of physical attractiveness in children. *Developmental Psychology,* 1973, *8,* 67–71.

Cavior, N., Miller, K., & Cohen, S. H. Physical attractiveness, attitude similarity, and length of acquaintance as contributors to interpersonal attraction among adolescents. *Social Behavior and Personality,* 1975, *3,* 133–141.

Centers, L., & Centers, R. Peer group attitudes toward the amputee child. *Journal of Social Psychology,* 1963, *61,* 127–132.

Clifford, M. Physical attractiveness and academic performance. *Child Study Journal,* 1975, *5,* 201–209.

Clifford, M., & Walster, E. The effect of physical attractiveness on teacher expectations. *Sociology of Education,* 1973, *46,* 248–258.

Corter, C., Trehub, S., Boukydis, C., Ford, L., Celhoffer, L., & Minde, K. Nurses' judgments of the attractiveness of premature infants. *Infant Behavior and Development,* 1978, *1,* 373–380.

Cortes, J. B., & Gatti, F. M. Physique and self-description of temperament. *Journal of Consulting Psychology,* 1965, *29,* 432–439.

Cross, J. F., & Cross, J. Age, sex, race, and the perception of facial beauty. *Developmental Psychology,* 1971, *5,* 433–439.

Davidson, M., McInnes, R., & Parnell, R. The distribution of personality traits in seven-year-old children: A combined psychological, psychiatric, and somatotype study. *British Journal of Educational Psychology,* 1957, *27,* 48–61.

DesLauriers, A. M., & Carlson, C. F. *Your child is asleep: Early infantile autism.* Homewood, Ill.: Dorsey, 1969.

Dion, K. Physical attractiveness and evaluation of children's transgressions. *Journal of Personality and Social Psychology,* 1972, *24,* 207–213.

Dion, K. Young children's stereotyping of facial attractiveness. *Developmental Psychology,* 1973, *9,* 183–188.

Dion, K. Children's physical attractiveness and sex as determinants of adult punitiveness. *Developmental Psychology,* 1974, *10,* 772–778.

Dion, K. The incentive value of physical attractiveness for young children. *Personality and Social Psychology Bulletin,* 1977, *3,* 67–70.

Dion, K., & Berscheid, E. Physical attractiveness and peer perception among children. *Sociometry,* 1974, *37,* 1–12.

Dion, K., Berscheid, E., & Walster, E. What is beautiful is good. *Journal of Personality and Social Psychology,* 1972, *24,* 285–290.

English, R., & Palla, D. A. Attitudes towards a photograph of a mildly and severely mentally retarded child. *Training School Bulletin,* 1971, *68,* 55–63.

Epstein, Y. M., Krufat, E., & Obudho, C. Clean is beautiful: Identification and preference as a function of race and cleanliness. *Journal of Social Issues,* 1976, *32,* 109–118.

Frodi, A. M., Lamb, M. E., Leavitt, L. A., Donovan, W. L., Neff, L., & Sherry, D. Fathers' and mothers' responses to the faces and cries of normal and premature infants. *Developmental Psychology,* 1978, *14,* 490–498.

Gellert, E., Girgus, J. S., & Cohen, J. Children's awareness of their bodily appearance: A developmental study of factors associated with the body percept. *Genetic Psychology Monographs,* 1971, *84,* 109–174.

Glueck, S., & Glueck, E. *Physique and delinquency.* New York: Harper, 1956.

Goffman, E. *Stigma: Notes on the management of spoiled identity.* Englewood Cliffs, N.J.: Prentice-Hall, 1963.

Goodman, N., Richardson, S. A., Dornbusch, S. M., & Hastorf, A. H. Variant reactions to physical disabilities. *American Sociological Review,* 1963, *28,* 429–435.

Gray, J. D., Cutler, C. A., Dean, J. G., & Kempe, C. H. Prediction and prevention of child abuse and neglect. *Child Abuse and Neglect,* 1977, *1,* 45–58.

Haley, E. G., & Hendrickson, N. J. Children's preferences for clothing and hair styles. *Home Economics Research Journal,* 1974, *2,* 176–193.

Hess, E. H. Ethology and developmental psychology. In P. Mussen (Ed.), *Carmichael's manual of child psychology* (Vol. 1). New York: Wiley, 1970.

Hildebrandt, K. A. *Adult responses to infant cuteness.* Unpublished master's thesis, Michigan State University, 1976.

Hildebrandt, K. A. *Parents' perceptions of their infants' physical attractiveness.* Paper presented at the International Conference on Infant Studies, New Haven, Conn., April 1980.

Hildebrandt, K. A., & Fitzgerald, H. E. Gender bias in observers' perceptions of infants' sex: It's a boy most of the time! *Perceptual and Motor Skills,* 1977, *45,* 472–474.

Hildebrandt, K. A., & Fitzgerald, H. E. Adults' responses to infants varying in perceived cuteness. *Behavioural Processes,* 1978, *3,* 159–172.

Hildebrandt, K. A., & Fitzgerald, H. E. Adults' perceptions of infant sex and cuteness. *Sex Roles,* 1979, *5,* 471–481. (a)

Hildebrandt, K. A., & Fitzgerald, H. E. Facial feature determinants of perceived infant attractiveness. *Infant Behavior and Development,* 1979, *2,* 329–339. (b)

Hildebrandt, K. A., & Fitzgerald, H. E. Mothers' responses to infant physical appearance. *Infant Mental Health Journal,* 1981, *2,* 56–61.

Iwawaki, S., & Lerner, R. M. Cross-cultural analyses of body behavior relations: I. A comparison of body build stereotypes of Japanese and American males and females. *Psychologia: An International Journal of Psychology in the Orient.* 1974, *17,* 75–81.

Iwawaki, S., & Lerner, R. M. Cross-cultural analyses of body behavior relations: III. Developmental intra- and inter-cultural factor congruence in the body build stereotypes of Japanese and American males and females. *Psychologia,* 1976, *19,* 67–76.

Iwawaki, S., Lerner, R. M., & Chihara, T. Development of personal space schemata among Japanese in late childhood. *Psychologia,* 1977, *20,* 89–97.

Johns, N. Family reactions to the birth of a child with a congenital abnormality. *Medical Journal of Australia,* 1971, *1,* 277–282.

Kagan, J. Body build and conceptual impulsivity in children. *Journal of Personality,* 1966, *34,* 118–128.

Kalick, S. M. Toward an interdisciplinary psychology of appearances. *Psychiatry,* 1978, *41,* 243–253.

Kehle, T. J., Bramble, W. J., & Mason, J. Teachers' expectations: Ratings of student performance as biased by student characteristics. *Journal of Experimental Education,* 1974, *43,* 54–60.

Kleck, R. E., Richardson, S. A., & Ronald, L. Physical appearance cues and interpersonal attraction in children. *Child Development,* 1974, *45,* 305–310.

Klein, M., & Stern, L. Low birth weight and the battered child syndrome. *American Journal of Disabilities in Childhood,* 1971, *122,* 15–18.

Lambert, C. N., Hamilton, R. C., & Pellicore, R. J. The juvenile amputee program: Its social and economic value. In G. T. Aitken (Ed.), *The child with an acquired amputation.* Washington, D.C.: National Academy of Sciences, 1972.

Langlois, J. H., & Downs, A. C. Peer relations as a function of physical attractiveness: The eye of the beholder or behavioral reality? *Child Development,* 1979, *50,* 409–418.

Langlois, J. H., & Stephan, C. Effects of physical attractiveness and ethnicity on children's behavioral attributions and peer preferences. *Child Development,* 1977, *48,* 1694–1698.

LaVoie, J. C., & Adams, G. R. Teacher expectancy and its relation to physical and interpersonal characteristics of the child. *Alberta Journal of Education Research,* 1974, *20,* 122–132.

LaVoie, J. C., Andrews, R. Facial attractiveness, physique and sex role identity in young children. *Developmental Psychology,* 1976, *12,* 550–551.

Lerner, R. M. The development of stereotyped expectancies of body build–behavior relations. *Child Development,* 1969, *40,* 137–141.

Lerner, R. M. "Richness" analyses of body build stereotype development. *Developmental Psychology,* 1972, *7,* 219.

Lerner, R. M. The development of personal space schemata toward body build. *Journal of Psychology,* 1973, *84,* 229–235.

Lerner, R. M. *Concepts and theories of human development.* Reading, Mass.: Addison-Wesley, 1976.

Lerner, R. M., & Gellert, E. Body build identification, preference, and aversion in children. *Developmental Psychology,* 1969, *1,* 456–462.

Lerner, R. M., & Iwawaki, S. Cross-cultural analyses of body behavior relations: II. Factor structure of body build stereotypes of Japanese and American adolescents. *Psychologia,* 1975, *18,* 83–91.

Lerner, R. M., & Korn, S. J. The development of body build stereotypes in males. *Child Development,* 1972, *43,* 908–920.

Lerner, R. M., & Lerner, J. Effects of age, sex, and physical attractiveness on child-peer relations, academic performance, and elementary school adjustment. *Developmental Psychology,* 1977, *13,* 585–590.

Lerner, R. M., & Pool, K. B. Body build stereotypes: A cross-cultural comparison. *Psychological Reports,* 1972, *31,* 527–532.

Lerner, R. M., & Schroeder, C. Physique identification, preference, and aversion in kindergarten children. *Developmental Psychology,* 1971, *5,* 538.

Lerner, R. M., Karabenick, S. A., & Meisels, M. Effects of age and sex on development of personal space schemata towards body build. *Journal of Genetic Psychology,* 1975, *127,* 91–101.

Lerner, R. M., Venning, J., & Knapp, J. R. Age and sex effects on personal space schemata toward body build in late childhood. *Developmental Psychology,* 1975, *11,* 855–856.

Lerner, R. M., Iwawaki, S., & Chihara, T. Development of personal space schemata among Japanese children. *Developmental Psychology,* 1976, *12,* 466–467.

Longacre, J. J. *Rehabilitation of the facially disfigured.* Springfield, Ill.: Charles C Thomas, 1973.

Marwit, K. L., Marwit, S. J., & Walker, E. Effects of student race and physical attractiveness on teacher's judgments of transgressions. *Journal of Educational Psychology,* 1978, *70,* 911–915.

McCandless, B. R. *Children: Behavior and development* (2nd ed.). New York: Holt, Rinehart & Winston, 1967.

Podol, J., & Salvia, J. Effects of visibility of a prepalatal cleft on the evaluation of speech. *Cleft Palate Journal,* 1976, *13,* 361–366.

Power, T. G., Hildebrandt, K. A., & Fitzgerald, H. E. Adults' responses to smiling and crying infants. *Infant Behavior and Development,* 1982, *5,* 33–44.

Rich, J. Effects of children's physical attractiveness on teacher evaluations. *Journal of Educational Psychology,* 1975, *5,* 599–609.

Richardson, S. A. The effect of physical disability on the socialization of a child. In D. A. Goslin (Ed.), *Handbook of socialization theory and research.* Chicago: Rand McNally, 1969.

Richardson, S. A. Age and sex differences in values toward physical handicaps. *Journal of Health and Social Behavior,* 1970, *11,* 207–214.

Richardson, S. A. Children's values and friendships: A study of physical disability. *Journal of Health and Social Behavior,* 1971, *12,* 253–258. (a)

Richardson, S. A. Handicap, appearance, and stigma. *Social Science and Medicine,* 1971, *5,* 621–628. (b)

Richardson, S. A., & Emerson, P. Race and physical handicap in children's preference for other children: A replication in a Southern city. *Human Relations,* 1970, *23,* 31–36.

Richardson, S. A., & Friedman, M. J. Social factors related to children's accuracy in learning peer group values toward handicaps. *Human Relations* 1973, *26,* 77–87.

Richardson, S. A., & Royce, J. Race and physical handicap in children's preferences for other children. *Child Development,* 1968, *39,* 467–480.

Richardson, S. A., Goodman, N., Hastorf, A. H., & Dornbusch, S. M. Cultural uniformity in reaction to physical disabilities. *American Sociological Review,* 1961, *26,* 241–247.

Richardson, S. A., Ronald, L., & Kleck, R. E. The social status of handicapped boys in a camp setting. *Journal of Special Education,* 1974, *8,* 143–152.

Richman, L. C. Behavior and achievement of cleft palate children. *Cleft Palate Journal,* 1976, *13,* 4–10.

Richman, L. C. The effects of facial disfigurement on teachers' perceptions of ability in cleft palate children. *Cleft Palate Journal,* 1978, *15,* 155–160. (a)

Richman, L. C. Parents and teachers: Differing views of behavior of cleft palate children. *Cleft Palate Journal,* 1978, *15,* 360–364. (b)

Rist, R. C. Student social class and teacher expectations: The self-fulfilling prophesy in ghetto education. *Harvard Educational Review,* 1970, *40,* 411–451.

Rosenthal, R., & Jacobson, L. *Pygmalion in the classroom.* New York: Holt, Rinehart & Winston, 1968.

Ross, M. B., & Salvia, J. Attractiveness as a biasing factor in teacher judgments. *American Journal of Mental Deficiency,* 1975, *80,* 96–98.

Rubin, J. Z., Provenzano, F. J., & Luria, Z. The eye of the beholder: Parents' views on sex of newborns. *American Journal of Orthopsychiatry,* 1974, *44,* 512–519.

Sallade, J. B. A comparison of the psychological adjustment of obese vs. nonobese children. *Journal of Psychosomatic Research,* 1973, *17,* 89–96.

Salvia, J., Sheare, J. B., & Algozzine, B. Facial attractiveness and personal-social development. *Journal of Abnormal Child Psychology,* 1975, *3,* 171–178.

Shasby, G., & Kingsley, R. F. A study of behavior and body type in troubled youth. *The Journal of School Health,* 1978, *48,* 103–107.

Sheldon, W. H. *The varieties of human physique.* New York: Harper, 1940.

Sheldon, W. H. *The varieties of temperament.* New York: Harper, 1942.

Siperstein, G. N., & Gottlieb, J. Physical stigma and academic performance as factors affecting children's first impressions of handicapped peers. *American Journal of Mental Deficiency,* 1977, *81,* 455–462.

Staffieri, J. R. A study of social stereotype of body image in children. *Journal of Personality and Social Psychology,* 1967, *7,* 101–104.

Staffieri, J. R. Body image stereotypes of mentally retarded. *American Journal of Mental Deficiency,* 1968, *72,* 841–843.

Staffieri, J. R. Body build and behavioral expectancies in young females. *Developmental Psychology,* 1972, *6,* 125–127.

Starr, P., Chinsky, R., Canter, H., & Meier, J. Mental, motor, and social behavior of infants with cleft lip and/or cleft palate. *Cleft Palate Journal,* 1977, *14,* 140–147.

Styczynski, E., & Langlois, J. The effects of familiarity on behavioral stereotypes associated with physical attractiveness in young children. *Child Development,* 1977, *48,* 1137–1141.

Sugarman, D. B. *Perceiving physical attractiveness: Making things relative.* Paper presented at the meeting of the Eastern Psychological Association, Philadelphia, April 1979.

Tisza, V. B., & Gumpertz, E. The parents' reaction to the birth and early care of children with cleft palate. *Pediatrics,* 1962, *30,* 86–90.

Trnavsky, P. A., & Bakeman, R. *Physical attractiveness: Stereotype and social behavior in preschool children.* Paper presented at the meeting of the American Psychological Association, Washington, D.C., September 1976.

Walker, R. N. Body build and behavior in young children: I. Body build and nursery school teachers' ratings. *Monographs of the Society for Research in Child Development,* 1962, *27*(Whole No. 84), 2–94.

Walker, R. N. Body build and behavior in young children: II. Body build and parents' ratings. *Child Development,* 1963, *34,* 1–23.

Woods, T. Comments on the dynamics and treatment of disfigured children. *Clinical Social Work Journal,* 1975, *3,* 16–23.

Young, R. D., & Avdzej, A. Effects of obedience–disobedience and obese-nonobese body type on social acceptance by peers. *The Journal of Genetic Psychology,* 1979, *134,* 43–51.

5

Development of the Father–Infant Relationship

MICHAEL W. YOGMAN

1. INTRODUCTION

Until recently, our theories and empirical studies of infant social development largely ignored the father. The father's role with young infants was considered to be mainly indirect, supporting the mother, who was biologically adapted to be the infant's caregiver. This chapter will review recent studies that suggest that the father's role with young infants is far less biologically constrained than once thought. In what way biologically based sex differences constrain the social interactions of fathers with their infants must await future research, but such constraints now seem far more subtle than once believed. Wide variability in the behavior and roles of the two parents challenges many of the stereotypes of the father as incompetent or uninvolved with the infant and leaves ample opportunity for wide variations in the way parents and their infants relate to each other.

Transformations in Western societies have had a dramatic influence on family life (Keniston, 1977; National Academy of Sciences, 1976) and provide background for understanding why fathers are becoming more involved with their infants. Increasingly, young parents are geographically isolated from family and friends and the early months of infancy are often stressful. During this time, the father's role seems increasingly important as a readily available source of support

---

Michael W. Yogman, M.D. • Department of Pediatrics, Harvard Medical School, and Child Development Unit, Children's Hospital Medical Center, Boston, 333 Longwood Avenue, Boston, Massachusetts 02115. The generous support of the Robert Wood Johnson Foundation, the Carnegie Corporation, and the National Institute of Mental Health is gratefully acknowledged. Parts of this work were conducted at the Mental Retardation Research Center, Children's Hospital Medical Center, Boston, Massachusetts.

for the mother. In the increased proportion of families in which mothers of young children return to work (44% of mothers of children less than age 6 in one recent survey, Collins, 1979), fathers may of necessity play a more active role with their infants. The increased interest in father's role with infants in medical practice, the courts, the social sciences, and government policy may, in part, reflect these transformations. Whereas the changing definitions of women's roles in society may influence the role fathers play in the family, changes in men's roles may reflect an independent historical shift, probably stimulated in part by the change in women's roles. Perhaps men are seeking an increased opportunity for emotional contact with the infants in their families because the pressures of their workplace offer such limited opportunities for emotional expression. These shifts in adult roles are certain to influence the nature of the male transition to parenthood. The focus here will be on the nature of the father–infant relationship.

This chapter acknowledges that fathers can and do form a significant relationship with their infants and attempts to describe the similarities and differences between the patterns of father–infant interaction and mother–infant interaction. In an attempt to understand paternity in biological as well as psychological terms, the first two sections will review historical, cross-cultural, phylogenetic, and anthropological data on the father's role in infancy. The third section will describe in some detail recent studies which elucidate the normal development and quality of the father's relationship with his infant during the first two years of life, including research by the author on the development of father–infant interaction during the first six months of life. A conceptual basis for understanding these developmental changes in the father–infant relationship is presented in section 4, in which the process of fathering fosters the infant's differentiation and individuation. The application of the principles of cybernetic theory to information transfer among infant, mother, and father as a useful way of studying family social interaction will also be addressed. Section 5 considers studies of the impact on the father–infant relationship of environmental stresses, individual differences of the infant, physical illness of the infant, and some clinical applications. Finally, the social policy implications of research on fathers and infants will be discussed.

2. HISTORICAL BACKGROUND

Throughout history and in virtually all cultures, children have lived in proximity to and have often been cared for by both male and female adults. The relationship of male adults to children is considerably more diverse than that of female adults. In some cultures, the "social" father may be a relative of the mother rather than the biological father (Malinowski, 1927). Other taboos may restrict what a father can do or determine when he can interact with a baby. In this chapter, I will suggest that cultural and social values and economic and political conditions both determine and reflect the father's specific role with infants. At the same time, regardless of differences in the specific role of fathers, fathering as a process serves a more general psychological function within most societies and cultures—the fostering of autonomy and the enhancement of individuation.

Writings about the history of fathers' roles in ancient cultures describe overall family structure but provide little information on the father–infant relationship. For example, though we know that the ancient Hebrews practiced polygamy while Greeks and Romans practiced monogamy and that male adults were dominant over women, comments about infants were limited to reports of infanticide (Queen & Adams, 1952). In ancient China and in Arab countries, Confucian and Islamic ideas both stressed reverence and love for the father as a function of his age and sex (Lang, 1946; Goode, 1963). Although it seems clear that patriarchy or male dominance represented a common theme of major religious and philosophical traditions, we know little about the nature of the father–infant relationship. Better information is available on the history of this relationship in Western industrialized nations, particularly in Great Britain. According to Stone's (1977) social history of the family in England, the nuclear family as we now know it was formed by the year 1500. Family life at that time was characterized by rigid patriarchy, high rates of infant and maternal mortality, and harsh discipline of children. Infant care practices consisted regularly of wet nursing and swaddling of infants (Shorter, 1975). Marriage was very much a bargaining process and was capitalized on to secure title or property rights.

By 1700, shifts occurred in philosophical thinking about the importance of the individual and early experience (Locke, 1727). Stone (1977) argues that there began to be evidence of deeper personal relationships within marital and parent–child relationships. This shift seems to have steadily influenced the evolution of the contemporary Western nuclear family. To what extent economic, demographic, and political forces have contributed to changing the character of relationships within families is open to question, but it is certain that industrial and technological developments played a role in stabilizing this pattern.

Currently, at least in the United States, a new phase of family relationships appears to have begun, perhaps in reaction to the excesses of industrialized growth and technology (separation of work from home). This phase is characterized by social values that place increasing significance on individual fulfillment, on emotional involvement, and on sexual equality. It is probably based on a population shift from rural to urban settings, on the availability and efficacy of new forms of birth control, on increased opportunities for leisure, and on the women's movement with accompanying demands for social, economic, and political equality. Fathers as well as mothers may be trying to reassert control over their own lives in the face of an onslaught of technology and experts telling them what to do. The Carnegie Commission has documented some of these changes. Over the past 200 years, families have become less self-sufficient in passing on the essential skills of life and work to children. As these functions eroded, various agencies or agents of society have taken their place: the school, physicians, nutritionists, and others, leaving parents with a diminished sense of authority and control over the rearing of their own children. In this sense, the experts have taken the place of a more natural social support system, leaving parents in the role of what has been termed the "weakened executive" (Keniston, 1977). It is in the context of these changes in the modern history of Western Europe and the United States that new definitions of men's roles, and fathers' roles in particular, are emerging.

3. PHYLOGENETIC AND ANTHROPOLOGICAL BACKGROUND

In an effort to understand the increased involvement of fathers with their infants, attempts have been made to relate paternal involvement with offspring to the evolutionary transition from ape to early hominid (Australopithecus). Because the offspring of man were slower to mature and were dependent on caregivers for a longer time than offspring of other species, the amount of time and energy required to raise a baby was so great that selection favored situations in which all adult members of the species (males and females) were involved in caring for the young, either through kin relationships or direct protection, provision, and care of offspring (Zihlmann, 1978). The evidence for this hypothesis relating paternal involvement to principles of evolutionary biology is not persuasive.

Attempts to review phylogenetic evidence of the role of fathers have shown no simple evolutionary trend (Kaufman, 1970; Mitchell, 1969; Redican, 1976; Rypma, 1976; Spencer-Booth, 1970). While in most nonprimate species mothers care for infants, in others caretaking is shared by both males and females (herring gull, fox, penguin) and in some male caretaking predominates (stickleback, midwife toad, ostrich, seahorse) (Rypma, 1976). The range of activities for highly involved males includes defending territory, nest-building, keeping a vigil over the eggs and keeping the brood together in the stickleback (Tinbergen, 1952), to the predigestion and regurgitation of food for the infant in the wolf, to protection, play, and transport in the coyote and fox (Rypma, 1976).

Among the primates the range is also wide. The marmoset, a New World monkey (see Figure 1), assists during birth, premasticates food during the first week, and carries the infant at all times, except during nursing in the first three months (Hampton, Hampton, & Landwehr, 1966). The barbary macaque on Gibraltar begins in the newborn period to elicit social chatter and later encourages beginning locomotion, which functions to orient the infant toward interaction with other group members (Burton, 1972). Male adoptions of orphans have been reported among baboons, chimps, and macaques (Hrdy, 1976). However, male primates are by no means universally nurturant. Male tree shrews, bush babies, and langurs are known to be hostile to infants of their own species and are capable

Zoological Society of San Diego

FIGURE 1. Marmoset father with infants. From *The Evolution of Primate Behavior* by Alison Jolly, New York: Macmillan, 1972. Copyright 1972 by Alison Jolly. Reprinted by permission.

of infanticide, while feral male chimps show little interest in and have little contact with their infants.

Variability in male caretaking exists not only between species, but among individuals within species. Hormonal influences seem to affect the behavior of adult males toward their young. Androgen levels of Japanese male monkeys have a yearly cycle and are lowest during the birth season when paternal behavior is highest (Alexander, 1970). Studies on adult rhesus males demonstrate the influences of environment on the expression of affection and aggression toward their young. Both in the wild and when living in groups in captivity, rhesus males are either aggressive or indifferent to young. However, in the laboratory when the female rhesus is removed from the group, the males do more caretaking and engage in more intense affectionate and reciprocal play with their infants (Redican & Mitchell, 1973). Furthermore, when adult male rhesus monkeys were housed in laboratory-created nuclear family social environments, they played with infants far more than did mothers (Suomi, 1977). Even males reared in social isolation are capable of learning to form affectionate social relationships (Gomber & Mitchell, 1974), although they are still less nurturant toward their young than females raised under comparable conditions (Chamove, Harlow, & Mitchell, 1967).

Since the ecology of the social group seems more important than phylogeny in influencing male care of infants, attempts have been made to suggest environmental or ecological factors which favor male caretaking across species. Trivers (1972) used the concept of *parental investment* to describe the influence of natural selection on parent–infant behavior as an evolutionary mechanism which increases survival of the offspring. Variables said to be associated with higher male parental investment include: monogamous social organization with prolonged pair bonding, the defense of territory, familiarity with mother, close kinship ties, permissive mother–infant interaction, and relative isolation from other conspecifics (Mitchell, 1969; Redican, 1976; West & Konner, 1976).

Anthropological data provide further insights into the way social and cultural variables influence paternal care of and involvement with infants. In cultures such as the Arapesh, fathers play an active and joint role with mothers during pregnancy as well as in caring for infants after birth (Howells, 1969). !Kung San (Bushmen) fathers, representative of the earliest hunter–gatherer societies, were found to be affectionate and indulgent, often holding and fondling their infants, although they provided little of the routine care compared with mothers (West & Konner, 1976). Fathers from the Lesu village in Melanesia who live in monogamous nuclear families and are gardeners are reported to play with their infants for hours.

More commonly, fathers compared with mothers played only a minor role with infants in a study of 80 nonindustrialized cultures (Barry & Paxson, 1971). In this study, a significant negative correlation ($r = -.48$) was found between cultures with a high degree of father–infant proximity and cultures with an emphasis on the early development of a child's autonomy. In cultures where father–infant contact is minimal, there are other intervening variables: Bedouin males are war-like and authoritarian and the Thonga of South Africa prefer to involve the mother's family with her infant and restrict father's involvement with taboos (West & Konner, 1976).

Analysis of social organization in different cultures suggests that males have a closer relationship with their infants when families are monogamous, when both parents live together in isolated nuclear families, when women contribute to subsistence by working, and when men are not required to be warriors (West & Konner, 1976; Whiting & Whiting, 1975).

Taboos and rituals both restrict and enhance the father's role in many cultures. The couvade ritual has been most widely discussed. In its traditional form, the father takes to bed during the woman's pregnancy, labor, and delivery, as a means of sharing in the experience. The remnant of this ritual in modern cultures is evidenced by the couvade syndrome in which men experience psychosomatic symptoms during their wives' pregnancies (Trethowan, 1972; Trethowan & Conlon, 1965). After reviewing anthropological evidence, Rivière (1976) has presented a novel interpretation of the couvade which assigns the father the responsibility for creating the spiritual existence of the infant, while the mother gives birth to the physical person. A ritual among the Ainu carries out a similar function: the mother gives the child its body and the father its soul. The body is acquired during pregnancy, the soul is acquired during the twelve days following birth. Father spends the first six days after birth in a friend's hut and the remainder in his own hut. At the end of this period, the infant is considered a complete person with body and soul (VanGennep, 1960).

In summary, phylogenetic and anthropological evidence underscores the diversity of the father–infant relationship across species and cultures. Fathers are involved with infants and play a significant role in many species and cultures. Biological or hormonal influences on male parenting have not been well studied. The limited data suggest that ecological factors may be more powerful determinants of male parenting than biological factors. Moreover, the conditions of modern Western culture may help explain the current increased involvement and interest by fathers in infant care.

4. INFANTS AND THEIR FATHERS: NORMAL ADAPTATION

In the past ten years, empirical studies based on direct observation of father–infant interaction and interviews with fathers have allowed us to begin to organize these data in a framework that acknowledges the simultaneous development of the infant and his parents (both father and mother) within a broader social world that also includes grandparents, friends, siblings, peers, and strangers. Previous conceptualizations of infant social development focused almost exclusively on the mother–infant relationship, whether the importance of the mother was tied to her gratification of instinctual drives in psychoanalytic theory or to her association with the feeding experience in social learning theory. Although "attachment" theory conceptualized the infant as active rather than passive in seeking the caregiving and love necessary for survival, this theory also tended to minimize the role of the father in the infant's first year of life (Bowlby, 1969; Ainsworth, 1973). No theory acknowledged a meaningful direct role for fathers until the child entered the oedipal period and began to identify with the father (Freud, 1923, 1925) or until the father could play a clear instrumental role in the family such as teaching his child to throw a ball (Parsons & Bales, 1955).

Sociocultural shifts which have occurred in the last 10 years have legitimized the study of the father–infant relationship, and a number of comprehensive reviews of studies of the father–infant relationship have appeared (Earls & Yogman, 1979; Lamb, 1975, 1976c, 1979; Lewis & Weinraub, 1976; Lynn, 1974; Nash, 1965; Parke, 1979).

First, secular changes in infant care documented by Bronfenbrenner (1961, 1976) have resulted in fathers' playing a more active role. These secular changes include an increase in the number of families in which both parents work and caretaking is shared (Howells, 1973) and, while rare in absolute terms, an increase in single-parent families in which the father is the primary caretaker (Mendes, 1976; Orthner, Brown, & Ferguson, 1976). They also include a shift in the father's role in the traditional family from a more hierarchical authoritarian one to a more individualized and flexible one, in which the father is encouraged to have more direct contact with his infant (Benedek, 1970).

Second, our understanding of the mother–infant relationship has been modified by findings that the feeding experience is not as critical as once thought (Harlow, 1958), that social responsiveness and stimulation are key dimensions contributing to psychological development (Rheingold, 1956), that infants could be attached to fathers who were not primary caretakers (Schaffer & Emerson, 1964), and by suggestions that studies of maternal deprivation were, more accurately, studies of parental deprivation (Green & Beall, 1962).

Interest in the father–infant relationship has come at a time when the field of infancy research has enormously expanded. Studies have demonstrated a wide range of perceptual, cognitive, and social competencies that are either present at birth or acquired by infants in the first few weeks (Appleton, Clifton, & Goldberg, 1975; Bower, 1977; Brazelton, 1973; Bruner, 1973; Kessen, 1970; Lewis & Brooks, 1975; Lipsitt, 1976; Stone, Smith, & Murphy, 1973; Wolff, 1963). Many of these competencies represent biological preadaptations of the infant which function to elicit caregiving from adults and insure the infant's survival. Not only, then, does the infant help the adult to develop appropriate responses (Bell, 1968; Lewis

& Rosenblum, 1974), but the processes of mutual recognition and regulation by both infant (Cassell & Sander, 1975) and parent (Klaus & Kennell, 1976) begin much earlier than attachment theory suggests. This shift in thinking about infants as capable of influencing caregivers means that any theory of the father–infant relationship must account for bidirectional influences of both the infant on the father and the father on the infant. A theory of the father–infant relationship must also recognize that these reciprocal influences can be direct as well as indirect (mediated through the mother or another family member) so that the family is often the meaningful unit of analysis. In many ways the development of the father–infant relationship is similar to the mother–infant relationship in that infants can elicit competent loving caregiving from both male and female adults. Furthermore, there are similarities in the developmental transitions of adulthood that mothers and fathers experience in becoming parents. In other ways, however, it seems that the father–infant relationship is unique and complementary to the mother–infant relationship.

In order to simplify this review, let us separately examine what empirical studies have told us about fathers and infants during five different developmental periods: preconceptual, prenatal, neonatal, one to six months, and six to 24 months. Almost all of the studies reviewed here were done in the United States in the last 10 years, so that caution must be exercised in generalizing the findings beyond contemporary American society.

4.1. Preconceptual and Prenatal Periods

The decision to have a child represents a major transition in adult development. The availability of contraception means that many prospective parents now choose whether to have children and when to have them. While many families are postponing the age of having a first child, the factors influencing that decision and the consequences of delaying the age of childbearing have not been studied. Whether or not males actively make a choice is probably influenced by such variables as prior sex education, early experiences as a child, available role models of fathering, and the status of the current marital relationship. An example of how intricate this decision can be for a male is reported by Gurwitt (1976), who described the responses of a young man in the midst of making the decision to become a father while he was undergoing psychoanalysis. As this man approached the decision to become a father, he experienced great internal turmoil as he attempted to rework past and current relationships with mother, father, siblings, and wife and to resynthesize his sense of self. This father's drivenness at moving and settling into a new home had the symbolic quality, evidenced in his associations, of preparing a nest. Similarly, the intensity with which he completed and defended his doctoral dissertation represented his attempt to complete his own creation prior to embarking on the shared creation of a baby.

Although the classic studies of Bibring (1959) have given us detailed insights into the mother's prenatal experiences, no comparable studies exist for fathers. Gurwitt (1976) in his case report discusses two aspects of the male prenatal experience: first, the man's reactions to changes in his wife's physical and psychological status throughout the pregnancy and second, a reworking of significant relationships and events early in his life. As was true in Bibring's (1959) studies, Gurwitt finds it "remarkable" that by the time of the birth of the child the preceding turmoil had been covered by an "amnesic blanket." This turmoil has been called a crisis of paternal identity by Ross (1975), who suggests that the impending obligations of parenthood signal the end of childhood and a loss of one's parents as parents. This crisis involves conflicts of gender, generative, and generational identity and a reworking of the oedipal relationship with the man's own father.

As part of a larger study of family psychological adaptations to first pregnancies, Liebenberg (1973) interviewed 64 normal, expectant fathers on several occasions during their wives' pregnancy. Initially, most of the fathers were pleased about the pregnancy but worried about the accompanying increased emotional and financial responsibilities. Many of them identified with their wives and expressed envy at her ability to have a baby. Of the fathers in Liebenberg's study, 65% developed complaints similar to those of pregnant women: fatigue, nausea, backache, headaches, and vomiting. A few transiently gained 10 to 20 pounds, and several stopped smoking and switched from drinking coffee to milk. Many of these symptoms were alluded to earlier in this chapter in the discussion of couvade, although the incidence of symptoms in this study was much higher than Trethowan (1972) suggested. During the course of the pregnancy, 52% of the fathers were increasingly unavailable at home because of heavy work or class schedules, and yet 44% of the families managed to move to different dwellings. One can speculate that these manifestations reflect the man's attempt to do something creative in his work and at the same time to establish a "nest" as Gurwitt reported. Other studies have demonstrated the important influence the father has on the mother during the prenatal period. An association was found between a husband's responsiveness to his wife's pregnancy and her successful adaptation to it (Shereshefsky & Yarrow, 1973).

While there are few studies of men becoming fathers, the limited data suggest that the period prior to the birth of the baby is, at least superficially, similar for fathers and mothers. For both, it is a time requiring psychological readjustment as they integrate the roles of child and spouse with that of expectant parent, and financial readjustment as they reorganize their living space and prepare for increased financial responsibilities. However, the expectant father does not feel the physical presence of the fetus growing within him, and this lack may stimulate a father to search for alternative evidence of his productivity and creativity, for example, through increased attention to his work and the provision of financial security for his family. Perhaps expectant men are more able to express their hopes and fears about the subsequent child since they are reported to be more interested in babies than expectant women are (Feldman & Nash, 1978). Nevertheless, the struggle for the expectant male during the prenatal period is to remain emotionally available to his wife and at the same time meet his own needs for feeling responsible and productive and maintain his self-esteem.

4.2. Labor, Delivery, and the Newborn Period

Fathers are now increasingly encouraged by obstetric services to accompany their wives during labor and delivery. In fact, Anderson and Standley (1976) have shown that husband support lessens the degree of maternal distress during this time. A recent study has looked more directly at the experience of fathers during the immediate postnatal period. Thirty fathers of healthy firstborns in London were given a written questionnaire 48–72 hours after birth to describe their feelings toward their babies. Half the fathers chose to be present in the delivery room and witness the birth and half did not. While the questionnaire did not show significant differences between the two groups, it documented the powerful impact on the father of seeing and holding the newborn whether or not he was present at delivery (Greenberg & Morris, 1974). This study suggests that immediate physical contact between father and the newborn may be a more crucial variable in the developing relationship than the father viewing the birth process. A subsample of the fathers was also interviewed. These fathers emphasized their desire to touch, pick up, move, hold, and play with their newborn and were particularly impressed by the liveliness, reflex activity, and movements of the baby: "When she starts moving I go and pick her up and she starts moving in your hands and your arms and you can feel her moving up against you. It's like a magnet." While these descriptions may be particularly characteristic of fathers, other descriptions appear characteristic of either parent: feelings of extreme elation, relief that the baby is healthy, feelings of pride and increased self-esteem, and feelings of closeness when the baby opens his eyes (Robson & Moss, 1970). Fathers' descriptions of their newborns have also been reported to be more sex-typed than mothers' descriptions, as evidenced by postpartum interviews on day one, in which fathers rated sons as firmer, more alert, stronger, and hardier and daughters as softer, finer-featured, and more delicate (Rubin, Provenzano, & Luria, 1974).

While debate goes on about the existence of a sensitive period for maternal contact with newborns (Klaus, Jerauld, Kreger, McAlpine, Steffa, & Kennell, 1972), no comparable studies of early contact exist for fathers. However, in one study of a small sample of five fathers, the sequence in which fathers touch their newborn over the first three days of life has been shown to be the same as with mothers: first with fingertips and then with full palms and first on the extremities and later on the trunk (Abbott, 1975; Klaus, Kennell, Plumb, & Zuehlke, 1970),

TABLE I. Studies of Father–Newborn Interaction

Reference	Sample	Measures	Findings
Parke, O'Leary, & West (1972)	Mothers, fathers, and 1- to 3-day infants; middle class	Time-sampled observations in hospital	Parents similar on most measures; fathers held and rocked more.
Parke & O'Leary (1976)	Mothers, fathers, and 1- to 3-day infants; lower class	Time-sampled observations in hospital	Parents similar on most measures; fathers held more, offered more physical and auditory stimulation while mothers smiled more. Fathers explored babies and smiled more in presence of mothers than when alone.
Parke & Sawin (1975)	Mothers, fathers, and 1- to 3-day infants	Time-sampled observations during feeding	Similarities between parents: both equally sensitive to infant cues during feeding.
Parke & Sawin (1977)	Mothers, fathers, and 1- to 3-day infants	Time-sampled observations in hospital, feeding observation, toy play observation, questionnaire	Fathers provide greater visual and auditory stimulation during toy play; fathers perceive newborn as more perceptually competent and as needing more affection than does mother.
Pedersen (1975)	Mothers, fathers, and 4-week-old infants	Interview of father; mother–infant feeding observation	Father's support of mother associated with more alert, motorically mature baby and with mother's effectiveness during feeding situation.

although the fathers took longer before they displayed this progression. Even fathers whose babies were delivered by cesarean section displayed the same progression with increased eye-to-eye contact (Rödholm & Larsson, 1979). Whether this sequence is specific to parents or characteristic of all human adults generally is yet to be determined.

Studies by Parke and Sawin (1975, 1977) of father–newborn interaction in the postpartum period suggest that fathers and mothers are equally active and sensitive to newborn cues during the postpartum period. Current published studies are summarized in Table I, and in general the conclusions hold for middle-class as well as lower-class families and in both the dyadic (father–infant) and triadic (mother–father–infant) situation. These studies also suggest differences between father- and mother–newborn interaction: fathers held, rocked, and provided more auditory and physical stimulation to their infants. Also, fathers of newborns reported that their babies needed more stimulation and affection and were more perceptually competent than mothers did. Pedersen (1975), studying four-week-old babies, has shown that fathers had an indirect influence on their babies as well, mediated through support of mother, that resulted in a more effective mother–infant relationship. While these studies describe how fathers and mothers interact with their newborn babies, little is known about the baby's influence on the father during the newborn period. It seems reasonable to assume that a father's sleep rhythms are modified by the new baby's schedule of night waking. Considering Sander's studies (Sander, Julia, Stechler, Burns, & Gould, 1975) on the entrainment of biorhythms for mothers and infants, one wonders about similarities and differences in this process for fathers and infants as family schedules and periodicities become reorganized during the early weeks of postnatal life.

4.3. The First Six Months

During the first six months of life, infants become increasingly social as they begin to smile and vocalize. One might suspect that these socially responsive infants are good elicitors of social interaction with fathers as well as mothers. Together with colleagues at Boston Children's Hospital, I have studied the social interactions of fathers with their infants two weeks to six months of age. In contrast to functional tasks such as feeding and diapering, we studied unstructured face-to-face interaction because it placed maximal demands on the social capabilities of the participants. While face-to-face communication may occupy only a small proportion of an infant's day at home, videotaped interactions in the laboratory allowed us to elicit and study in a detailed way exchanges of expressive communication that may underlie the developing father–infant relationship. This method of studying early social interaction was developed by Brazelton, Tronick, Adamson, Als, and Wise (1975) and has been used to characterize mother–infant interaction as a mutually regulated reciprocal process.

We compared the face-to-face interaction of infants with fathers to their interaction with mothers and with strangers in order to study how infants differ in their patterns of expressive behavior during interaction with fathers as compared with mothers and strangers.

4.3.1. *Subjects*

The subjects were six first-born infants (three females, B, D, and F, and three males, A, C, and E) recruited in the newborn period as part of an ongoing study of social interaction, their mothers and fathers, and several adult strangers. Mothers were the primary caretakers in all families. The strangers were both male and female adults and varied in previous experience with infants.

Infants were all full-term, weighed more than 2900 gm at birth, and were delivered vaginally after uncomplicated pregnancies. Apgars were all greater than 7, and all were healthy neonates.

4.3.2. *Observations*

Each infant was studied during interaction with each parent and a stranger and seen weekly in a laboratory. A schematic representation of the laboratory can be seen in Figure 2 (Brazelton *et al.,* 1975). Two-minute face-to-face interactions were videotaped. The infant, when alert and calm, is seated in an infant seat

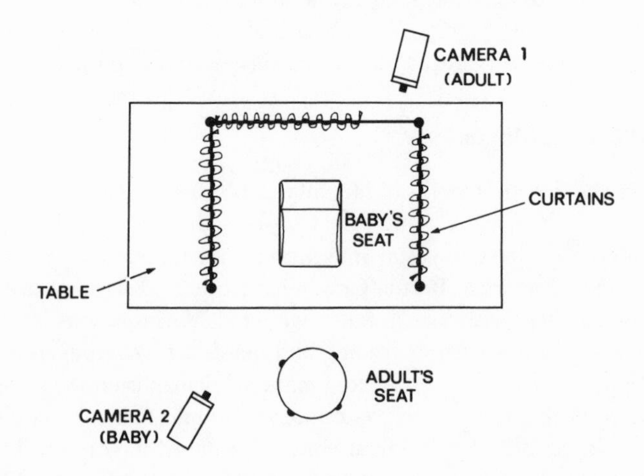

FIGURE 2. Schema of laboratory during observations of adult–infant interaction. From "Early Mother–Infant Reciprocity" by T. B. Brazelton, E. Tronick, L. Adamson, H. Als, and S. Wise, in *Parent-Infant Interaction,* edited by R. Hinde (Ciba Foundation Symposium No. 33). Amsterdam: Elsevier, 1975. Copyright 1975 by Elsevier and Associates Scientific Publishers. Reprinted by permission.

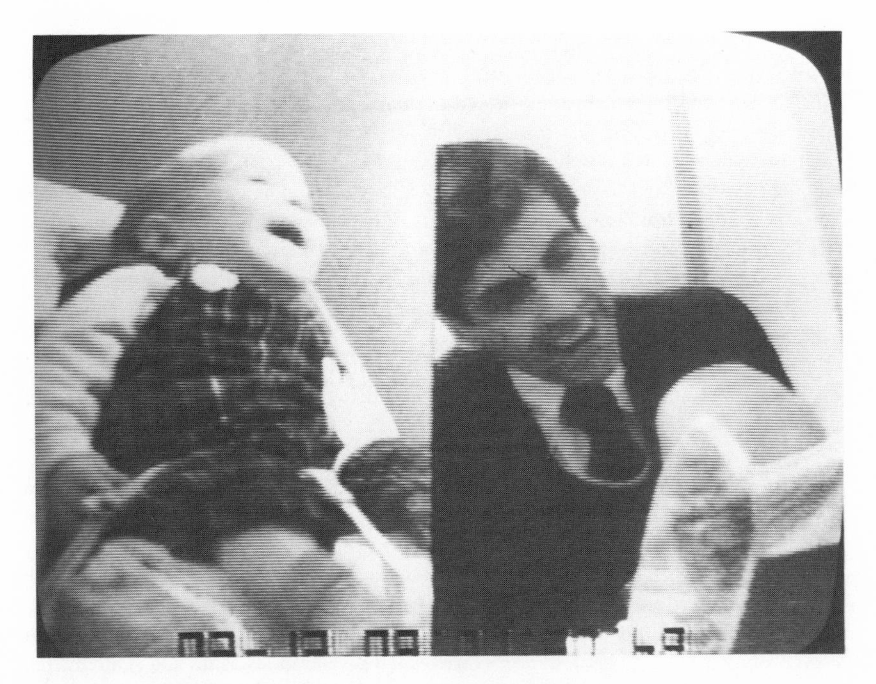

FIGURE 3. Picture of TV monitor during father–infant interaction.

placed on a table surrounded by curtains. The adult enters from behind the curtain, sits in front of the infant and is instructed to "play without using toys and without removing the infant from the seat." One video camera focuses on the infant, the other on the adult. The two images appear simultaneously on a split screen monitor (Figure 3), which shows a single frontal view of adult and infant along with a digital time display. Sound is simultaneously recorded.

Each session consisted of a total of seven minutes' recording: two-minute periods of play with mother, father, and a stranger, separated by 30 seconds of infant alone to assure that she or he was alert and comfortable. The 30-second period allowed standardization of infant state prior to the entrance of each adult. Adults waited outside the laboratory when not interacting with the infant so that they could not observe and be influenced by each other's behavior with the infant. The order of adults entering was counterbalanced.

4.3.3. Description of Session

Because the behaviors analyzed became more meaningful in context, a narrative description of a typical interaction between a 96-day-old infant and his father, his mother, and a stranger follows:

As father enters, the infant sits upright, becomes still, and watches the father intently, everything quiet for a short time. Then he presents a bright smile greeting, with a short, intense coo and an abrupt kick as if to punctuate the father's entrance. The father responds saying, "Are you glad to see me?" The infant's body posture remains upright, he remains attentive and absolutely still for extended periods of time; vocalizations are often laughs—short and intense—followed by long pauses. The limb movements that do occur are large, abrupt, and staccato in quality. The father begins with a greeting, then waits for the baby and continues in an almost adult-like flowing narrative vocalization. He uses animated facial expressions with shifts in eyebrows to amplify the infant's responses and sits at a medium distance from the infant. The father demonstrates rhythmic touching patterns, for example, walking up the baby's arm with his fingers or cycling the baby's legs, as part of a game. As the father leaves and says, "I gotta go, be right back," the infant watches him go and then turns to the side and shifts his facial expression from a bright smile to a more sober mien.

As the mother enters, saying "Hi," in a high-pitched voice, the infant returns her smile with a bright one of his own and then begins a dialogue of reciprocal, varied, and drawn-out vocalizations (oh's, coos, gurgles), infant and mother in turn initiating and terminating the dialogue. The infant's movements of limbs, fingers, and toes alternately cycle smoothly toward and away from the mother and underscore the rhythmic quality of the interaction. Mother and baby together reach a peak of excitement and involvement in the interaction and then slowly withdraw as if to allow time for recovery. They repeat this cycle of alternate and gradual accelerations and decelerations again and again during their interaction. The mother places her arms and hands on the baby's abdomen, containing and modulating the baby's movements. She sits very close to the infant and uses repetitive, high-pitched, imitative baby talk and animated facial expressions. As the mother leaves, the baby visually follows her out, turns his head to the side, and begins to squirm.

As the stranger enters with a bright smile, the infant stares, half smiles, abruptly moves all extremities, and then displays a more sober expression and begins biting his lower lip. The baby continues to make episodic, brief overtures with looks, smiles, and vocalizations as if cautiously attempting to elicit interaction with the stranger, but vocalizations and smiles are sparse. The stranger sits far away from the infant, makes only occasional tentative touches and either remains silent for longer periods or talks continuously through the baby's verbal response. As the stranger extends her finger, the baby falls forward in the infant seat. The process of withdrawal proceeds as the infant begins to play with the strap of the infant seat; he continues to make brief overtures looking up with visual checks, smiles, and vocalizations. Finally, he leans way over and begins to squirm and grunt as the stranger leaves and comments, "He promptly ignored me, didn't he?"

4.3.4. Analysis of Data

We have been able to use the tapes to answer several different questions:

a. Could infants three months of age in an interactive situation discriminate familiar adults, mother, and father, from each other and from unfamiliar strangers?

b. What are the similarities in mutual regulation and reciprocity between mother–infant and father–infant interaction?
c. What are the differences both in the quality of regulation and in the structure and content of games parents play with their infants?
d. What are the developmental changes which occur in father–infant interaction during the first six months of life?

Each of these questions required a different method of data analysis. The analysis of discrete behaviors will be discussed first, followed by analyses of more structural characteristics of social interaction, monadic phases, and social games.

4.3.4a. Differential Interaction: Analyses of Discrete Behavior. First, we set out to determine whether the three-month-old infants in our study interacted differently with mothers, fathers, and strangers as was suggested in the narrative account for one family. Infants visited the laboratory several times during the first two months of life, and the families were familiar with the setting. For each infant the session closest to the middle of the third month at which the infant was alert and comfortable during the 30-second period prior to interaction with each adult was selected. The mean age of the six sessions selected for this analysis was 80 days (SD = 10.4 days).

The video tapes were analyzed by using a microbehavioral scoring system to describe the behaviors each participant displayed second by second. The system consists of four categories of infant behavior, each with three mutually exclusive subdivisions, and five categories of adult behavior, each with four mutually exclusive subdivisions. Each subdivision of each category is defined in a scoring manual and labeled as shown in Table II. The scoring system includes, for the infant, categories of gaze patterns, limb movements, facial expressions, and vocalizations and, for the adult, categories of gaze patterns, body positions, facial expressions, vocalizations, and touching patterns. This system is a modification of one previously developed and used by Brazelton *et al.* (1975).

Two observers simultaneously analyzed the tapes at one seventh of normal speed, viewed each tape once for each category, and scored each behavior according to its occurrence throughout a one-second time period, not on single frames stopped one second apart. All decisions were joint ones and any disagreements between observers were resolved by reviewing the tape and coming to agreement. Test–retest reliability was maintained at greater than 85% agreement for all infant behaviors and greater than 80% agreement for all adult behaviors. In order to eliminate the possibility that the scorers were biased by knowing which adult was simultaneously present on the split screen with the infant, the tapes were rescored with the adult side of the split screen mechanically covered and the sound turned off, so that the scorers coding infant behaviors were unaware of which

TABLE II. Behavior Scored for Infant and Adult

Infant	Adult
Movement	Body position
1. Still: still	1. back, going away
2. Small: sporadic small and medium movements	2. sitting, neutral, sideways shifts
3. Large: medium and large cycling or abrupt movements, touches adult	3. medium close
	4. very close, large forward and back
Facial	Facial
1. Negative: cry, grimace, wary, lidded, frown	1. frown, serious
2. Neutral: neutral, soft, yawn	2. neutral
3. Positive: smile, coo, bright	3. bright, animated, smile
	4. play face, broad smile, large animation, coo face, kisses
Look	Look
1. Away: averted, away 6 seconds	1. predominantly away
2. Interrupted: interrupted look	2. interrupted look (predominantly toward)
3. Toward: concentrating, following, toward $\geq$ 10 seconds	3. toward $>$ 6 seconds
Vocalization	Vocalization
1. Negative: cry, protest, whimper	1. stern, angry, rapid, tense
2. None: none, isolated sound, grunt	2. none ($>$ 4 seconds)
3. Positive: coo, gurgle, laugh, repeated sound	3. narrative, noise, low burst-pause
	4. high burst-pause, verbal imitation
	Touch
	1. abrupt poke, jerk, readjust position
	2. none
	3. containment
	4. tapping

adult was playing with the infant. Agreement with initial scores was greater than 80% for all categories.

Results: Analysis of the video tapes enabled calculation of the amount of time each infant displayed each specific behavior with various adults during the laboratory session. Because of the large variability in the data, the Friedman two-way analysis of variance (ANOVA) by Ranks (Siegel, 1956) was used to test for differences in infant behavior with the three adults. Specific differences between the pairs (mother–father, father–stranger, mother–stranger) were explored using the Sign test (Siegel, 1956). Table III shows the data for infant behaviors which differed significantly with the three adults. The results of the ANOVA showed significant differences in the amount of time these six infants displayed negative facial expressions with mothers, fathers, and strangers ($\chi_r^2 = 6.58$, $df = 2$, $p \leq 0.05$). The Sign test indicated that infants showed negative expressions significantly more of the time with strangers than with either fathers ($p \leq 0.05$) or mothers ($p \leq$

0.05). No differences existed in the amount of time infants displayed negative expressions with mothers as compared with fathers. The ANOVA for the amount of time infants displayed positive expressions with the three adults showed no significant differences.

Table III also shows that infants differed in body movements depending on the adult interactant. All infants held their limbs still for significantly different amounts of time with the three adults ($\chi_r^2 = 7.00$, $df = 2$, $p \leq 0.05$). Analysis of pairs (Sign test) showed that infants remained still more often with fathers than with mothers ($p \leq 0.05$). The differences in facial expression and limb movement are shown graphically in Figure 4. While not significant, infants showed a trend to remain still more often with fathers than with strangers ($p \leq 0.1$). There were no other significant differences in limb movement or in infant looking time, vocalizations, or periods of silence with the three adults.

Discussion: In these six families, infants by 80 days of age displayed different patterns of interaction with fathers as compared with mothers and strangers. They differentiated unfamiliar adults from their familiar parents as evidenced by less frequent displays of negative facial expressions with both mothers and fathers than with strangers. Infant differentiation of mother from father was more subtle and was evidenced only by the fact that they remained still more with their fathers than with mothers. Infant looking and vocalizations were not different with mothers, fathers, and strangers during these sessions. Analysis of data from sessions with infants as young as six weeks of age has shown similar findings (Yogman, Dixon, Tronick, Adamson, Als, & Brazelton, 1976a,b).

Not only does the infant interact differently with father, mother, and stranger, but the adults also behave differently (Yogman *et al.*, 1976a,b). Our data indicate that within the ongoing context of an interaction the infant rapidly adapts his behaviors to the familiarity and actions of his partner even within a two-minute period. Carpenter (1974) has shown in a noninteractive setting that two-week-

TABLE III. Percent of Time Infants Display Behavior with Mother (M), Father (F), and Stranger (S)

	Negative facial expression			Still limb movement		
Infant	M	F	S	M	F	S
A	3.9	0	27	48.8	61.7	36.8
B	5.4	0	5.8	83.7	95.6	86.6
C	13.9	5.6	24.6	78.8	85.5	85.4
D	0	30.5	48.7	67.7	71.9	74.7
E	0	0	0	25.5	73.8	7.1
F	0	0	70.4	12.5	69.4	35.6
	χ_r^2 ($df = 2$) = 6.58, $p \leq 0.05$[a]			χ_r^2 ($df = 2$) = 7.00, $p \leq 0.05$[a]		

[a] χ_r^2 = Friedman two-way ANOVA by Ranks (Siegel, 1956).

old infants, as evidenced by their looking patterns, can visually discriminate between mother and stranger. However, neither Carpenter's results nor our data argue that the infant is able to recognize mother, father, and stranger if that is taken to mean that the infant has a non-stimulus-bound concept of these persons.

4.3.4b. Similarities in Mutual Regulation: Analysis of Monadic Phases. Our analysis of discrete behaviors comparing a single behavior of one partner of a dyad to a single behavior of the other partner was useful for questions about infant differentiation but seemed too static for understanding patterns of interaction. Each discrete behavior seemed to derive specific meaning only as part of a cluster of behaviors embedded in a sequence of such clusters interacting with clusters from the other member of the dyad.

Because we believed that the interactive messages and communicative meaning were carried in such clusters of substitutable behaviors, we adapted a method of data analysis developed by Tronick (1977; Tronick, Als, & Brazelton, 1980) and clustered the discrete behaviors into more meaningful units of analysis called monadic phases. For both the infant and the adult, *a priori* decision rules were

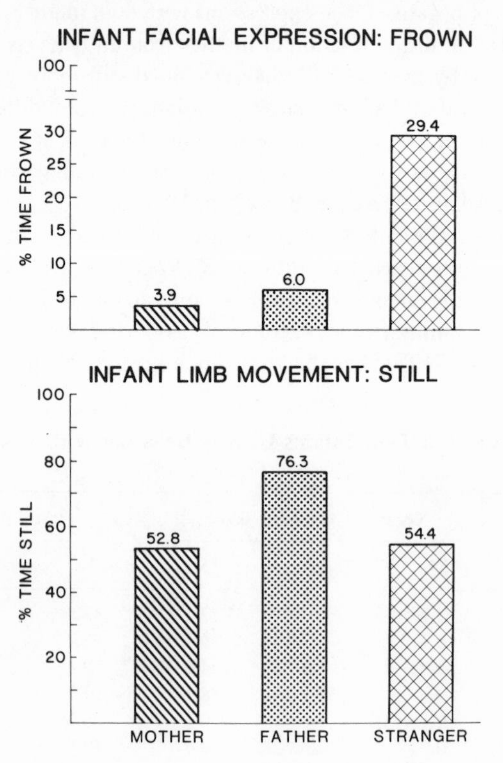

FIGURE 4. Percentage of time infants 3 months of age frown or remain still with mothers, fathers, and strangers.

created to translate each second-by-second display into one of the following monadic phases: Talk, Play, Set, Elicit, Monitor, Avert, and Protest/Avoid. Each of the monadic phases is made up of a set of substitutable second-by-second displays and is not defined by a single behavior. The choice of phases and rules for clustering were guided by the expectation that each phase would convey an affective message from one partner to the other.

The monadic phase analysis enabled us to look at both similarities and differences in the structural characteristics of mother–infant and father–infant interaction. For example, the monadic phase analysis derived from the session described earlier in this section is depicted graphically in Figure 5. For each second of interaction, infant and adult behavior has been translated into monadic phases, and the sequence of phases with adult and infant superimposed is shown graphically (infant with a solid line, adult with a broken line).

Looking first at the father–infant graph, one sees that both father and infant cycle through similar phases shifting from Set up to Play and Talk and then back down to Set again. Both father and infant spend more than 90% of the interaction in phases Set, Play, and Talk, the three most affectively positive phases.

Furthermore, the graph allows one to focus on the transitions between phases and to see that these transitions are jointly regulated. The small letters a through j below the graph mark transitions in which both partners move in the same direction within one to two seconds of each other. One can see that father and infant often change phases simultaneously during face-to-face interaction.

Comparing father–infant interaction on the top of Figure 5 with mother–infant interaction on the bottom, one notes that the similarities are striking. Mother and infant also cycle through similar affectively positive phases and on several occasions also change phases simultaneously as indicated by the small letters a through h directly below the graph. With both parents, the cycling between phases limits the duration of time spent in any one phase. With both mothers and fathers, this cycling appears to maintain the level of affective involvement of each partner within certain limits in a homeostatic fashion.

Below each of the graphs is another visual representation of the amount of meshing or mutual regulation which occurs during these dyadic interactions. During each second, parent and infant monadic phases may be related in one of three ways: match, conjoint, and disjoint.

Match occurs when the phases are identical; in conjoint, both partners are in adjacent phases, as when the adult is in Play and the infant is in Talk. Disjoint occurs when the partners are more than one phase apart, as when the adult is in Monitor and the infant is in Play.

The shaded boxes below the graphs depict the relationship between infant and parent phases during each second as match on top, conjoint in the middle and disjoint on the bottom. After a few seconds of disjoint states, the remainder of infant interaction with both parents consists of conjoint and match states.

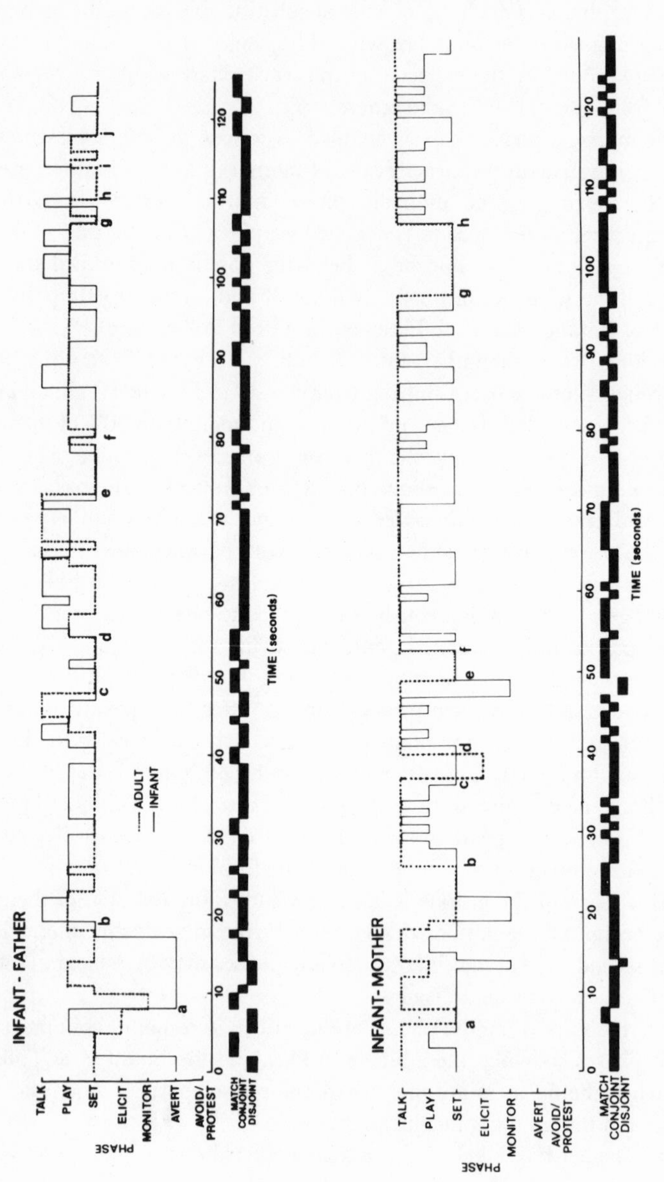

FIGURE 5. Monadic phases during session of 96-day-old infant with mother and father.

Looking more closely at similarities in dyadic states—match, conjoint, and disjoint—we calculated the relative proportion of time spent in each of these three dyadic states and the transitions between states during infant interaction with mother and father. Interactions were mostly conjoint with both parents: 56% of the time with mother and 60% of the time with father. There are also a large proportion of matching states with both parents, while disjoint states are rare, occurring less than 10% of the time. Furthermore, the transitions between dyadic states demonstrate the way partners achieve this meshing. Most of the second-to-second transitions either remain within conjoint or match states or cycle between the two. When the partners diverge to disjoint states, they readjust back to conjoint or match states within the next one or two transitions. These data suggest similarities in joint regulation and reciprocity displayed during dyadic interactions of infants with mothers and fathers. Similarities exist in the levels of affective involvement of the partners, the almost simultaneous timing of transitions between phases present during interactions with both parents, and in the quality of dyadic states and nature of transitions between dyadic states.

4.3.4c. Differences in Quality of Regulation: Analyses of Transitions and Games. Differences between father–infant and mother–infant interaction are also evident in these graphs. These differences exist in the kinds of transitions between phases which occur. Focusing just on episodes of infant Talk in Figure 5 and looking at the transitions the infant makes from that phase to other phases, one notes differences between infant interaction with mothers and fathers. We have characterized two different types of transitions from the phase of infant Talk. Transition A represents a shift from Talk to a lower phase and then back up to Talk on the subsequent transition. An example can be seen on the infant–mother graph between the small letters b and c (Figure 5). Transition B represents the same initial shift from Talk to a lower phase, but then a subsequent shift to a phase other than Talk instead of back to Talk. An example can be seen on the infant–father graph between the small letters b and c (Figure 5). The bar graphs in Figure 6 show the relative proportion of infant transitions A (left) and B (right) occurring during mother–infant and father–infant interaction.

With mothers, transitions are more likely to be Type A, whereas with fathers these transitions are more likely to be Type B. Furthermore, the mean duration of the interval between episodes of Talk is also longer with fathers (8.0 sec) than with mothers (2.8 sec). These data suggest that after infants talk with fathers, they are more likely to shift through lower phases and remain there for a longer time, while with mothers they are more likely to return to Talk. This difference in the quality of transitions between phases is characteristic of the more accentuated shifts from peaks of maximal attention to valleys of minimal attention that occur during infant interaction with fathers. This can be compared with the more gradual and modulated shifts that occur during mother–infant interaction.

Games: Further differences are evident in the temporal structure and specific

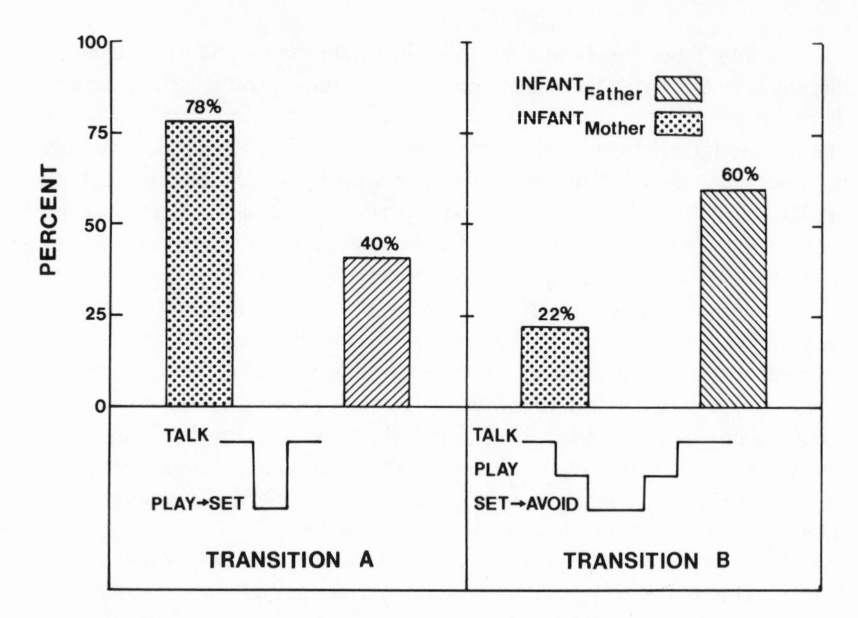

FIGURE 6. Transitions from infant talk during sessions with 96-day-old infant.

behavioral content of the games played by parents and infants. An "interactive game" is defined as: "A series of episodes of mutual attention in which the adult uses a repeating set of behaviors with only minor variations during each episode of mutual attention" (Stern, 1974).

Examples of the father–infant games included such activities as exercising an infant in a "pull-to-sit" game (for example, see Figure 7), repeatedly tapping the infant at three places (under the nose and at both corners of the mouth) while accompanying the taps with clicks of varying pitch, repeatedly buttoning the baby's lips (see Figure 8), and bicycling the baby's legs (Yogman, 1977). These games resulted in a very aroused, attentive infant and were seen with both male and female infants as young as 23 days of age.

We chose to look at verbal games and motor games in more detail. In order to describe accurately the components of these games, we reviewed the videotapes from all six babies in the initial study and described in detailed narrative form episodes that met Stern's definition. The sample for this analysis consists of nine sessions recorded for each infant with mother and father at 4, 6, 8, 10, 12, 14, 16, 20, and 24 weeks of age. Descriptions of the games which occurred during these 54 sessions were then categorized according to a system devised by Crawley, Rogers, Friedman, Iacobbo, Criticos, Richardson, and Thompson (1978) for motor games, which we then modified by adding verbal games and combinations of games and omitting gross body movement games since our setting precluded their occurrence. The categories we used and the results can be seen in Table IV.

First of all, although games occurred during most sessions, they were more likely to occur during sessions with fathers than with mothers. Mothers and infants played games during 75% of the 54 sessions, while fathers and infants played games during 87% of the sessions. Parents usually played more than one game per session (mean for mothers = 1.7 games/session; mean for fathers = 1.65 games/session). Of the games that did occur, pure tactile and pure verbal games and combinations of the two were quite common with both parents. Visual games in which the parent displays distal motor movements that may be observed by the infant and appear to be attempts to maintain the visual attention of the infant also were quite common, whether or not accompanied by verbal games. These visual games were more common with mothers than with fathers. They represented the most common mother–infant games (36% of all games played) and occurred in 46% of all mother–infant sessions (61% of sessions in which any games occurred). With fathers, these games represented only 20% of all games played, significantly lower than with mothers. The most common type of father–infant games were tactile games, representing 27% of all father–infant games. In contrast to mothers, fathers more often engaged in limb movement games in which their behavior attempted to arouse the infant. These limb movement games (whether or not accompanied by verbal games) occurred in 31% of all father–infant sessions

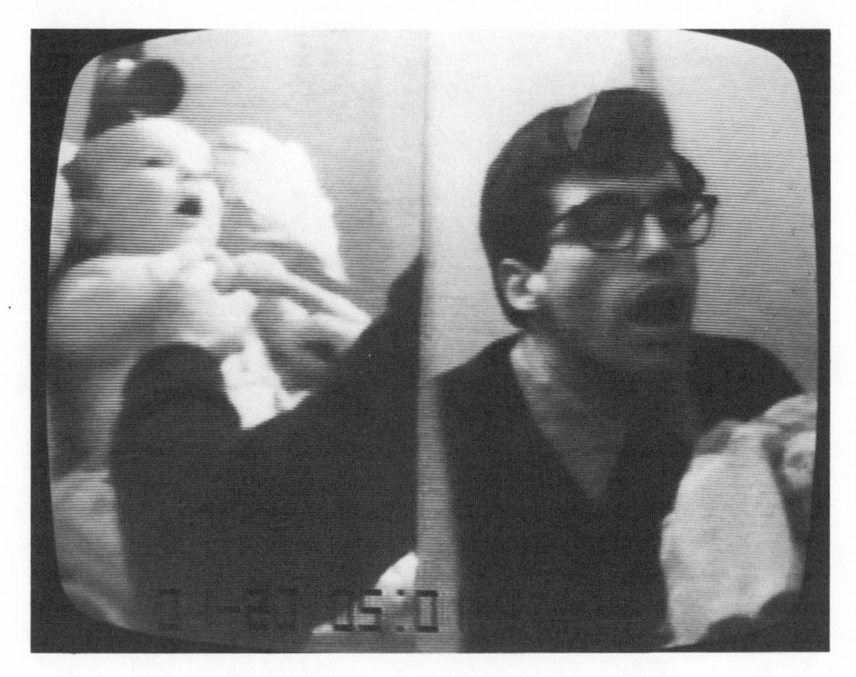

FIGURE 7. Father–infant (45 days) "pull to sit" game.

(36% of sessions in which any games occurred) and represented 21% of all father–infant games, while they occurred in only 7% of all mother–infant sessions and represented only 4% of mother–infant games. All of these differences were significant (*t*-test, Bruning & Kintz, 1968). Mothers also played frequent limb movement games, but these were a different category in which the infant assumed a conventional motoric role such as pat-a-cake, peek-a-boo, or waving.

The visual games more often played by mothers may represent a more distal attention-maintaining form of interactive play than the more proximal, idiosyncratic limb movement games played more often by fathers. Studies of the games parents play with eight-month-old infants show similar findings: mothers played more distal games, while fathers engaged in more physical games (Power & Parke, 1979). Stern (1974) has suggested that the goal of such games is to facilitate an optimal level of arousal in the infant in order to foster attention to social signals. The more proximal games of infants and fathers may serve to modulate the infant's attention and arousal in a more accentuated fashion than occurs during the more distal games of infants and mothers. Since accentuated temporal shifts may also increase the infant's arousal (Stern, 1979), further attention to the temporal structure of these games and their burst–pause patterns may provide additional evidence of differences between mother– and father–infant interaction.

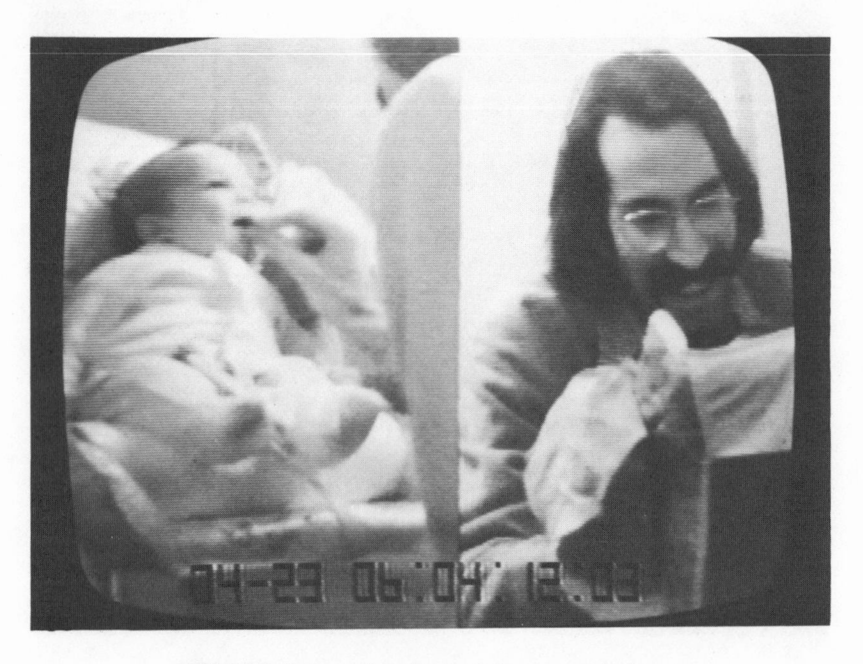

FIGURE 8. Father–infant (64 days) "button your lip" game.

TABLE IV. Types of Games Played by Parent and Infant

	% of all games		% of all sessions		% sessions any games occurred	
	Mother–infant	Father–infant	Mother–infant	Father–infant	Mother–infant	Father–infant
Tactile	26.0	27.0	37.0	31.4	48.7 [a]	36.1
Limb movement	4.3 [a]	21.3	7.4 [a]	31.4	9.7 [a]	36.1
Conventional limb movement	8.6 [a]	2.2	14.8 [a]	3.7	19.5 [a]	4.2
Visual	35.8 [a]	20.2	46.2	31.4	60.9 [a]	36.1
Conventional visual	2.1	3.3	3.7	1.8	4.8	2.1
Combination	8.6	5.6	9.2	7.4	12.2	8.5
Pure verbal	14.1	20.2	22.2	29.6	29.2	34.0

[a]$p \leq 0.05$ (t-test) (Bruning & Kintz, 1968).

4.3.4d. Developmental Changes. We next used the monadic phase analysis and the description of games to look in more detail at the ontogenesis of these patterns of father–infant communication. We wished to compare them to the developmental changes in mother–infant interaction described by Brazelton and Als (1979). We chose to illustrate the developmental trends in father–infant interaction by describing in detail the data from one family in which the infant is a healthy female. We analyzed two minutes of father–infant face-to-face interaction from sessions recorded at infant ages of one month (37 days), two months (56 days), three months (91 days), and four months (140 days). The data for the monadic phases of both father and infant during each second of interaction at each of these four sessions are depicted graphically in Figure 9. The percentage of time spent in each of the phases for both father and infant at each age is also shown in Table V.

The data suggest that as the infant's age increases, the range of the infant's affective displays with father expands and the infant spends a greater proportion of time in more affectively positive phases. For example, the percentage of time spent in the negative phase Protest is greatest at the youngest age, 37 days, while the percentage of time spent in either of the positive phases Play or Talk is greatest at the oldest age, 140 days. At 37 days, the infant's predominant phase is Monitor, and she spends over 90% of her time in Monitor and Set. She spends five seconds in Protest and only goes to a phase more positive than Set for a single, brief two-second interval after 90 seconds of interaction. At this time, the father alternates between Set and Play and remains about two phases higher than the infant, as if encouraging her without overwhelming her. None of the phase transitions for father and infant occur simultaneously.

By 56 days, while the infant's predominant phase is still Monitor (43.6%),

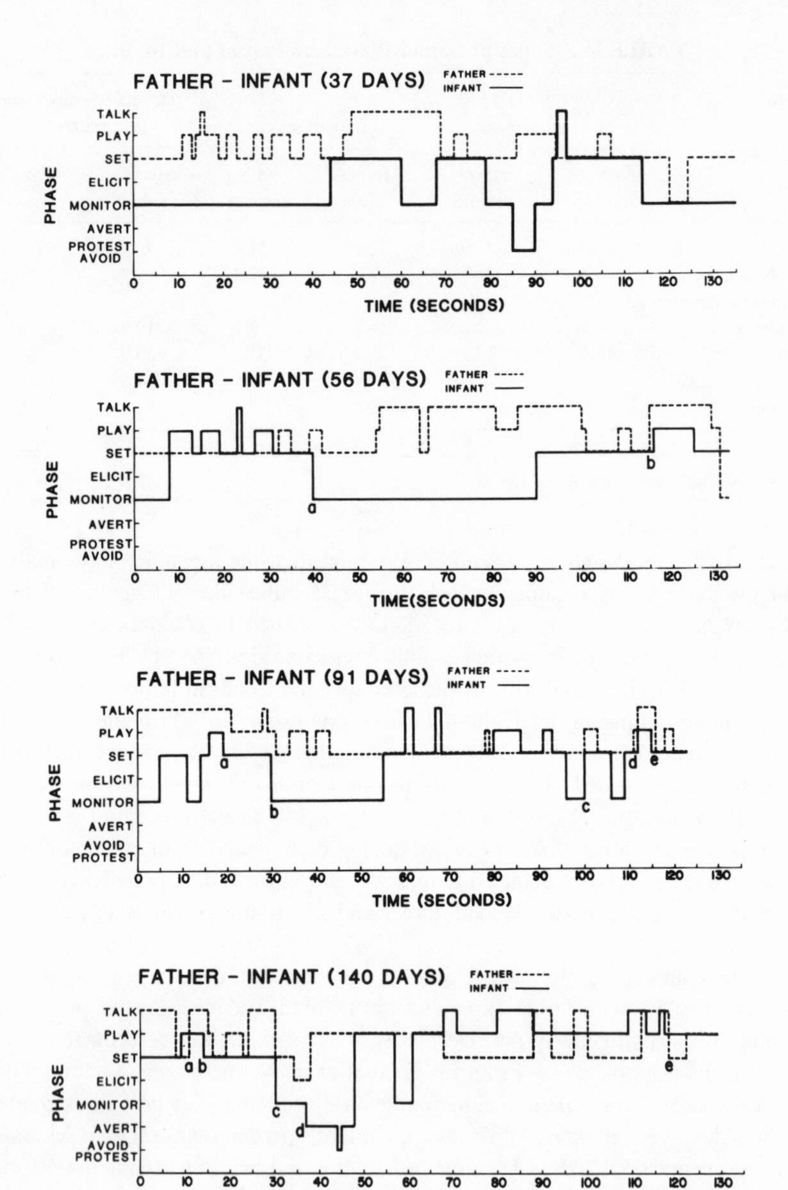

FIGURE 9. Developmental changes in father–infant interaction: monadic phases during sessions with infant aged 1, 2, 3, and 4 months.

she first enters the phase Play after just 8 seconds of interaction and spends over 17% of her time in either Play or Talk. She never enters Protest and spends over 80% of her time in either Set or Monitor. The father remains above Set, frequently in a higher phase than his infant, encouraging her expansion. On two occasions (a and b noted on graph), simultaneous transitions occur for father and infant. By 91 days, the infant's predominant phase is now Set (53.6%), and she reaches the phase Talk on two occasions and Play on several occasions, returning to Set in between. She now seems to be able to integrate smiling and vocalizing with the more neutral Set and she now spends over 65% of her time in these three most affectively positive phases. Simultaneous transitions occur on five occasions (a–e noted on graph). The father's phases now more frequently overlap those of the infant and he spends more of his time in Set than before as if he were now allowing the infant to lead more of the time.

Finally, by 140 days, the infant now spends half her time in the phase Play and over 80% of her time in the three positive phases. She enters the phases Play and Talk repeatedly and remains there for sustained periods. She now displays the full range of phases so that while she Protests for one second, she can also Avert (8.6% of time), conceivably as a means of taking a time-out from the interaction and preparing for the prolonged, intense involvement in Play and Talk. Simultaneous transitions occur on five occasions (a–e) in this session. The father's phases show considerable overlap with the infant's, and the two partners seem to alternate cycling up and down between Set, Play, and Talk as if taking turns.

Games which occurred between this father and infant seemed to encourage the infant's arousal and expansion of her range of affect. For example, at 37 days the father repeatedly tapped his infant, circling her mouth with his finger. He repeated this game several times while repeatedly verbalizing, "Come on." The

TABLE V. Developmental Changes in Structure of Father–Infant Interaction (Percentage of Time in Each Monadic Phase)

	Monadic phase						
	Protest	Avert	Monitor	Elicit	Set	Play	Talk
Infant with father							
Age 37 days	3.7	0	61.5	0	33.3	0	1.5
Age 56 days	0	0	43.6	0	39.1	16.5	.75
Age 91 days	0	0	32.5	0	53.6	11.4	2.4
Age 140 days	.78	8.6	8.6	0	19.5	50.0	12.5
Father with infant							
Age 37 days	0	0	0	0	58.5	25.9	15.5
Age 56 days	0	1.5	0	0	21.8	14.3	62.4
Age 91 days	0	0	0	0	60.1	18.7	21.1
Age 140 days	0	0	0	7.8	38.2	29.7	24.2

infant responded by alerting, stilling, opening her mouth and vocalizing "ahh," softly. These games are repeated as the infant grows older and expanded to incorporate conventional motor activities such as clapping and "pat-a-cake" as well as verbal accentuation ("ahh boo," "come on," and clicking noises).

This infant, while engaging in social interaction with her father, demonstrated an orderly sequence over the first five months of life in which her affective range expanded and her affective expressions became more differentiated. Furthermore, the interactions of father and infant showed a similar progression of mutually regulated and modulated social exchanges as has previously been described for mothers and infants (Als, Tronick, & Brazelton, 1979; Brazelton *et al.,* 1975; Brazelton & Als, 1979).

In sum, although the data described in this section are based on a few selected families, they support the hypothesis that fathers are capable of skilled and sensitive social interaction with young infants. These studies suggest that both similarities and differences exist between these mother–infant and father–infant interactions. Dyadic interaction of infants with both mothers and fathers appears mutually regulated, and in both cases partners build to a peak of attentional involvement and come down in an orderly and cyclical fashion. Furthermore, infants exhibit well-organized expressive displays that are affectively positive with both parents and that become more differentiated in a similar, orderly, developmental progression of modulated social exchanges. Through these mutually regulated reciprocal exchanges, both parents provide a responsive, protective environment that matches the infant's developmental capacities.

Along with these structural similarities, we have illustrated differences in the quality and content of play which we believe provide the anlage for different functional tracks of development. Both types of play offer the infant the opportunity to participate in "turn-taking" activities, through which the infant develops early notions of sharing control in an interactive situation. Through play, the infant learns the rules of culture and of family (Bruner, Jolly, & Silva, 1976). At least for the six families studied in this manner, interaction with fathers can be characterized as heightened and playful whereas the interactions with mothers appeared more smoothly modulated and contained.

Each of these differential tracks may serve a unique function in the infant's development. Together, they foster the development of a wider range of social skills than would develop if only one pattern were available. Both of these tracks appear to be available to infants as early as one month of age, and both depend on the parents' ongoing relationship with their infant. This relationship allows both parents to be aware of the physiological and psychological capacities of their infant and, in turn, both to support and to test the limits of those capabilities.

One possible explanation of the differences between mother–infant and father–infant interaction found in these studies and others (Clarke-Stewart, 1978; Lamb, 1976c) is that in families in which the father is out of the home at work all day, the infant may show a more excited response to his father because he is more novel. In an effort to assess the contribution of the familiarity of the parents to the quality of their interaction with their babies, I began to study families in which traditional caregiving roles were reversed: mothers returned to work in the first few weeks and fathers remained at home as primary caregivers. For this study, our data consist of naturalistic home observations as well as face-to-face interactions in the laboratory. While this study is still in progress, our observations of the quality of father–infant interaction during caretaking tasks suggest that the qualitative differences between mothers and fathers observed in the earlier studies are not based purely on the infant's familiarity with the parent. The primary caretaker fathers I have observed while bathing, diapering, and feeding their babies carry out these tasks with great skill and sensitivity but with the same vigorous, exciting quality seen in face-to-face play with secondary caretaker fathers. One infant's response to the father's vigorous rubdown after the bath was to chortle, cycle his legs, and engage the father in a limb movement game identical in structure to that seen in the earlier study. One study of four-month-old infants in a laboratory face-to-face interaction with primary and secondary caretaker fathers and with primary caretaker mothers has shown similar findings: increased game playing and poking by both groups of fathers (Field, 1978). In this study, primary caregivers (mothers and fathers) engaged in more smiling and baby talk with their infants.

4.3.5. Studies of Father–Infant Interaction at Home

While these laboratory studies document the social capacities of fathers and young infants, we know very little about the actual experiences of fathers and infants at home during these first few months, a period of transition and adaptation to the new baby. Both laboratory- and home-based studies are summarized in Table VI. Pedersen, Andersen, and Cain (1977) have done home observations with infants aged five months and find that the father's behavior with his infant is closely related to the quality of his relationship with his wife even though their perceptions of the baby's temperament may differ widely. Marital tension and conflict were associated with less competent maternal feeding and with the display of negative affect during mother–infant interaction (Pedersen *et al.,* 1977). Given Pedersen's earlier (1975) findings that the baby's alertness and motor maturity at four weeks of age are associated with the husband's support of his wife, it appears that the baby's behavior may influence the relationship between the parents at five months as well. It is likely that during these early months there is wide variability in the kinds and amount of paternal caregiving and interaction with their infants. In one study, fathers did not appear to talk to their infants very much during this period, since on tape recordings collected in the home fathers talked to their infants (aged two weeks to three months) for a mean of 37.5 seconds a day (Rebelsky &

TABLE VI. Studies of Father–Infant Interaction (1–6 Months)

Reference	Sample	Measures	Findings
Yogman, Dixon, Tronick, Adamson, Als, & Brazelton (1976a,b)	Mothers, fathers, strangers, and infants (2 weeks to 6 months)	Behavioral description of face-to-face interaction in the laboratory	Infants by 2 months displayed differential interactive patterns with mothers, fathers, and strangers.
Yogman (1977)	Mothers, fathers, and infants (3 months)	*A priori* clusters of expressive behavior during face-to-face interaction in the laboratory	Similarities: Both mother–infant and father–infant interactions are reciprocal and jointly regulated. Differences: Father–infant interaction characterized by more physical games.
Field (1978)	Primary caretaker mothers, primary and secondary caretaker fathers, infants (4 months)	Face-to-face interaction in the laboratory, behavioral description	Both groups of fathers did more game-playing and poking.
Pedersen, Anderson, & Cain (1977)	Mothers, fathers, and infants (5 months)	Direct observation at home	Behavior of father with infant linked with husband–wife relationship.
Pedersen, Rubenstein & Yarrow (1978)	Mothers and infants (5 months)	Maternal report on fathers, Bayley test	Increased male–infant social responsiveness during Bayley test associated with increased paternal involvement.

Hanks, 1972). More recent studies suggest that the father's verbalizations may not be a very useful index of father–infant interaction.

4.4. Attachment and the Second Year

The study of the father–infant relationship with infants aged 6 to 24 months focuses primarily on the development of attachment as Bowlby (1969) and Ainsworth (1973) have defined it. These studies have asked questions such as: do infants greet, seek proximity with, and protest on separation from fathers as well as mothers? Such studies provide conclusive evidence, as summarized in Table VII, that infants are attached to fathers as well as to mothers. By 7–8 months of age, when the home environment is relatively low in stress, infants are attached to both mothers and fathers and prefer either parent over a stranger (Lamb, 1975, 1977a, 1978a,b). During the second year, most studies also show attachment to both mother and father (Clarke-Stewart, 1978, 1980; Kotelchuck, 1976; Lamb, 1977b), although in a more stressful setting in the laboratory some studies have shown that infants between 12 and 18 months prefer mothers (Lamb, 1976a,b; Ban & Lewis, 1974; Cohen & Campos, 1974). The issue of preference for mother or father depends on what measure is used, but independent estimates by Kotelchuck (1973) in Boston and Schaffer and Emerson (1964) in Scotland are similar: between 12 and 21 months 51–55% of infants show maternal preferences, 19–25% show paternal preferences, and 16–20% show joint preferences. Lamb's data (1978a, 1979) suggest that it is mainly boys who prefer their fathers because their fathers more actively engage them.

Clinically important and yet not well studied are the influences on father–infant attachment. Pedersen and Robson (1969) report that the father's investment in caretaking as well as the level of stimulation and play he provides are positively correlated with the attachment greeting behaviors in 8- to 9-month old babies. Infants whose fathers participate highly in caregiving show less separation protest and cry less with a stranger than infants whose fathers are less involved (Kotelchuck, 1975; Spelke, Zelazo, Kagan, & Kotelchuck, 1973). Qualitative aspects of infant attachment to mothers and fathers studied using Ainsworth's "strange situation" suggested that infants could develop a secure attachment with the father in spite of an insecure attachment with the mother (Lamb, 1978b).

Given that infants are attached to both mothers and fathers, how are their relationships different? First, mothers and fathers interact differently: fathers engage in more play than caretaking activities with six-month-olds (Rendina & Dickerscheid, 1976) and more often pick up their infants (eight months) to play physical, idiosyncratic, rough-and-tumble games, while mothers are more likely to hold infants, engage in caregiving tasks, and either play with toys or use conventional games such as peek-a-boo (Lamb, 1977a). By age $2\frac{1}{2}$ when parents were

TABLE VII. Studies of the Father–Infant Relationship (6 Months–2 Years)

Reference	Sample	Measures	Findings
Schaffer & Emerson (1964)	Mothers, fathers, and infants (12, 18 months)	Parental interview	Increased separation protest with mother at 12 months, $\frac{3}{4}$ of infants attached to fathers as well by 18 months.
Kotelchuck (1973, 1976); Spelke, Zelazo, Kagan, & Kotelchuck (1973); Lester, Kotelchuck, Spelke, Sellers, & Klein (1974)	Mothers, fathers, and infants 6–12 months in U.S. and Guatemala	Separation protest in lab	No differences when mother or father leaves.
Pedersen & Robson (1969)	Mothers and first-born infants (9 months)	Maternal interview	Increased paternal investment in caretaking and stimulation during play associated with increased greeting for father.
Lamb (1975)	Mothers, fathers, and infants (8 months)	Strange situation in the home	Attachment to both parents; mother holds infant more for caregiving, father holds infant more for play, and infant responded more positively to father when held for play.
Lamb (1976a,b)	Mothers, fathers, and infants (12, 18 months)	Strange situation and free play in the lab	Mother preferred only under high stress.
Lamb (1977b)	Mothers, fathers, and infants (24 months)	Strange situation in the lab	Attachment to both parents, no preference.
Willemsen, Flaherty, Heaton, & Ritchey (1974)	Mothers, fathers, and infants (12 months)	Separation protest in the lab	No difference with mothers and fathers.
Cohen & Campos (1974)	Mothers, fathers, and infants (10, 13, 16 months)	Proximity-seeking, separation protest	No difference in separation protest, increased proximity-seeking with mother.

Ban & Lewis (1974)	Mothers, fathers, and infants (12 months)	Free play in lab, proximal and distal attachment behaviors	Increased proximal behaviors with mother, male infants looked more at fathers.
Lewis & Weinraub (1974)	Mothers, fathers, and infants (12, 24 months)	Free play in lab, proximal and distal attachment behaviors	No differences between infant behaviors with mothers and fathers.
Lamb (1977a,b)	Mothers, fathers, and infants (7–8 months, 12–13 months, 15, 18, 21, 24 months)	Structured and unstructured home observation	No preference in attachment behaviors during first year; increased attachment to father during second year; fathers do more physical, rough-and-tumble play, mothers do more conventional play using a toy.
Clarke-Stewart (1978, 1980)	Mothers, fathers, and infants (15, 20, 30 months)	Structured and naturalistic home observation	No difference in infant attachment to mother and father (20 months); fathers' play is briefer, more physical and arousing, and they enjoyed play more and were better able to engage child; infant is more involved, excited, and interested during play with father.
Power & Parke (1979)	Mothers, fathers, and infants (8 months)	Semi-structured play	Fathers played more physical games; mothers played more distal games.
Lamb (1978b)	Mothers, fathers, and infants (12 months)	Strange situation in the lab, attachment typologies	Quality of attachment was independent with each parent.
Frodi, Lamb, Leavitt, & Donovan (1978)	Mothers and fathers of 18-month-olds	Adult physiological responses to infant cry	Both parents equally sensitive to infant smile and cry.

asked to engage the child in specific play activities, fathers were better able to engage the child in play. Father's play with his child was more likely to be proximal (as was described for younger infants), social, physical, arousing, and briefer in duration, and fathers reported that they enjoyed it more than mothers (Clarke-Stewart, 1978, 1980). Infants at eight months responded more positively to play with fathers than mothers (Lamb, 1977a) and by age $2\frac{1}{2}$ not only preferred to play with fathers, but were judged to be more involved and excited with them (Clarke-Stewart, 1978). It is fascinating to note that fathers' physical play with their infants correlates most highly with mothers' verbal stimulation (.89) and toy play (.96) (Clarke-Stewart, 1980). These are maternal behaviors found in other studies to be part of a pattern of "optimal maternal care" (Clarke-Stewart, 1973). Although there may be qualitative differences in the relationship of mothers and fathers with their infants, parents of nine-month-olds were equally sensitive to infant smiles and cries (Frodi, Lamb, Leavitt, & Donovan, 1978).

All of these studies have been based on the assumption that the quality of the father–infant relationship is more important than the quantity given some yet-to-be-defined lower limit. In natural observations, the mother is clearly the predominant partner (Clarke-Stewart, 1980). Reports of father involvement vary widely from a mean of 8 hours per week spent playing and 26 hours per week available at home with awake nine-month-old babies (Pedersen & Robson, 1969) to 3.2 hours per day ($\frac{1}{3}$ that of mother) spent with the infant (Kotelchuck, 1976), to 30 minutes per day spent alone with the infant ($\frac{1}{12}$ that of mother) (Pedersen, Yarrow, Anderson, & Cain, 1979). When one looks at specific activities (feeding, cleaning, play), fathers also spend much less time doing each than mothers do, although a greater proportion of fathers' time with the baby is spent in play (37.5% vs. 25.8%) (Kotelchuck, 1975). Even though fathers do not spend as much time with their babies as mothers do, at least 25–50% of fathers are involved in some caretaking responsibilities (Kotelchuck, 1975; Newson & Newson, 1963; Richards, Dunn, & Antonis, 1977). Almost 90% of fathers often played with their infants, while fathers were somewhat likely to feed their infants and least likely to change diapers and bathe the baby (Kotelchuck, 1975; Richards *et al.,* 1977; Newson & Newson, 1963). The degree to which fathers participate in caregiving and in play with their infants is related to their sex role classification: increased participation is associated with androgynous sex roles (Russell, 1978). Finally, the actual time parent and infant actually spend interacting may be substantially less than simply the total amount of time they spend engaged in any of these specific activities. Differences in the time mothers and fathers spend interacting with their infants may be less than they first appear.

In summarizing these studies, infants seem to be attached to both parents although when stressed prefer their mothers. Fathers engage in more physical play and their infants, particularly males, respond more positively to play with their fathers.

4.5. Influences on Later Development

The influence of the father–infant relationship on later cognitive, social, and emotional development of the infant has not been well studied. In a prenatal and postpartum interview study of the influence of maternal and paternal expectations and roles, Fein (1976) suggests that the most effective male postpartum adjustment was related not simply to high or low paternal involvement but rather to a coherent role that met both the needs of the father himself, the mother, and the baby.

The influence of the father–infant relationship on the infant during these early months is suggested by the report that, at least for males, increased father involvement at home is associated with greater infant social responsiveness at five months of age during a Bayley test (Pedersen, Rubenstein, & Yarrow, 1979). Concurrent predictions of infant Bayley scores at 16 and 22 months were related to the father's positive perceptions of the child and his ability to engage the child in play and to anticipate independence on the part of the child. Predictions of concurrent social competence were related to the father's verbal and playful behavior and, for girls, his expectation of independence (Clarke-Stewart, 1978, 1980). Boys in particular have been found to be more autonomous when both parents are warm and affectionate (Baumrind & Black, 1967). The use of cross-lagged correlations led Clarke-Stewart (1980) to suggest that the mother's warmth, verbal stimulation, and play with toys with infants at 15 months of age were related to higher infant Bayley scores at 30 months which, in turn, influenced fathers to engage in play more often, to expect more independence, and to perceive their children more positively.

Most attempts to assess the impact of the father–infant relationship on later development have looked at father-absent families and the relationship to sex role identification. Father-absence, particularly prior to age five (Mischel, 1970), has been shown to influence masculine sex role adoption and cognitive style among boys (Biller, 1970, 1976; Carlsmith, 1964; Hetherington, 1966) and heterosexual roles among girls (Johnson, 1963). Since these studies have been criticized for confounding both the underlying reason for the father's absence and its effect on the mother (Herzog & Sudia, 1973), the focus has shifted to understanding the differential relationship of fathers with sons and daughters during infancy. These studies show that not only do fathers vocalize and play more with sons than daughters but that this is especially true for first-born sons (Parke & O'Leary, 1976; Parke & Sawin, 1975, 1977; Parke, Power, Tinsley, & Hymel, 1979). The preference seems true not only in the United States but in Israel (Gerwirtz & Gerwirtz, 1968) and among the !Kung San bushmen (West & Konner, 1976). Studies of infant preferences show that one-year-old male infants look more at fathers than at mothers during free play in the laboratory (Ban & Lewis, 1974), remain closer and vocalize more to fathers than to mothers during a more stressful laboratory procedure (Spelke *et al.*, 1973), and that by 20 months of age, male infants show a specific play preference for fathers (Clarke-Stewart, 1980). Belsky (1979) attempted to determine the direction of this preference in home observations and attributes the preferences of male 15-month-olds for their fathers primarily to the behavior of their fathers.

In sum, these studies of the normal adaptation of infants and their fathers during the first two years have supported the idea that parent and infant development proceed in parallel. The pregnancy experience seems to be a time of developmental transition for fathers as well as mothers. After birth, both the father and infant are available for meaningful social interaction although there appear to be qualitative differences between mother–infant and father–infant interaction. Communication between the parents also appears to have a major influence on the parents' relationship with their baby. By the end of the first year, an enduring attachment exists between fathers and infants, even though fathers may spend less time interacting with their infants than mothers do. Finally, interactions between fathers and infants commonly involve physical play. These studies have barely begun to unravel the complexity of the influence of the father–infant relationship on the child's later cognitive and socioemotional development.

5. THEORETICAL CONSIDERATIONS

These studies of fathers and infants are beginning to form a basis for theorizing about the transition to parenthood in adult male development, about the nature of the father–infant relationship, and about the influences of paternal involvement on infant personality development.

First of all, studies of father–infant interaction suggest that fathers can have a meaningful and direct relationship with their infants right from birth. Fathers undergo significant psychological changes during pregnancy when they share the anticipation of the birth of their baby with mothers, and they describe feelings of "engrossment" at the time of birth. Similarities and differences between the experiences of expectant mothers and expectant fathers still need to be studied. One can speculate about the relationship between the transition to parenthood and the successful negotiation of Erikson's (1959) psychosocial crisis of adulthood: generativity *vs.* self-absorption. For the father the opportunity for increased investment in an infant offers an opportunity for a profound kind of generativity and seems both biologically and psychologically adaptive. Biologically, increased male "parental investment" (Trivers, 1972, as discussed in the introduction) may represent adaptation by the species for infant care in the current ecological setting. Such an adaptation may better insure the survival of offspring and thereby maximize parents' contributions to the gene pool. Although the biological components of male parental investment are much less evident than the more obvious female role during birth and nursing, they may be no less real, but more subtle. Hormonal studies of male adults during the perinatal period would be useful in this regard. Perhaps when men are expected to participate in caregiving and do so, their androgen levels may decrease.

Increased emotional investment in infants by fathers may be psychologically adaptive if the father views the infant as a joint creation of both parents and if this supports his desire for greater participation in the emotional life of his family to balance the often conflicting demands of ambition and work. Perhaps most obvious, the father's direct relationship with the infant is valuable for the mother in that it offers her both the support and time out that an extended family may have offered in the past. As families have grown smaller, it seems easier for a mother to have another person or other persons to share the caretaking responsibilities.

The fact that both parents can have a direct, sensitive, and responsive relationship with their infant suggests that highly invested human adults are capable of forming a consistent, loving relationship sensitive to an infant's emotional state and immature attentional capacities. The fact that either parent can learn to perform certain caretaking tasks may be a reflection of both the infant's eliciting capabilities and the human adult's capability to learn. As Rossi (1977) points out, however, while what men and women do is not entirely genetically determined, the biological contributions shape what is learned and there are differences in the ease with which the different sexes can learn certain things.

Therefore, it seems unlikely that the relationships of fathers and mothers with their infants are identical and redundant. The biological fact of pregnancy, the nine-month period of physical symbiosis, sets the stage for an intimacy between mother and baby from birth and a facilitated process of entrainment that a father can only develop over a period of time. While there is little evidence that human systems are unmodifiable, it is likely to require more learning for the father to achieve a similar level of intimacy with his baby. In our studies, fathers were able to achieve mutually regulated, reciprocal interactions with their infants by three months of age, but the fathers offered the infant a qualitatively different experience than that with the mother.

In spite of different research strategies and samples, the studies have shown considerable consistency in describing the nature of father–infant interaction. In the newborn period fathers held and rocked their babies more than mothers and were more stimulating (Parke & Sawin, 1975). By three months of age, fathers and infants in our studies engaged in more proximal, arousing, idiosyncratic games, whereas mothers and infants were more likely to engage in smoothly modulated, soothing distal games, especially verbal ones. Others have found that fathers are more likely to play physical, arousing, idiosyncratic games with their infants at one and two years of age while mothers are more likely to play conventional games (Clarke-Stewart, 1980; Lamb, 1975). Furthermore, infants respond to their fathers with more excitement. In sum, while fathers have been shown to be sensitive to infant cues (Parke & Sawin, 1975, 1977) and skilled interactants with young infants (Yogman, 1977), consistent and rather stable differences in the quality of behavioral regulation have been demonstrated between father–infant and mother–infant interaction. Fathers seem more likely to develop a heightened, arousing, and playful relationship with their infants (Clarke-Stewart, 1980; Parke, 1979; Yogman, 1977) and to provide a more novel and complex environment (Pedersen *et al.,* 1979).

Although it is clear that cultural and social influences and the amount of time spent with the infant influence these interactions, it may also be true that these qualitative differences in the way fathers and infants interact are based on sex-linked biological predispositions. When reinforced by cultural stereotypes, both parents can be viewed as playing direct but complementary roles with their infants in which attempts by either to substitute entirely for the other require prolonged periods of learning. The interaction of biological predispositions and cultural influences could be elucidated by studying the influences on the father's role in several modern societies; for example, a comparison of the father's role with infants in France, Germany, Japan, and the Soviet Union might help us understand the influences of political and economic forces on the transmission of values within a culture.

Keeping in mind the impact of differential parent behavior, it is also important to consider the influence of the infant on the father. For example, Rosenblatt's (1969) studies of rats showed that maternal behavior was under the control of both a short-term hormonal system surrounding parturition and a second longer-term postpartum system that relies on the presence of the infant to elicit maternal behavior. Given our understanding of the competencies of the human newborn, it is likely that the infant's capabilities to elicit and shape parenting are more long-lasting than either biological or hormonal influences in parents. Perhaps future research will more clearly delineate the infant's role in shaping his parents to play different but complementary roles, that are both stable and expectable from each parent.

Given the qualitative differences found between mother–infant and father–infant interaction and given the increased involvement of fathers with young infants, it seems useful to consider the influences of paternal involvement on infant personality development. The optimal caregiving environment for the infant during the first two years of life serves two primary functions: (1) it offers a nurturant, protective environment which matches the infant's developmental capacities and (2) it offers an environment which challenges the infant's capabilities, facilitates the differentiation of new skills, and gradually encourages him to become independent and to explore and learn about the world as an autonomous individual. Such terms as *attachment* (Bowlby, 1969), *autonomy* (Erikson, 1959), and *symbiosis, hatching,* and *separation–individuation* (Mahler, Pine, & Bergman, 1975) have been used to describe these processes. One wonders whether the father's more stimulating and arousing pattern of interaction with his infant is in any way related to the acquisition and differentiation of new skills by the infant or to the development of independence, exploration, or autonomy. One also wonders whether an infant who is securely attached to both father and mother is in some way less vulnerable to stress since the relationship with one parent could compensate for any difficulty in the relationship with the other.

The developmental process for both parents and infant in such a system may be conceptualized in terms of Werner's developmental theory as one of increasing differentiation and integration of skills (Langer, 1970) and in terms of Sander's (1979) theory of systems regulation. Shortly after birth, the infant's biorhythms become entrained with those of the caregiver (Sander *et al.,* 1975) and the infant becomes more stable physiologically. These accomplishments are shaped by social interactions with mothers, which are modulated and contained, in keeping with the infant's physiological immaturity, and by other interactions characterized by the more intense, arousing stimulation which commonly occurs with fathers. In these early months, interactions with mother may prolong the concurrent state of the infant, often resulting in prolonged periods of quiet alertness, while interactions with father may elicit repeated changes of state and may be associated with greater proprioceptive and kinesthetic stimulation. These social interactions become a way of learning not only about basic trust in the external world engendered by synchronous interaction, but also learning about one's own mechanisms for internal control and homeostasis when stressed (Brazelton, Koslowski, & Main, 1974).

Interactive games during this period also provide differential experiences for the infant. One can speculate about the developmental significance of these games. Conventional games, more common with mothers, may allow the establishment and consolidation of rules of interchange that provide the foundation for later language development, while more arousing physical games, more common with fathers, may differentiate into alternate forms of social play, eventually incorporating objects and later leading into further instrumental activities (Tronick, 1977; Yogman, 1977).

By the end of the first year, the infant is beginning to explore his surroundings while maintaining focused attachment relationships with both his mother and father. Although we know that many infants show comparable attachment behaviors toward both parents, we know much less about the way either parent encourages the child's exploration during physical play. Finally, by around 18 months, infants' interactions with parents provide differential experiences: (1) the opportunity to be closer to their mothers (rapprochement) (Mahler *et al.,* 1975) and (2) the opportunity to use their fathers as the third member of the family triangle to facilitate their separation and individuation: "There must be an I, like him, wanting her" ("intrapsychic triangulation") (Abelin, 1971, 1975). Autonomy seems to appear as a gradually evolving phenomenon, following the progression of competence in the infant and the infant's feelings of mastery and control in an increasingly novel and stressful situation.

At various stages of development, the father–infant relationship may complement the infant's relationship with the mother and facilitate the development of autonomy by providing a range of novel, arousing, and playful experiences for the infant. As yet, we know little about the salience of specific aspects of the infant's experience with fathers as compared with mothers (differences in kinds of stimulation, quality of voice or of handling) or about how early perceptions are influenced by repeated experience with various caretakers. It is interesting to speculate about the relationship between the recent trend for fathers to be increasingly involved with their infants and the evolving cultural objectives of child-rearing to prepare children for more complex roles in society. The greater social emphasis on individualism and self-sufficiency in personal growth is demanding that children be capable of autonomy at increasingly earlier ages (as an example, consider the growth of day-care and preschool education in Western societies). One hopes that future studies will elucidate the influences of paternal involvement with infants on personality development.

The triadic model of mother and father playing complementary roles in rearing their baby provides one model for a nuclear family. The fact that single parents, and in particular, single women, also raise autonomous children means that parents can and do play dual roles with their children, but in no way does it imply that playing dual roles is an easy task.

It is interesting to think of this triad in cybernetic terms in which information exchange occurs between all three dyads within a mutually regulated feedback system (Ashby, 1956). The strength of such a system is that the larger triadic system provides some overall stability while the feedback system within any one of the dyads can be transiently disrupted. This carries with it the opportunity for differentiation, separation, and individuation within a dyad while a stable matrix for developmental organization is maintained in the larger system.

Flexibility in such a system is related to how loosely coupled the subsystems are. When these are loosely coupled, the system can allow the temporary independence of subsystems while the larger system remains stable (Sander, 1979). In such a situation, the infant can differentiate, individuate, and become autonomous within an organizing system which balances his arousal and excitement with the opportunity for recovery and balances his exploration of the world with a secure home base to learn about himself. Not only can the infant learn from two other people in a triad, but the system allows any two members to readjust and cope with stress or disruption in the other member.

Finally, such a model not only may help us understand a family's capacity to cope with stress, but also allows us to conceptualize maladaptation within a family (Brazelton, Yogman, Als, & Tronick, 1979). Pathological symbiotic affective relationships within a family might be thought of as tightly coupled subsystems in which whatever occurs in one dyad is replicated in the other. Such a system is less flexible and might interfere with optimal individuation for the infant. Conversely, uncoupled subsystems might provide no consistent organization and be more characteristic of depriving family environments.

6. CLINICAL APPLICATIONS

Given the paucity, until recently, of any theoretical or empirical base for understanding a father's role in early infancy, it is not surprising that fathers have often been excluded, ignored, and patronized in the management of the clinical problems of infants. By discussing a few clinical problems, I would like to suggest that improved outcomes for families may result from the systematic inclusion of fathers in pediatric management of these problems.

6.1. Infants at High Risk

The birth of a high-risk infant confronts fathers as well as mothers with a major stress. Preterm birth is a good example. Initially, fathers are often required to play a primary role in decision-making either because the infant has been transferred to a different hospital or because of the physical incapacity of the mother. In one study in progress, the father's behavior at this initial visit appears to be associated with subsequent contacts during the first month. Fathers who touched and talked to their newborns and visited longer than 15 minutes were more likely to call and visit more frequently during the subsequent four weeks (Johnson & Gaiter, 1980). While studies such as this are beginning to describe fathers' behavior during early visits, we know much less about long-term influences. On the basis of a few families referred for psychiatric help, Herzog (1979) has expressed concern about early increased paternal involvement with the infant at the expense of maternal attachment. Fathers may need permission to express their feelings of fear and grief during the perinatal period even if it violates the stereotype of masculinity. Zilboorg (1931) has examined the interaction between psychodynamic and sociocultural influences on postpartum reactions in men and women and suggests that men are more likely than women to develop symptoms of extreme regression, psychosis, or paranoia because depression is a socially unacceptable symptom in males. According to Zilboorg, society instructs men to view symptoms such as passivity, impotence, and crying as a threat to their self-image. Therefore, professionals working with fathers in a premature nursery may need to view bizarre and regressive behavior as less pathological than in other situations and as analogous to maternal depression.

Once high-risk infants leave the hospital, fathers may become even more highly involved in caregiving or, on the other hand, they may become less directly involved. In a study of families followed after cesarean section delivery, home observations when infants were five months of age suggested that fathers were more responsive to infant distress than mothers were (Pedersen, Zaslow, Cain, & Anderson, 1980). In a cohort of families with preterm infants whom we are following at Children's Hospital as part of a longitudinal study, fathers of preterm infants report that they do more caregiving (bathing and diapering of their infants) than reported by fathers of a comparison group of term infants. Whether the father plays a directly supportive role or a more indirect role of encouraging and supporting the mother through a difficult adjustment, the father's capability and availability to offer support to his family during this crisis may be critical to long-term outcome. Any high-risk event represents a stress for a family, and increased paternal involvement may represent an example of the way families mobilize support systems to cope with stresses.

Professionals must be alert to signs of maladaptive paternal roles: complete emotional withdrawal from his infant and spouse or an excessively competitive primary attachment with the baby which excludes the spouse. Heightened paternal feelings of competition with his spouse have been shown to occur with term infants, particularly if the father has a close relationship with the baby and the mother is nursing. Husbands of nursing mothers described feelings of inadequacy, envy, and exclusion, and the competition may actually undermine the mother's attempts at breast-feeding unless these feelings are addressed (Lerner, 1979).

Given the difficulties of social interaction with a premature infant whose state and motor organization may be labile (Field, 1979; Goldberg, 1978) and whose cry seems more aversive to father and mother (Frodi, Lamb, Leavitt, Donovan, Neff, & Sherry, 1978), the tendency of many fathers to excite, play with, and vigorously stimulate their infant may stress an already vulnerable infant, interfere with social interaction, and lead a father to withdraw. Describing the baby's cues (particularly when they are poorly readable) may be of great help to a father in this situation and may offer him a more adaptive alternative to complete withdrawal. Unfortunately, most intervention programs for high-risk infants do not specifically address the father's role with these infants (Bricker & Bricker, 1976). The expectation that most fathers will not participate quickly becomes a self-fulfilling prophecy unless the father's participation is actively encouraged.

6.2. Sudden Infant Death

Paternal responses to sudden, unexplained infant death are another instance in which pediatricians must be sensitive to the father's grief reaction as well as to the mother's. In a study of 46 such families (Mandell, McAnulty, & Reece, 1980), in which both parents were interviewed, several identifiable mourning patterns seemed more characteristic of men: (1) the necessity to keep busy with extra jobs or increased work loads, (2) a feeling of diminished self-worth, (3) self-blame because of lack of "care" involvement, and (4) the overwhelming limited ability to ask for help. Health care providers often seemed unwittingly to promote masculine stoicism and managerial functions which may serve to obstruct the full expression of grief. While mothers often request help, their presenting problems were often their concerns about the fathers who would not ask for help. On the basis of these interviews, the authors suggest that if fathers are given an opportunity to express feelings and a validation of their feelings, they will constructively utilize support. The task for the health professional is to respond to masculine expressions of distress as well as the more overt feminine ones.

6.3. Handicapped Infants

The paternal response to the birth of a retarded or defective infant is similar in some ways to his response to the sudden loss of a child. When informed, fathers are less emotional and expressive than mothers and ask questions about future problems (Price-Bonham & Addison, 1978). Gath's case-control study of families experiencing the birth of an infant with Down's syndrome suggests that the birth resulted in severe degrees of marital tension and strain during the subsequent 18 months (Gath, 1974). The study suggested that the source of marital tension was commonly sexual dissatisfaction, which may have been a reflection of the fact that the father mourns as well as the mother after having produced a defective offspring.

Fathers may also serve as a protective factor in a family undergoing a stressful situation. Studies of infant outcome when the mother is retarded or psychotic must look not only at the mother's coping capacities as a parent but at the father's direct role as a caregiver. I have followed many families in which the father's capacity to buffer external stresses was critical to the infant's healthy development.

6.4. Differences in Infant Temperament

When an infant is temperamentally irritable or difficult, the father as well as the mother may find difficulty interacting with the infant. However, a father unsure of himself with a young infant is likely to blame himself for these difficulties and either withdraw or become more stimulating, overwhelming an already hypersensitive infant. We recently described father–infant interaction in a family in which the infant had difficulties with state regulation and was especially irritable at birth (Yogman, Dixon, Tronick, Als, Adamson, & Brazelton, 1976c). During the early weeks, this infant was either difficult to rouse or after minimal interaction began fussing. He went on to develop facial eczema at three months of age. By videotaping face-to-face interactions and reviewing and discussing the videotapes with the parents, we may have helped father as well as mother to adapt to this infant's temperamental characteristics.

During the first two months after birth, neither parent was able to engage this infant successfully in sustained reciprocal social interactions. The infant remained predominantly fretful or uninvolved. By three months of age, brief periods of reciprocal vocalization and smiling began to occur and, just as the parents began to feel successful, their infant developed a facial rash. In response to the itching and discomfort, the infant again became fussy. Both parents responded by providing even more stimulation. After a difficult period lasting a few weeks, the mother began to slow down, modulate her behavior, and search for ways to help the baby comfort himself. The father, however, became even more intrusive and his interactions with the infant were characterized by frequent infant gaze aversion and arching. Meanwhile, the baby got appropriate medical intervention and the rash began to subside. Finally, by the time the infant was seven months of age, the father had found less intrusive ways of interacting with his baby. He became less over-stimulating, and he and his infant established their own style of reciprocity and communication which now often included toys. On follow-up visits, the infant developed into an autonomous, competent two-year-old. This father and mother both had to learn to adapt to the stresses of this infant's temperament and facial rash.

6.5. Behavioral Problems of Infants and Toddlers

When a pediatrician is asked for help with the behavioral problems of infants and toddlers, meeting with the father as well as the mother may be crucial to understanding these problems and helping the family find solutions. For example, helping a family whose baby is persistently fussy may require establishing a strong alliance between the parents that supports both parents and avoids placing responsibility or blame on either. Sleep problems are another example. Difficulties in bedtime settling may be exacerbated if the father arrives home just before bedtime, elicits the predictable excited reaction from the baby, and then attempts to convince the baby to go to sleep. Persistent waking at night can create substantial marital tension. Issues such as who goes to the baby, do the parents take the baby into bed, and the impact of sleep loss on the parent's job may become major sources of conflict between parents and result in a fixed symptom for the infant. By directly talking with fathers as well as mothers about these problems, the pediatrician can diffuse some of the tension while remaining the child's advocate.

Although scheduling pediatric visits when working parents can come is often difficult, it is surprising how willing fathers are to modify their own schedules in order to come to the pediatrician when they are concerned about their infant and when they believe that the pediatrician is really willing to listen to them. When parents are having difficulties with their child's negativism or with discipline and control during the second year, discussions with the father as well as the mother may be crucial. Toddlers are unusually capable of highlighting any inconsistencies between mothers and fathers in discipline and limit-setting. Unless parents have an opportunity to discuss their feelings about the child's behavior with each other, they may find themselves competing for the child's favor and exacerbating the problems. By discussing the situation with fathers and mothers together in a few extended counseling visits, the pediatrician can be very helpful in either alleviating the problems or helping the parents to accept a referral for more intensive therapy.

6.6. Child Abuse

The clinical importance of the father's role with young infants is perhaps best illustrated by the literature on child abuse. Green (1979; Green, Gaines, & Sandgrund, 1974), on the basis of studies of a cohort of 60 abused children in New York, reports that 43% of the children were abused by their fathers or father surrogates and 14% were jointly abused by mothers and fathers. Gil (1970) concurs, reporting that when fathers were present in the home, they were involved in two thirds of the abusive incidents. He suggests that personality disorders of parent, environmental stresses, and child characteristics are all potentiating factors, just as with abusing mothers (Newberger, Reed, Daniel, Hyde, & Kotelchuck, 1977). Other reviews (Rigler & Spinetta, 1972) suggest that substitute or social fathers (in most cases, a boyfriend) more commonly abuse children conceived from a prior relationship of the mother than biological fathers abuse their own children.

Given the sampling biases, these studies merely serve to emphasize the importance of intervening with fathers as well as mothers in these situations. Recent evidence that links varying manifestations of domestic violence—spouse abuse, child abuse—reinforces the need both to understand and to intervene with fathers to prevent child abuse whether the father's role be direct or indirect.

6.7. Teenage Pregnancy

Given the current concerns about the increase in teenage pregnancies, it is surprising that more attention has not been devoted to the role of adolescent fathers. Perhaps professionals have been reluctant to study adolescent fathers because of an early report (Lorenzi, Klerman, & Jekel, 1977) that when the adolescent mother finds a permanent relationship with the putative father a second pregnancy occurs within 18 months of the first and that this pregnancy carries a higher biological risk for the infant than did the first. Very little is known about the emotional status or the motives for reproduction in adolescent males. Increasing our knowledge of fatherhood should contribute to improved services and may direct the way toward effective intervention in the high rates of unwanted pregnancy in this age group.

In following teenage mothers with infants, I have noted the great influence of the father and his family on both mother and baby. While impermanence of relationships between adolescent parents is certainly common, it is rather surprising to find how often the parents desire a permanent relationship. Several studies have documented that 40–50% of clinic samples of adolescent mothers have maintained a relationship with a single male, the putative father, for a period of more than two years following the first pregnancy (Platts, 1968; Ewer & Gibbs, 1975). Many of these fathers sustain an investment in their infant long after their emotional relationship with the baby's mother has ended. In fact, the father's parents often play an active role as substitute caregivers or even primary caregivers if the mother gives the child up. Pediatricians may find themselves in the middle of this complex web if the mother begins to amplify symptoms or illnesses in the baby as a way of maintaining a relationship with the father. An optimal role for the pediatrician in this situation must be to remain the advocate for the child, and I believe that this often requires direct communication with the father as well as the mother. Unfortunately, such contact is often difficult to arrange, a problem very few programs have addressed. Given the continuing involvement of these fathers with their infants, our exclusion of fathers and our failure to provide educational and social support for paternal roles leave fathers uninformed about infant care, infant competencies, and infant needs. To be truly effective in reaching adolescent males, such education must be offered as part of a school curriculum (Sawin & Parke, 1976).

6.8. Child Custody

The more active role in child-rearing played by fathers has begun to influence child custody decisions in divorce cases. Courts have begun to award contested custody to fathers (38% of cases in one Minneapolis study, Friedman, 1973), and joint custody arrangements are becoming more common. I believe that pediatricians must become forceful advocates for the child's interests in these situations and must make that role explicit with parents. The difficulty often lies in coordinating the child's needs for consistent, stable, secure caregiving with the needs to maintain the already established relationship with both parents. Whenever possible, parental counseling and mediation serve the child's interests more than a contested court struggle.

In the few limited follow-up studies comparing the effects of different custody arrangements on older children, parental cooperation appears to be related more to outcome than is any specific custody arrangement (Abarbanel, 1979; Warshak & Santrock, 1979).

Single parents, both mothers and fathers, are increasingly common in the United States, and the demands on them to play dual roles often require that the pediatrician be a supportive listener, educator, and, at times, interpreter for the child. Not only are fathers becoming single parents as a result of divorce or the death of a spouse, but single males are becoming adoptive parents as well (Levine, 1976). In the 1975 United States census, 324,000 males were single parents. Although it is estimated that only 1% of children are reared by primary caretaker fathers in this country, more than 15% of children under age 14 are cared for by fathers while the mothers work, and pediatricians must increasingly deal with the concerns of these fathers as well.

In sum, these are just a few examples of the importance of understanding the father–infant relationship for more effective management of pediatric problems.

7. IMPLICATIONS AND RECOMMENDATIONS FOR SOCIAL POLICY

Many fathers seem both competent as caregivers and committed to the care of their infants. Social policies have been slow to acknowledge a meaningful direct role for fathers in the infant's life. In situations wherein the culture and family group support fathers' direct involvement with infants, our programs and policies ought to facilitate and legitimize this participation so that fathers will both execute these tasks more effectively and view their behavior as role-consistent. When culture and family discourage the direct involvement of fathers with their infants, programs can still encourage paternal support of the mother. The changes in obstetric and neonatal services which encourage the father's presence during labor and delivery and in the nurseries have gone far toward reversing the separation and exclusion of fathers from their infants. Postpartum support groups for fathers described by Rizzuto (1978) document fathers' unmet needs for discussing their concerns and fears without being labeled aberrant. Several educational programs have shown that brief interventions with fathers in the perinatal period could influence their attitudes, caregiving skills, and knowledge of infant capabilities for as long as three months (Parke, Hymel, Power, & Tinsley, 1980). In Sweden, fathers who received simple instructions on bathing, changing, and feeding in the perinatal period were found to have higher degrees of infant caretaking activity as recorded on a maternal questionnaire six weeks after discharge (Johannesson, 1969). Simple interventions with fathers of older children (instructing them to play with their 12-month-old sons 50 minutes a day) were also effective. One month later, infants in the intervention group showed greater degrees of proximity-seeking to their fathers in a free play context than fathers in the nonintervention group (Zelazo, Kotelchuck, Barber, & David, 1977). More prolonged interventions have been shown to improve the infant's competence as well (Dickie & Carnahan, 1979). Changing our stereotypes about the relationship of fathers and infants probably requires educating adolescents about infants as part of a high school curriculum in parenting.

In addition to institutionally based interventions, employment and economic policies have a major impact on the capacities of fathers as well as mothers to care for their babies the way they wish. Zaleznik (1979) has discussed the enormous pressures on male and female business executives that compete with family responsibilities. Many fathers complain about the way in which inflexible work schedules prevent them from spending adequate time with their infants. Sweden has now offered men the option of taking paternity leaves to help care for their babies (Palme, 1972). Suggestions for changes in the United States to support fathers as well as mothers have included flexible work schedules and part-time employment. None of these programs acknowledges the difficulties parents face when an infant develops even a minor illness. The effectiveness of such changes in work schedule also requires a concomitant shift in the values of the work place, such that people selecting those options are not viewed as less effective contributors.

8. CONCLUSION

While much has recently been written about fathers and infants, our knowledge of the social and emotional significance of paternal involvement with young infants is still quite limited. Therefore, this concluding section will attempt both to summarize this chapter and to pose future questions for research.

Whereas a cultural stereotype exists for mothering, no such entity exists for fathering. When a man becomes a parent, the options available to him are exceedingly varied. While the biological component of parenting may well prove to be weaker in men than women, recent social and cultural changes have supported men who want to be directly involved with their infants right from birth. The studies described in this chapter suggest that fathers can successfully and reciprocally interact with their young infants.

While sociocultural changes have enabled both mothers and fathers to play similar roles if they wish, biologically based predispositions to adult sex differences may shape the quality of parental behavior. These sex differences appear to start in infancy. Males are more active and more sensitive to physical stimuli and females are more verbally productive. These differences are then supported by cultural expectations (Macoby & Jacklin, 1974). Fathers in the studies described here interacted with their babies with a more arousing, more physical style than did mothers. While the absence of rigidly prescribed parenting roles can be liberating for men as well as for women, parents may also find this situation stressful if they must redefine their role in each situation they face. Ultimately, a better understanding of adult development and the transition to parenthood for both men and women is needed. Studies have hypothesized that increased contact of fathers with their infants may aid adult maturation (Heath, 1978). Regardless of the degree of the father's involvement, events such as the mother's return to work may spark another kind of developmental transition in male parenting.

If this apparent trend of increased paternal investment in pregnancy as well as with their young infants continues, important influences on subsequent generations might occur. For example, male infants today are increasingly reared in a culture more free of sex role stereotypes than previously. How will these infants act when they become fathers? As increased options become available to men as well as women and stereotypes fade, it will be possible to study a range of interactive styles in adult males to see if they relate to caregiving and play activities.

Increased paternal involvement with young infants may have important influences on the father himself and may provide critical support for the mother, particularly if she is stressed. Influences on the infant are less well understood. If the infant and caregiver together are viewed as part of a biological system, then increased paternal involvement may influence the regulation of that system and alter the structure of the infant's social experiences. Fathers seem to provide infants with more intense, arousing, differentiating experiences, and such experiences, along with others, may shape the infant's temperament. It is tempting to speculate that infants may also be predisposed to seek an appropriate balance of both arousing and well-modulated, contained stimulation from the caregiving system. Both the expectations and goals of adult caregivers and the needs of the infant may interact to maintain the regulation of the system even as social and cultural expectations are shifting.

Finally, given our limited knowledge, any efforts to change clinical care and social policy toward fathers should aim to increase the options available to fathers as well as mothers rather than promoting specific alternative prescriptions. At the same time, any renegotiation of caregiving tasks with regard for adult equality should remain sensitive to the implications for infants and their needs. Ideally, many families will continue to seek and to find an optimal balance between adult fulfillment from work and the responsibilities and emotional rewards of parenthood—the sense of delight, renewal, and creativity from observing one's infant grow up.

Acknowledgments

I would like to express my appreciation to the families who participated in these studies and shared a special time in their lives with me. Much of the research upon which this chapter was based was done in collaboration with Dr. Suzanne Dixon and with the encouragement and assistance of Edward Tronick, Ph.D., Lauren Adamson, Ph.D., Heidelise Als, Ph.D., Barry Lester, Ph.D., and T. Berry Brazelton, M.D. Thanks also go to Carey Halsey for technical assistance. Finally, I'd like to thank Ms. Karin Madrid for her care and patience in preparing this manuscript.

9. REFERENCES

Abarbanel, A. Shared parenting after separation and divorce: A study of joint custody. *American Journal of Orthopsychiatry,* 1979, *49*, 320–328.

Abbott, L. *Paternal touch of the newborn: Its role in paternal attachment.* Thesis submitted to Boston University, School of Nursing, 1975.

Abelin, E. L. The role of the father in the separation–individuation process. In J. B. McDevitt & C. F. Settlage (Eds.), *Separation–individuation.* New York: International Universities Press, 1971.

Abelin, E. L. Some further observations and comments on the earliest role of the father. *International Journal of Psychoanalysis,* 1975, *56*, 293–302.

Ainsworth, M. D. S. The development of infant–mother attachment. In B. Caldwell & H. Ricciuti (Eds.), *Review of child development research* (Vol. 3). Chicago: University of Chicago, 1973.

Alexander, B. K. Paternal behavior of adult male Japanese monkeys. *Behaviour,* 1970, *36*, 270–285.

Als, H., Tronick, E., & Brazelton, T. B. Analysis of face-to-face interaction in infant–adult dyads. In M. Lamb, S. Suomi, & G. Stephenson (Eds.), *The study of social interaction.* Madison, Wisc.: University of Wisconsin, 1979.

Anderson, B. J., & Standley, K. *A methodology for observation of the childbirth environment.* Paper presented to the American Psychological Association, Washington, D.C., September 1976.

Appleton, T., Clifton, R., & Goldberg, S. The development of behavioral competence in infancy. In F. Horowitz (Ed.), *Review of child development research* (Vol. 4). Chicago: University of Chicago, 1975.

Ashby, W. R. *An introduction to cybernetics.* London: Methuen, 1956.

Ban, P. L., & Lewis, M. Mothers and fathers, girls and boys: Attachment behavior in the one-year-old. *Merrill-Palmer Quarterly,* 1974, *20*, 195–204.

Barry, H., & Paxson, L. M. Infancy and early childhood: Cross-cultural codes. *Ethnology,* 1971, *10*, 467.

Baumrind, D., & Black, A. E. Socialization practices associated with dimensions of competence in preschool boys and girls. *Child Development,* 1967, *38*, 291–327.

Bell, R. A reinterpretation of the direction of effects in studies of socialization. *Psychological Review,* 1968, *75*, 81–95.

Belsky, J. Mother–father–infant interaction: A naturalistic observational study. *Developmental Psychology,* 1979, *15*, 601–607.

Benedek, T. Fatherhood and providing. In E. Anthony & T. Benedek (Eds.), *Parenthood: Its psychology and psychopathology.* Boston: Little, Brown, 1970.

Bibring, G. L. Some considerations of the psychological processes in pregnancy. *Psychoanalytic Study of the Child,* 1959, *14*, 113.

Biller, H. Father absence and the personality development of the male child. *Developmental Psychology,* 1970, *2*, 181–201.

Biller, H. The father and personality development: Paternal deprivation and sex-role development. In M. Lamb (Ed.), *The role of the father in child development.* New York: Wiley, 1976.

Bower, T. G. R. *A primer of infant development.* San Francisco: Freeman, 1977.

Bowlby, J. *Attachment and loss* (Vol. 1). New York: Basic Books, 1969.

Brazelton, T. B. *Neonatal Behavioral Assessment Scale.* Monograph No. 50, Spastics International Medical Publications. Philadelphia: Lippincott, 1973.

Brazelton, T. B., & Als, H. Four early stages in the development of mother–infant interaction. In A. J. Solnit, R. Eissler, A. Freud, M. Kris, P. B. Neubauer (Eds.), *The psychoanalytic study of the child* (Vol. 34). New Haven: Yale University, 1979.

Brazelton, T. B., Koslowski, B., & Main, M. The origins of reciprocity. In M. Lewis & L. A. Rosenblum (Eds.), *The effect of the infant on its caregiver.* New York: Wiley, 1974.

Brazelton, T. B., Tronick, E., Adamson, L., Als, H., & Wise, S. Early mother-infant reciprocity. In R. Hinde (Ed.), *Parent-infant interaction.* Ciba Foundation Symposium No. 33. Amsterdam: Elsevier, 1975.

Brazelton, T. B., Yogman, M. W., Als, H., & Tronick, E. The infant as a focus for family reciprocity. In M. Lewis & L. Rosenblum (Eds.), *The child and its family.* New York: Plenum Press, 1979.

Bricker, W. A., & Bricker, D. A. The infant, toddler, and preschool research and intervention project. In T. D. Tjossem (Ed.), *Intervention strategies for high risk infants and young children.* Baltimore: University Park Press, 1976.

Bronfenbrenner, U. The changing American child: A speculative analysis. *Journal of Social Issues,* 1961, *17,* 6.

Bronfenbrenner, U. Who cares for America's children. In V. Vaughn & T. B. Brazelton (Eds.), *The family: Can it be saved?* Chicago: Yearbook Medical Publishers, 1976.

Bruner, J. Organization of early skilled action. *Child Development,* 1973, *44,* 1-11.

Bruner, J., Jolly, A., & Sylva, K. (Eds.). *Play.* New York: Basic Books, 1976.

Bruning, J. L., & Kintz, B. L. *Computational handbook of statistics.* Glenview: Scott, Foresman, 1968.

Burton, F. D. The integration of biology and behavior in the socialization of Macaca Sylvana of Gibraltar. In F. Poirier (Ed.), *Primate socialization.* New York: Random House, 1972.

Carlsmith, L. Effects of early father absence on scholastic aptitude. *Harvard Education Review,* 1964, *34,* 3-21.

Carpenter, G. Mother's face and the newborn. *New Scientist,* 1974, *61,* 742-744.

Cassell, T., & Sander, L. *Neonatal recognition processes and attachment: The masking experiment.* Paper presented to Society for Research in Child Development, Denver, March 1975.

Chamove, A., Harlow, H. F., & Mitchell, G. D. Sex differences in the infant-directed behavior of preadolescent rhesus monkeys. *Child Development,* 1967, *38,* 329-335.

Clarke-Stewart, K. A. Interactions between mothers and their young children. *Monographs of the Society for Research in Child Development,* 1973, *38.*

Clarke-Stewart, K. A. And daddy makes three: The father's impact on mother and child. *Child Development,* 1978, *49,* 466-478.

Clarke-Stewart, K. A. The father's contribution to children's cognitive and social development in early childhood. In F. A. Pedersen (Ed.), *The father-infant relationship: Observational studies in a family setting.* New York: Holt, Rinehart, & Winston, 1980.

Cohen, L. J., & Campos, J. J. Father, mother, and stranger as elicitors of attachment behaviors in infancy. *Developmental Psychology,* 1974, *10,* 146-154.

Collins, G. A new look at life with father. *The New York Times Magazine,* June 17, 1979, pp. 30-65.

Crawley, S., Rogers, P., Friedman, S., Iacobbo, M., Criticos, A., Richardson, L., & Thompson, M. Developmental changes in the structure of mother-infant play. *Developmental Psychology,* 1978, *14,* 30-36.

Dickie, J., Carnahan, S. *Training in social competence: The effect on mothers, fathers, and infants.* Paper presented at the biennial meeting of the Society for Research in Child Development, San Francisco, March 1979.

Earls, F., Yogman, M. W. The father-infant relationship. In J. Howells (Ed.), *Modern perspectives in the psychiatry of infancy.* New York: Bruner Mazel, 1979.

Emde, R., Gaensbauer, T. J., & Harmon, R. J. Emotional expression in infancy. *Psychological Issues X: Monograph 37,* 1976.

Erikson, E. Identity and the life cycle. *Psychological Issues,* 1959, *1,* 65-74.

Ewer, P., & Gibbs, J. Relationship with putative father and use of contraception in a population of black ghetto adolescent mothers. *Public Health Reports,* 1975, *90,* 417.

Fein, R. A. Men's entrance to parenthood. *Family Coordinator,* 1976, *25,* 341-351.

Feldman, S. S., & Nash, S. C. The effect of family formation on sex stereotypic behavior: A study of responsiveness to babies. In W. B. Miller & L. F. Newman (Eds.), *The first child and family formation*. Chapel Hill, N.C.: Carolina Population Center, 1978.

Field, T. M. Interaction behaviors of primary versus secondary caretaker fathers. *Developmental Psychology,* 1978, *14*, 183–184.

Field, T. M. Interaction patterns of preterm and fullterm infants. In T. M. Field (Ed.), *Infants born at risk: Behavior and development*. New York: S. P. Medical and Scientific Books, 1979.

Freud, S. *New introductory lectures on psychoanalysis*. New York: Norton, 1923.

Freud, S. The passing of the Oedipus complex. In *Collected papers* (Vol II). London: Hogarth, 1925.

Friedman, L. Fathers don't make good mothers, said the judge. *The New York Times,* January 28, 1973.

Frodi, A. M., Lamb, M., Leavitt, L., & Donovan, W. L. Fathers' and mothers' responses to infant smiles and cries. *Infant Behavior and Development,* 1978, *1*, 187–198.

Frodi, A., Lamb, M., Leavitt, L., Donovan, W., Neff, C., & Sherry, D. Fathers' and mothers' responses to the faces and cries of normal and premature infants. *Developmental Psychology,* 1978, *14*, 490–498.

Gath, A. The impact of an abnormal child upon the parents. *British Journal of Psychiatry,* 1974, *125*, 568.

Gerwirtz, H. B., & Gerwirtz, J. L. Visiting and caretaking patterns for kibbutz infants. *American Journal of Orthopsychiatry,* 1968, *38*, 427–443.

Gil, D. G. *Violence against children*. Cambridge: Harvard University, 1970.

Goldberg, S. Prematurity: Effects on parent-infant interaction. *Journal of Pediatric Psychology,* 1978, *3*, 137–144.

Gomber, J., & Mitchell, G. D. Preliminary report on adult male isolation-reared rhesus monkeys caged with infants. *Developmental Psychology,* 1974, *10*, 298.

Goode, W. J. *World revolution and family patterns*. Glencoe, Ill.: Free Press, 1963.

Green, A. H. Child-abusing fathers. *Journal of the American Academy of Child Psychiatry,* 1979, *18*, 270–282.

Green, A. H., Gains, R., & Sandgrund, A. Child abuse. *American Journal of Psychiatry,* 1974, *131*, 882–886.

Green, M., & Beall, P. Paternal deprivation—A disturbance in fathering—A report of nineteen cases. *Pediatrics,* 1962, *30*, 91–99.

Greenberg, M., & Morris, N. Engrossment: The newborn's impact upon the father. *American Journal of Orthopsychiatry,* 1974, *44*, 520–531.

Gurwitt, A. R. Aspects of prospective fatherhood. *Psychoanalytic Study of the Child,* 1976, *31*, 237–271.

Hampton, J. K., Hampton, S. H., & Landwehr, B. T. Observations on a successful breeding colony of the marmoset, oedipomidas oedipus: *Folia Primatologia,* 1966, *4*, 265–287.

Harlow, H. F. The nature of love. *American Psychologist,* 1958, *13*, 673–685.

Heath, D. What meaning and effects does fatherhood have for the maturing of professional men? *Merrill-Palmer Quarterly,* 1978, *24*, 265–278.

Herzog, E., & Sudia, C. E. Children in fatherless families. In B. M. Caldwell & H. N. Ricciuti (Eds.), *Review of child development research* (Vol. 3). Chicago: University of Chicago, 1973.

Herzog, J. Disturbances in parenting high risk infants: Clinical impressions and hypotheses. In T. M. Field (Ed.), *Infants born at risk*. New York: Spectrum, 1979.

Hetherington, E. M. Effects of paternal absence on sex-typed behaviors in negro and white preadolescent males. *Journal of Personality and Social Psychology,* 1966, *4*, 87–91.

Howells, J. G. Fathering. In J. G. Howells (Ed.), *Modern perspectives in international child psychiatry*. Edinburgh: Oliver & Boyd, 1969.

Howells, M. Employed mothers and their families. *Pediatrics,* 1973, *52*, 252–263.

Hrdy, S. B. Care and exploitation of nonhuman primate infants by conspecifics other than the mother. In J. R. Rosenblatt, R. A. Hinde, E. Shaw, & C. Beer (Eds.), *Advances in the study of behavior* (Vol. 6). New York: Academic Press, 1976.

Johannesson, P. *Instruction in child care for fathers*. Dissertation at University of Stockholm, 1969.

Johnson, A., & Gaiter, J. L. *Father–infant bonding in the intensive care nursery*. Work in progress, Children's Hospital National Medical Center, 1980.

Johnson, M. M. Sex role learning in the nuclear family. *Child Development*, 1963, *34*, 315–333.

Jolly, A. *The evolution of primate behavior*. New York: Macmillan, 1972.

Kaufman, C. Biologic consideration of parenthood. In E. J. Anthony & T. Benedek (Eds.), *Parenthood: Its psychology and psychopathology*. Boston: Little, Brown, 1970.

Keniston, K. (Ed.). *All our children*. New York: Harcourt Brace Jovanovich, 1977.

Kessen, W. Human infancy. In P. Mussen (Ed.), *Carmichael's manual of child psychology*. New York: Wiley, 1970.

Klaus, M., & Kennell, J. *Maternal–infant bonding*. St. Louis: Mosby, 1976.

Klaus, M., Kennell, J., Plumb, N., & Zuehlke, S. Human maternal behavior at the first contact with her young. *Pediatrics*, 1970, *46*, 187–192.

Klaus, M., Jerauld, R., Kreger, N., McAlpine, W., Staffa, M., & Kennell, J. Maternal attachment: Importance of the first postpartum days. *New England Journal of Medicine*, 1972, *286*, 460–463.

Kotelchuck, M. *The nature of the infant's tie to his father*. Paper presented to the Society for Research in Child Development, Philadelphia, March 1973.

Kotelchuck, M. *Father-caretaking characteristics and their influence on infant–father interaction*. Paper presented to American Psychological Association, Chicago, September 1975.

Kotelchuck, M. The infant's relationship to the father: Experimental evidence. In M. Lamb (Ed.), *The role of the father in child development*. New York: Wiley, 1976.

Lamb, M. E. Fathers: Forgotten contributors to child development. *Human Development*, 1975, *18*, 245–266.

Lamb, M. E. Twelve-month-olds and their parents: Interaction in a laboratory playroom. *Developmental Psychology*, 1976, *12*, 237–244. (a)

Lamb, M. E. Effects of stress and cohort on mother and father-infant interaction. *Developmental Psychology*, 1976, *12*, 435–443. (b)

Lamb, M. E. The role of the father: An overview. In M. E. Lamb (Ed.), *The role of the father in child development*. New York: Wiley, 1976. (c)

Lamb, M. E. Father–infant and mother–infant interaction in the first year of life. *Child Development*, 1977, *48*, 167–181. (a)

Lamb, M. E. The development of mother–infant and father–infant attachments in the second year of life. *Developmental Psychology*, 1977, *13*, 637–648. (b)

Lamb, M. E. The father's role in the infant's social world. In J. E. Stevens, Jr., & M. Mathews (Eds.), *Mother/child, father/child relationships*. Washington, D.C.: National Association for the Education of Young Children, 1978. (a)

Lamb, M. E. Qualitative aspects of mother- and father–infant attachments. *Infant Behavior and Development*, 1978, *1*, 265–275. (b)

Lamb, M. E. Paternal influences and the father's role: A personal perspective. *American Psychologist*, 1979, *34*, 938–943.

Lang, O. *Chinese family and society*. New Haven: Yale University, 1946.

Langer, J. Werner's comparative organismic theory. In P. Mussen (Ed.), *Carmichael's manual of child psychology* (Vol. 1). New York: Wiley, 1970.

Lerner, H. Effects of the nursing mother–infant dyad on the family. *American Journal of Orthopsychiatry*, 1979, *49*, 339–348.

Lester, B. M., Kotelchuck, M., Spelke, E., Sellers, J. J., & Klein, R. E. Separation protest in Guatemalan infants: Cross-cultural and cognitive findings. *Developmental Psychology,* 1974, *10,* 79–85.

Levine, J. A. *Who will raise the children? New options for fathers.* Philadelphia: Lippincott, 1976.

Lewis, M., & Brooks, J. Infant's social perception: A constructivist view. In L. B. Cohen & P. Salapatek (Eds.), *Infant perception: From sensation to cognition* (Vol. II). New York: Academic Press, 1975.

Lewis, M., & Rosenblum, L. *The effect of the infant on its caregiver.* New York: Wiley, 1974.

Lewis, M., & Weinraub, M. Sex of parent x sex of child: Socioemotional development. In R. C. Friedman, R. M. Richart, & R. L. Vander Wiele (Eds.), *Sex differences in behavior.* New York: Wiley, 1974.

Lewis, M., & Weinraub, M. The father's role in the infant's social network. In M. E. Lamb (Ed.), *The role of the father in child development.* New York: Wiley, 1976.

Liebenberg, B. Expectant fathers. In P. Shereshefsky & L. Yarrow (Eds.), *Psychological aspects of a first pregnancy and early postnatal adaptation.* New York: Raven, 1973.

Lipsitt, L. *Developmental psychobiology: The significance of infancy.* Hillsdale, N.J.: Lawrence Erlbaum, 1976.

Locke, J. Essay on human understanding. In *The Works.* London: Butterworth, 1727.

Lorenzi, M. E., Klerman, L. V., & Jekel, J. F. School-age parents: How permanent a relationship? *Adolescence,* 1977, *12,* 13–23.

Lynn, D. *The father: His role in child development.* Monterey, Calif.: Brooks/Cole, 1974.

Macoby, E. E., & Jacklin, C. N. *The psychology of sex differences.* Stanford, Calif.: Stanford University, 1974.

Mahler, M. S., Pine, F., & Bergman, A. *The psychological birth of the human infant.* New York: Basic Books, 1975.

Malinowski, B. *The father in primitive psychology.* New York: Norton, 1927.

Mandell, F., McAnulty, E., & Reece, R. Observations of paternal response to sudden unexplained infant death. *Pediatrics,* 1980, *65,* 221–225.

Mendes, H. A. Single fathers. *Family Coordinator,* 1976, *25,* 439–445.

Mitchell, G. D. Paternalistic behavior in primates. *Psychological Bulletin,* 1969, *71,* 399–417.

Mischel, W. Sex typing and socialization. In P. Mussen (Ed.), *Carmichael's manual of child psychology.* New York: Wiley, 1970.

Nash, J. The father in contemporary culture and current psychological literature. *Child Development,* 1965, *36,* 261–297.

Newberger, E., Reed, R. B., Daniel, J. H., Hyde, J., & Kotelchuck, M. Pediatric social illness: Toward an etiologic classification. *Pediatrics,* 1977, *60,* 178–185.

Newson, J., & Newson, E. *Patterns of infant care in an urban community.* New York: Penguin, 1963.

Orthner, D., Brown, T., & Ferguson, D. Single-parent fatherhood: An emerging life style. *Family Coordinator,* 1976, *25,* 429–439.

Palme, O. The emancipation of man. *Journal of Social Issues,* 1972, *28,* 237–246.

Parke, R. Perspectives on father–infant interaction. In J. D. Osofsky (Ed.), *Handbook of infancy.* New York: Wiley, 1979.

Parke, R., & O'Leary, S. Father–mother–infant interaction in the newborn period: Some findings, some observations, and some unresolved issues. In K. Riegel & J. Meacham (Eds.), *The developing individual in a changing world. Vol. II. Social and environmental issues.* The Hague: Mouton, 1976.

Parke, R., & Sawin, D. *Infant characteristics and behavior as elicitors of maternal and paternal responsibility in the newborn period.* Paper presented to Society for Research in Child Development, Denver, March 1975.

Parke, R., & Sawin, D. *The family in early infancy: Social interactional and attitudinal analyses.* Paper presented to Society for Research in Child Development, New Orleans, March 1977.

Parke, R., O'Leary, S. E., & West, S. Mother–father–newborn interaction: Effects of maternal medication, labor, and sex of infant. *Proceedings of the American Psychological Association,* 1972, 85–86.

Parke, R., Power, T. G., Tinsley, B., & Hymel, S. The father's role in the family system. *Seminars in Perinatology,* 1979, *3,* 25–34.

Parke, R., Hymel, S., Power, T., & Tinsley, B. Fathers and risk: A hospital based model of intervention. In D. B. Sawin, R. C. Hawkes, L. O. Walker, & J. H. Penticuff (Eds.), *Psychological risks in infant environment transactions.* New York: Bruner/Mazel, 1980.

Parsons, T., & Bales, R. *Family, socialization and interaction process.* Glencoe, Ill.: Free Press, 1954.

Pedersen, F. A. *Mother, father, and infant as an interactive system.* Paper presented to the American Psychological Association, Chicago, September 1975.

Pedersen, F. A., & Robson, K. S. Father participation in infancy. *American Journal of Orthopsychiatry,* 1969, *39,* 466–472.

Pedersen, F. A., Anderson, B. J., & Cain, R. L. *An approach to understanding linkages between the parent–infant and spouse relationships.* Paper presented at the Society for Research in Child Development, New Orleans, March 1977.

Pedersen, F. A., Yarrow, L. J., Anderson, B. J., & Cain, R. L. Conceptualization of father influences in the infancy period. In M. Lewis & L. Rosenblum (Eds.), *The child and its family.* New York: Plenum Press, 1979.

Pedersen, F. A., Zaslow, M., Cain, R. L., & Anderson, B. J. *Cesarean childbirth: The importance of a family perspective.* Paper presented to International Conference on Infant Studies, New Haven, April 1980.

Pedersen, F. A., Rubenstein, J., & Yarrow, L. J. Infant development in father-absent families. *Journal of Genetic Psychology,* 1979, *135,* 51–61.

Platts, K. A public agency's approach to the natural father. *Child Welfare,* 1968, *47,* 530.

Power, T. G., & Parke, R. D. *Toward a taxonomy of father–infant and mother–infant play patterns.* Paper presented to the Society for Research in Child Development, San Francisco, March 1979.

Price-Bonham, S., & Addison, S. Families and mentally retarded children: Emphasis on the father. *The Family Coordinator,* 1978, *27,* 221–230.

Queen, S. A., & Adams, J. B. *The family in various cultures.* Chicago: Lippincott, 1952.

Rebelsky, F., & Hanks, C. Fathers' verbal interactions with infants in the first three months of life. *Child Development,* 1972, *42,* 63.

Redican, W. K. Adult male–infant interactions in nonhuman primates. In M. E. Lamb (Ed.), *The role of the father in child development.* New York: Wiley, 1976.

Redican, W. K., & Mitchell, G. D. A longitudinal study of paternal behavior in adult male rhesus monkeys: 1. Observations on the first dyad. *Developmental Psychology,* 1973, *8,* 135–136.

Rendina, I., & Dickerscheid, J. D. Father involvement with first-born infants. *Family Coordinator,* 1976, *25,* 373–379.

Rheingold, H. The modification of social responsiveness in institutional babies. *Monographs of the Society for Research in Child Development,* 1956, *21.*

Richards, M. P. M., Dunn, J. F., & Antonis, B. Caretaking in the first year of life: The role of fathers' and mothers' social isolation. *Child Care, Health and Development,* 1977, *3,* 23–26.

Rigler, D., & Spinetta, J. The child-abusing parent: A psychological review. *Psychological Bulletin,* 1972, *77,* 296.

Rivière, P. G. The couvade: A problem reborn. *Man,* 1976, *9,* 423–435.

Rizzuto, A. *Intervention with fathers in the perinatal period: Father's support group.* Paper presented at the annual meeting of the Seminar in the Development of Infants and Parents, Boston, November 1978.

Robson, K., & Moss, H. Patterns and determinants of maternal attachment. *Journal of Pediatrics,* 1970, *77,* 976–985.

Rödholm, M., & Larsson, K. Father-infant interaction at the first contact after delivery. *Early Human Development,* 1979, *3,* 21-27.

Rosenblatt, J. S. The development of maternal responsiveness in the rat. *American Journal of Orthopsychiatry,* 1969, *39,* 36-56.

Ross, J. M. The development of paternal identity: A critical review of the literature on nurturance and generativity in boys and men. *Journal of American Psychoanalytic Association,* 1975, *23,* 783-817.

Rossi, A. A biosocial perspective on parenting. *Daedalus,* 1977, *106,* 1-32.

Rubin, J. Z., Provenzano, F. J., & Luria, Z. The eye of the beholder: Parent's views on sex of newborns. *American Journal of Orthopsychiatry,* 1974, *44,* 512-519.

Russell, G. The father role and its relation to masculinity, femininity, and androgyny. *Child Development,* 1978, *49,* 1174-1181.

Rypma, C. B. The biological bases of the paternal responses. *Family Coordinator,* 1976, *25,* 335-341.

Sander, L. The regulation of exchange in the infant-caretaker system and some aspects of the context-content relationship. In M. Lewis & L. Rosenblum (Eds.), *Interaction, conversation and the development of language.* New York: Wiley, 1979.

Sander, L., Julia, H., Stechler, J., Burns, P., & Gould, J. *Some determinants of temporal organization in the ecological niche of the newborn.* Paper presented to the International Society for the Study of Behavioral Development, Guildford, England, July 1975.

Sawin, D., & Parke, R. Adolescent fathers: Some implications from recent research on paternal roles. *Educational Horizons,* 1976, *55,* 38-43.

Schaffer, H. R., & Emerson, P. E. The development of social attachments in infancy. *Monographs of the Society for Research in Child Development,* 1964, *29.*

Shereshefsky, P., & Yarrow, L. *Psychological aspects of a first pregnancy and early postnatal adaptation.* New York: Raven, 1973.

Shorter, E. *The making of the modern family.* New York: Basic Books, 1975.

Siegel, S. *Nonparametric statistics.* New York: McGraw-Hill, 1956.

Spelke, E., Zelazo, P., Kagan, J., & Kotelchuck, M. Father interaction and separation protest. *Developmental Psychology,* 1973, *9,* 83-90.

Spencer-Booth, Y. The relationships between mammalian young and conspecifics other than mothers and peers: A review. In D. Lehrman, R. Hinde, & E. Shaw (Eds.), *Advances in study of behavior,* 1970, *3,* 119-139.

Stern, D. N. The goal and structure of mother-infant play. *Journal of the American Academy of Child Psychiatry,* 1974, *13,* 402-421.

Stern, D. N. Temporal expectancies of social behaviors in mother-infant play. In E. Thoman (Ed.), *The origins of the infant's responsiveness.* New York: Erlbaum, 1979.

Stone, L. *The family, sex, and marriage in England, 1500-1800.* New York: Harper & Row, 1977.

Stone, L. J., Smith, H., & Murphy, L. B. *The competent infant.* New York: Basic Books, 1973.

Suomi, S. Adult male-infant interactions among monkeys living in nuclear families. *Child Development,* 1977, *48,* 1215-1270.

Tinbergen, N. The behavior of the stickleback. *Scientific American,* 1952, *187,* 22-38.

Toward a national policy for children and families. Washington, D.C.: National Academy of Sciences, 1976.

Trethowan, W., & Conlon, M. F. The couvade syndrome. *British Journal of Psychiatry,* 1965, *111,* 57.

Trethowan, W. The couvade syndrome. In J. Howells (Ed.), *Modern perspectives in psycho-obstetrics.* Edinburgh: Oliver & Boyd, 1972.

Trivers, R. L. Parental investment and sexual selection. In B. Campbell (Ed.), *Sexual selection and the descent of man, 1871-1971.* Chicago: Aldine, 1972.

Tronick, E. *The ontogenetic structure of face-to-face interaction and its developmental functions.* Paper presented to the Society for Research in Child Development, New Orleans, March 1977.

Tronick, E., Als, H., & Brazelton, T. B. Monadic phases: A structural descriptive analysis of infant–mother face-to-face interaction. *Merrill-Palmer Quarterly,* 1980, *26,* 3–24.

VanGennep, A. *The rites of passage.* Chicago: University of Chicago, 1960.

Warshak, R., & Santrock, J. W. *The effects of father and mother custody on children's social development.* Paper presented at the Society for Research in Child Development, San Francisco, March 1979.

West, M. M., & Konner, M. J. The role of the father: An anthropological perspective. In M. E. Lamb (Ed.), *The role of the father in child development.* New York: Wiley, 1976.

Whiting, B., & Whiting, J. *Children of six cultures.* Cambridge: Harvard University, 1975.

Willemsen, E., Flaherty, D., Heaton, C., & Ritchey, F. Attachment behavior of one year olds as a function of mother vs. father, sex of child, session and toys. *Genetic Psychology Monographs,* 1974, *90,* 305–323.

Wolff, P. Observations on the early development of smiling. In B. Foss (Ed.), *Determinants of infant behavior* (Vol. II). New York: Wiley, 1963.

Yogman, M. W. *The goals and structure of face-to-face interaction between infants and fathers.* Paper presented to the Society for Research in Child Development, New Orleans, March 1977.

Yogman, M. W., Dixon, S., Tronick, E., Adamson, L., Als, H., & Brazelton, T. B. *Father–infant interaction.* Paper presented at American Pediatric Society—Society for Pediatric Research, May 1976. (a)

Yogman, M. W., Dixon, S., Tronick, E., Adamson, L., Als, H., & Brazelton, T. B. *Development of social interaction of infants with fathers.* Paper presented to the Eastern Psychological Association, New York, April 1976. (b)

Yogman, M. W., Dixon, S., Tronick, E., ALs, H., Adamson, L., and Brazelton, T. B. *Parent-infant interaction under stress: The study of a temperamentally difficult infant.* Paper presented to the American Academy of Child Psychiatry, Toronto, October 1976. (c)

Zaleznik, A. Isolation and control in the family and work. In T. B. Brazelton & V. Vaughn (Eds.), *The family: Can it be saved?* New York: Science and Medicine Publishing, 1979.

Zelazo, P. R., Kotelchuck, M., Barber, L., & David, J. *Fathers and sons: An experimental facilitation of attachment behaviors.* Paper presented to the Society for Research in Child Development, New Orleans, March 1977.

Zihlmann, A. L. Motherhood in transition: From ape to human. In W. B. Miller & L. F. Newman (Eds.), *The first child and family formation.* Chapel Hill, N.C.: Carolina Population Center, 1978.

Zilboorg, G. Depressive reactions to parenthood. *American Journal of Psychiatry,* 1931, *10,* 927.

Author Index

Subject Index